STUDIES IN ANALYTIC PHILOSOPHY
Quentin Smith, *Series Editor*

Philosophy of Religion, Physics, and Psychology

Essays in Honor of Adolf Grünbaum

Edited by
Aleksandar Jokić

Prometheus Books

59 John Glenn Drive
Amherst, New York 14228-2119

Published 2009 by Prometheus Books

Inquiries should be addressed to
Prometheus Books
59 John Glenn Drive
Amherst, New York 14228–2119
VOICE: 716–691–0133, ext. 210
FAX: 716–691–0137
WWW.PROMETHEUSBOOKS.COM

13 12 11 10 09   5 4 3 2 1

Library of Congress Cataloging-in-Publication Data

Adolf Grünbaum Symposium (2002 : Santa Barbara City College).
    Philosophy of religion, physics, and philosophy : essays in honor of Adolf
Grünbaum / edited by Aleksandar Jokić.
        p. cm.
    Includes bibliographical references and index.
    ISBN 9781591023692
    1. Religion—Philosophy—Congresses. 2. Physics—Philosophy—Congresses.
3. Psychology—Philosophy—Congresses. 4. Grünbaum, Adolf—Congresses.
I. Grünbaum, Adolf. II. Jokić, Aleksandar. III. Title.

BL51.A423 2002
210—dc22

2006001250

I dedicate this book very affectionately to Harvey and Leslie Wagner, gracious benefactors and very dear, warm friends.

—Adolf Grünbaum

CONTENTS

PREFACE:
THE ORIGINS OF THE BOOK

ALEKSANDAR JOKIĆ

T his book is the product of a longer project that officially got under way with a symposium honoring Professor Adolf Grünbaum's contributions to contemporary philosophy held on October 18–19, 2002, at Santa Barbara City College, and hosted by the college's Center for Philosophical Education (CPE). By this time Professor Grünbaum had been an "old friend" of CPE, having participated two years earlier at another CPE conference on the subject of modern cosmology, titled "In the Beginning." It was at this earlier event that the idea about the Adolf Grünbaum Symposium was announced.

The papers presented at the symposium covered an array of issues in philosophy of religion and philosophy of science. Evidence of Grünbaum's vast philosophical influence was shown in virtually all papers by some of the world's best-known philosophers of science. The highlight of the conference was the keynote address by Grünbaum, who gave an autobiographical sketch of his intellectual journey, beginning with a poignant account of his experiences as a preteen Jew confronting anti-Semitism in Germany. He spoke of his early queries about religious belief, repulsed by his observation of the suffering brought about in the name of one or another deity throughout history, and of what struck him as intolerable inconsistencies in sacred literature concerning the nature of God and God's relationship with mankind.

This book preserves much of the format of the symposium; it offers a chapter by each participant at the conference and Grünbaum's autobiographical-philosophical narrative. However, similarities between the two end there since the book includes a much-revised and expanded autobiographical chapter by Grünbaum, and all authors have significantly reworked and substantially revised their conference talks for the book. Bas van Fraassen has written an entirely new article for this book. Additionally, a reprint of Grünbaum's "The Poverty of Theistic Cosmology" further enriches the book.

When CPE was created in 1997 it was an attempt to chart new territory for professional academic philosophy. CPE was envisaged as a forum where high-end research—a project not usually associated with community colleges—through regular and varied mini-conferences involving prominent philosophers would be projected beyond the narrow confines of traditional academia. The Adolf Grünbaum Symposium is yet another one among those CPE efforts that has culminated in a book.

Working on this book with so many prominent philosophers and thinkers has been a tremendously rewarding experience for me. I am thankful to Dr. Steven Humphrey, president of the CPE executive board, for his support of this project from the start and his participation as both presenter at the symposium and author of a chapter in the book. Thanks are also due to Joseph White and James Chesher, CPE's other directors. Finally, the opportunity to work with Adolf Grünbaum will no doubt have positive and long-lasting effects on me as an inspiration well beyond this project.

PROLOGUE:
AN AUTOBIOGRAPHICAL-
PHILOSOPHICAL
NARRATIVE*

ADOLF GRÜNBAUM

I was born in Cologne/Rhine in Germany in 1923. When I entered a German high school ("gymnasium") there ten years later, the Nazis came to power. My parents were members of the Jewish community and belonged to a modern synagogue. This house of worship stood doctrinally between Orthodox Judaism, which I regarded even then as a form of distressingly obsessive ritualism, and Reform Judaism, which goes back to the eighteenth-century philosopher Moses Mendelssohn and emulated Christian styles of worship, such as featuring organ music.

To prove the biological inferiority of Jews, the Nazis decreed that the physical education of Jewish children in school be greatly curtailed or discontinued. My athletic development was also stunted psychologically by the close association between Nazi political mass rallies and athletic parades. In early 1938, my parents, my two younger siblings, Suzanne (Susi) and Norbert, and I left Cologne for the *life-saving* United States, only eight months before the *Kristallnacht* pogrom conducted by the Nazis.

---

*Professor Joseph White, the organizer of the Santa Barbara conference in October 2002, asked me to give, as the keynote address, a retrospect on my formative early experiences, as well as an outline of my later intellectual development and philosophical odyssey. This chapter is a very considerable expansion of that keynote address.

We emigrated with the crucial help of relatives who had already immigrated to the United States from Germany. Once I had the great good fortune of living in the United States, I began to realize that athletics could be good, clean fun and indeed valuable for body-building or good health. Yet, as soon as I entered college, I was alienated by the sheer idolatry accorded to sports teams on college campuses in the United States and by the role of their victories in the loyalty and donations of their alumni to their alma mater. Indeed, I venture the guess that during a prolonged, major baseball strike, the incidence of clinical depression in the adult male population at large would measurably increase. Yet the same might be said of soccer addicts in Europe.

As for my boyhood in Germany, I must say that there wasn't much fun in it. To avoid the rising hostility of some of the Christian boys in the neighborhood, it was best not to play on our street (*Rubensstrasse* in German, named after the famous Flemish painter Peter Paul Rubens) or in the neighborhood. The increasingly ominous remarks as to what Hitler might do next, which I overheard in the conversations of my parents, soon robbed me of boyish playfulness and innocence. Cologne, with its famous cathedral, was predominantly Roman Catholic, and initially its Christian population and some twelve thousand Jewish citizens seemed to live together harmoniously, so far as I knew then.

But I vividly recall that when a band of young thugs beat me in the street, they told me that "the Jews have killed our savior" ("*die Juden haben unseren Heiland getötet*"). At the time, I did not know yet that this inveterate charge of wicked deicide—which nowadays is hardly a thing of the past in some quarters—is *theologically incoherent* within the Christian redemptive scenario, as I learned in the early 1960s from my then secretary Elizabeth McMunn (cf. Matt. 26:54). According to that doctrine of salvation, Jesus *had* to die on the cross to purchase forgiveness for the sins of everyone else! And he was crucified.

Indeed, the well-known, formidable liberal Roman Catholic scholar Gary Wills wrote in the year 2000: "For one thing, 'the Jews' did not kill Christ, even if one could make the (uncertain) case that *certain* Jews were more responsible than the Romans who actually executed him." More importantly, Wills goes on to declare: "But there is a deeper theological reason why believers in Christian theology should never have considered one part of mankind [i.e., Jews] the killers of Jesus. Since he died for all sins, the only racial solidarity expressed in his suffering is that of the sinful human race, the joint cause and beneficiary of the redeeming death."[1] Incidentally, "Mother" Teresa reportedly denied analgesics to the terminal cancer patients in her hospice in Calcutta on the deplorable grounds that Jesus' sufferings on the cross had also been unrelieved by medication.

My experience with the charge of deicide as a curse upon the Jewish people suggests an explanation of why, among Jews, there may be a higher incidence of *dissident intellectuals*, such as Karl Marx, Sigmund Freud, and Albert Einstein, than among those who do not encounter comparable rejection and adversity in their society.

Yet any such *natural* motivational explanation would be anathema to the mid-twentieth century playwright Clare Booth Luce, who was the US ambassador to Italy (1953) and Brazil (1959), a congresswoman from Connecticut (1943), a prominent power broker in the Republican party, a member of the Congressional Intelligence Oversight Committee, and the wife of Henry Luce, publisher of *Time*, *Life*, and *Fortune* magazines. In 1947, *McCall's* magazine featured a series of articles by her, which were then reprinted in *Reader's Digest*, in which she offered a *theological* explanation of *non*conformity among Jews.

Commenting on the presumed ideological ills of the twentieth century, Clare Booth Luce attributed them to the theory of relativity, psychoanalysis, and Marxism, and to the mentality of their creators Albert Einstein, Sigmund Freud, and Karl Marx, respectively. All three were of Jewish parentage. And she claimed that their Jewish origin played a crucial role in the spawning of their theories.

In 1946, after the tragic death of her only child at a young age, Mrs. Luce renounced Protestantism and became a Roman Catholic under the tutelage of Bishop Fulton J. Sheen, a charismatic TV personality at the time. In the wake of this conversion, she noted that each of her three ideological villains rejected the divinity of Jesus. As she saw it, just this rejection was the *real cause* of the presumed intellectual mischief they wrought to the detriment of Western society: *As Jews, she claimed, they suffered from unfulfilled and forlorn messianic longings*, although—as it happens—believing Jews expect the Messiah, prophesied in their Bible, no less than devout Christians expect Jesus' Second Coming.

This gaping psychic void, she claimed, is destined to keep haunting the Jewish people, generating ever-new intellectual monstrosities, until and unless they embrace the Christian formula for spiritual salvation. Mind you, she does *not* identify the *adversity* suffered by Israelites in the Christian world as a contributing cause of Jewish innovative intellectual challenge to the conventional wisdom. Instead, she regards a *fundamentally religious deficit* to be the crucial factor in the proclivity of Jewish intellectuals to break with established traditions. It is essential to bear in mind the difference between these two proposed explanations.

In my view, it stands to reason that the assorted barriers traditionally erected in much of Western society by the non-Jewish majority to the full emancipation of the Jewish minority were an important causal factor in creating the skeptical frame of mind conducive to radical Jewish intellectual dissent and innovation. Paradoxically, Mrs. Luce did not complain about the theological ravages of Darwin's theory of evolution. After all, he was not Jewish.

Let us be clear that Einstein, Freud, and Marx denied the divinity of Jesus, *not* because they believed in Judaism, but because they rejected theism generically, be it Jewish, Christian, or Muslim for that matter. In the case of Einstein and Freud, however, this religious apostasy coexisted with their publicly declared sociocultural allegiance to the Jewish community, whereas Marx, whose father was an opportunistic convert to Protestantism, inveighed against "Jewish capitalists" in self-hating fashion. And I must mention very clearly that, in my own case as well, my atheism coexists harmoniously with my strong cultural identification with some, though only *some*, of my Jewish patrimony.

Yet you may ask: What was Mrs. Luce's bill of particulars against the allegedly wicked theories of the messianically unfulfilled triumvirate? What deviltry does she lay at the doors of Einstein and Freud? As any ordinary student of Einstein's theory of relativity knows, there is *nothing whatsoever* in it pertaining to the relativity or absoluteness of *morals*. Yet Mrs. Luce made the illiterate, pseudointellectual claim that Einstein had "relativized morals." Furthermore, she supplied no evidence that even a *vulgar* misunderstanding of relativity theory had lent legitimacy to the permissive society. And even if there were such evidence, the purveyors of the demagogic *caricature* of his theory would be liable for it, but not Einstein's supposed forlorn messianic longings.

Her reasoning is no better in the case of Freud. Although I have argued extensively that the core of his psychoanalytic theory is ill supported by the evidence, Freud deserves some credit, I think, for exposing the hypocritical neglect of the sexual problems prevalent at his time. Yet Mrs. Luce, who ironically herself lived the life of a sexual libertine, saddles him with culpability for *creating* these problems, though he only insisted on lifting the veil of dishonesty from them. It is a gauge of the need for candor in Freud's era that when male homosexuality was outlawed in England under Queen Victoria, lesbianism was not: To Her Majesty's mind, female homosexuality was such an abomination that she could not even acknowledge its existence by advocating its prohibition.

Mrs. Luce offers no cogent evidence that Freud's ideas *exacerbated* the sexual problems of modern society, let alone that belief in the divinity of Jesus would have prevented him from developing his sexual etiology of the neuroses. As we know, the Christian sex taboo did not prevent the sex scandals in the Roman Catholic clergy that have come to light in more recent years. Apparently, Mrs. Luce was not aware that, for decades in Freud's own time, the Swiss Lutheran clergyman Oskar Pfister was an ardent proponent of the use of psychoanalysis in pastoral work. Indeed, the Jesuit William Meissner, a professor at the Jesuit Boston College, is a dedicated practicing Freudian psychoanalyst.

Nor is it a grave social malady that the *vulgarized popularization* of Freudian theory has led some people to talk much fatuous and dreary psychobabble, as portrayed by the film director and actor Woody Allen. But even if the cultural impact of psychoanalysis had turned into a social disorder, Freud's theoretical edifice would not deserve to be blamed for it.

In any case, I can report that, although I am an acolyte of Einstein's theory, I don't have the least unfulfilled messianic longings. In my view, all such yearnings are inspired by delusional wish fulfillment, just as the belief in personal immortality, which I consider *multiply* untenable in the face of the evidence. I do not long in the least for the Jewish Messiah, who is allegedly still to come. Nor do I pine for Jesus' supposed Second Coming with its terrifying promise of Armageddon. This expected epiphany provides the basis, as the late Jerry Falwell told us with his usual smug grin, for the unreserved support given to the Likud party in Israel by the religious Right in this country.

So much for Mrs. Luce's disregard of adversity as the midwife of challenges to conventional wisdom.

I am about to relate how, in my early teens, I came to abandon the biblical theism in which I was reared, in favor of atheism. But a half century later—in a series of publications which I shall cite—I set forth my critique of such religious belief *systematically* by reference to some major theists.

One of my first feelings of rebellion against the God of the Old Testament, and later the God of the New Testament, was the moral outrage I felt in the face of the story in Genesis 22 about the first patriarch, Abraham. His son Isaac was his favorite, because that son was supposedly born of divine promise, being the offspring of Abraham's aged and heretofore barren wife Sarah. Once Abraham had thereby become a true believer in God, he had to pass the crucial test of his faith: He was expected to sacrifice Isaac willingly to God.

I found it morally outrageous that anyone should be asked to sacrifice an innocent person as a test of his fealty to God. Thus, I deemed the readiness of an Almighty God to accept such a homicidal sacrifice, even if it were offered *spontaneously*, to be ethically monstrous.

Furthermore, I could not avoid the conclusion that such a conception of God was modeled psychologically on experience with *capricious monarchs*: Their ire is dangerous, so that they need to be propitiated, or even bribed by sacrifices and offerings that please them, and by endless praise. But they also can be implored for crucial intercessions or special favors, much like nowadays the king or princes in Saudi Arabia.

Note the revealing frequency with which the Almighty is called "the King of the Universe," and his realm is denominated as "the *Kingdom* of God." Why should *royal status potentiate* the august dignity of a God whose complete supremacy is already guaranteed by being almighty? My moral indignation was not lessened by the tale in Genesis 22 of Isaac's miraculous deliverance from death-by-sacrifice. Later I learned that, in the Koran, the same story about Abraham is told about his son Ishmael, the offspring of his wife Hagar. The important Islamic festival Eid al-Adha commemorates Abraham's willingness to sacrifice Ishmael to God and comes at the end of the annual period of pilgrimage to Mecca, the Muslim holy city.

I also remember a formative episode from Hebrew class during a reading of the Torah, with surreptitious use of Martin Luther's excellent German translation as a pony. I soon learned that *Yehovah* (in Germanically pronounced Hebrew) was supposed to be God's *true name* and that Torah scribes went through ablutions before putting it on parchment. As I knew, no good Jew was allowed to pronounce that name, a prohibition smacking of a semantic confusion between the name and the object. Thus an array of substitute names were coined, such as *Adonai* (the Lord) and *Hashem*, which literally meant in Hebrew "the name," suggesting—in the logician Frege's later terminology—that instead of God's name being *used*, it was merely being *mentioned*, presumably to betoken reverence. For example, there is a ritual titled, in Hebrew, *Kiddush hashem*, meaning "sanctification of *the name*." I was also aware that when an observant Jew used the German word *Gott* for God *in writing*, it was spelled "G-tt," rather than "Gott." Many years later I saw in English the use of the spelling "G-d," in lieu of "God," as in some articles in a Jewish newspaper.

It struck me that the whole nomenclature fetish was redolent of the hoary superstition of word *magic*, much like the doctrine of the "evil eye." Thus, I decided to rebel in Hebrew class and to vent my disdain for the prohibition against saying *Yehovah*. So when my turn came to read a Torah passage containing that hallowed word, I blithely pronounced it out loud. That turned the Hebrew teacher, a Dr. Stein, simply apoplectic. Pounding the table thunderously, he expostulated that what I had done is just about the worst thing a Jew can do!

If looks could kill, the daggers in his eyes would surely have dispatched me, wherefore I dread to think what he would have done if he had known that my deed was brazenly mischievous, rather than ignorantly inadvertent. Alas, Dr. Stein's God allowed him to perish in the Holocaust, along with millions of others, being the God who countenanced Abraham's readiness to sacrifice Isaac as a test of his fealty.

When still a boy, I found it even more abhorrent morally that the homicidal sacrifice of an innocent person received its apotheosis in Christianity: God, the supposedly loving father, permitted his supposedly incarnated only son to be sacrificed on the cross in order to achieve redemption for the entire human race, including forgiveness for the purported *original sin* of each of us. In this scenario, the transgressions committed by each and every one of us are forgiven, *not* by our own efforts to make amends, if possible, to those whom we have wronged, and by asking their forgiveness; instead, atonement for the transgressions of any and all of us is purchased by the excruciating death of a person who obviously did not commit them, *even if that person was willing to suffer crucifixion for that purpose*, as Jesus reportedly was.

Likewise morally repugnant to me is the claim that when we achieve extrauterine existence as neonates, we are already guilty of *original sin*, except for the Virgin Mary, who—according to the Roman Catholic doctrine of the Immaculate Conception—was *not* saddled with original sin. It is psychologically significant, I think, à propos of original sin, that Augustine called attention to the following biological fact, although it is *morally irrelevant*: We are each born between feces and urine (*"Inter faeces et urinam nascemur"*).

I see no merit in the complaint of historical myopia against my moral objections to the divine covenants, injunctions, and eschatologies in both the Old and New Testaments: I am accused of being unmindful of the historical contexts and ancient times at which these notions were enunciated. But the charge of historical parochialism against my moral dismay boomerangs, precisely because the complaint does not cohere with the religious claim that the biblical norms are *eternally valid, absolute verities, rather than time-bound*! Indeed, proponents of divinely revealed ethical commandments have complained that admittedly time-bound ethical prescriptions invite pernicious ethical relativism.

The notion of God the Father as the capricious monarch, who may need to be implored for favors to do the right thing, also crops out in the essential role assigned to *intercessions* by the Virgin Mary or by saints. Thus, Pope John Paul II extolled the role of Mary as supposed *mediatrix*, and he intensified Marian piety to the point of engaging in truly bizarre causal reasoning.

On May 13, 1981, a Turk named Mehmet Ali Ağca attempted to assassinate this pope in Rome. Oddly, the pontiff considered it highly significant that May 13, 1981, was the anniversary of the supposed first apparition of the Virgin near Fatima, Portugal, in 1917. On May 13 of that year, three Portuguese children tending sheep reportedly told of having seen a vision of a lady in a cove. According to the children, the lady, dressed in a white gown and veil, told them to come there on the thirteenth day of each month until the following October, when she would tell them who she was. On October 13, she said that she was the Lady of the Rosary and told the children to recite the rosary every day.[2]

I cannot help but wonder why the Virgin chose to appear to highly suggestible and impressionable children, rather than, far *more convincingly*, to, say, an avowed atheist or even to a savvy Roman Catholic psychiatrist familiar with the malleability of human perception and memory, and with the confabulatory bending of memories by expectations.

Yet after the Turk's homicidal attempt of May 13, 1981, John Paul II considered it simply evident that his life had been spared because a "motherly hand," that is, the Virgin Mary, had interceded to save him from assassination.[3] But why did he suppose that God would have let him be killed *without* that purported intercession?

In my writings on fallacious causal inferences, I called attention to the error of inferring a *causal connection* between two events from a mere *thematic kinship* between them. Therefore, I dubbed this inferential error "the thematic affinity fallacy."[4] Alas, the pontiff committed just that fallacy when he took the *anniversary relation* between two events, which had both occurred on a May 13, to qualify as evidence for a *causal connection* between them.

Unaware of his blunder, on May 13, 1982, exactly a year after the unsuccessful assassination attempt, the pontiff went to Fatima to thank Mary for having saved his life, but leaving unexplained why it required *Mary's intercession* to prevent his death, rather than just the autonomous action of the loving Almighty God the Father himself.

Not to be outdone in astute causal inference by the pope, Nancy Reagan opined that John Hinckley's bullet missed her husband's heart because God sat on the president's shoulder, as it were. Alas, in her public religious tribute, Mrs. Reagan overlooked very *callously* that the same God allowed another one of Hinckley's bullets to rip off a major portion of the brain of Jim Brady, the president's press secretary, whose wife then became the nemesis of the National Rifle Association.

Decades after my early ethical disenchantment with theism, I was flabbergasted by the theistic apologias for the Holocaust offered by the Jewish philosopher Martin Buber in his "eclipse of God" doctrine, and far worse, I was appalled by the outright justification of the Holocaust championed by the Chief Orthodox Rabbi of the United Kingdom, Lord Immanuel Jakobovitz, who enjoyed the esteem of his highly placed friend Margaret Thatcher to the point of owing his knighthood and peerage to her.

In a 1987 article in the *Times* (London), this rabbi deemed the Nazi Holocaust to be divine punishment for the religious apostasy of the German Jews who founded and practiced assimilationist Reform Judaism. This "idol of individual assimilation," he declared almost gleefully, "was eventually melted down and incinerated in the crematoria of Auschwitz."[5] On Jakobovitz's obscene theodicy, the SS men who implemented the "final solution" can exculpate themselves as merely having been the instruments of the God of Moses.

Unencumbered by contrary facts, Jakobovitz ignored that the Nazis also incinerated devoutly Orthodox Jews from all over central Europe, not just the assimilationist German Jews. The God whom Jakobovitz depicted strikes me as a sadistic satanic monster deserving of cosmic loathing, rather than of worship and love. Yet Chief Rabbi Jakobovitz had no monopoly on the view that the Holocaust was justly deserved on Jewish religious grounds.

In a 1980 book titled *Faith and Science*, the late ultraorthodox Brooklyn rabbi Menachem Schneerson, who had studied engineering at the Sorbonne in Paris, obscenely informed us that, by permitting the Holocaust, God cut off the gangrenous arm of the Jewish people. Such was the compassion felt for his coreligionists by this devout purveyor of piety! Incidentally, his followers hailed him as the soon-to-be Jewish Messiah, and according to a *New York Times* advertisement, they expected him to be resurrected shortly after his death. They are still waiting.

Thus, on Schneerson's outrageous view, those who perished in the crematoria were just detritus sloughed off the Jewish body politic by their God. I have discussed these matters in a forty-page article, "The Poverty of Theistic Morality," which was published in 1995.[6] In that paper, I offered a vigorous defense of secular humanism, which has been embraced by my good friend John Moossy, an academic neuropathologist, in his forthcoming autobiography.

My central thesis therein was twofold: (1) theism and atheism are equally sterile as a theoretical foundation for concrete norms of ethical conduct; and (2) motivationally, belief in either theism or atheism is *far too crude* a touchstone to correlate with civilized moral conduct on the personal, social, or national level. In the parlance of the received androcentric idiom, the brotherhood of man does *not* depend on the fatherhood of God, either normatively or motivationally. It is time, I claim, that this major lesson be heeded widely in word and deed, especially by those who are at the levers of power and *who vociferously deny it* (p. 240).

Examples are George W. Bush, a born-again Christian, and the sanctimonious, theocratic Jewish senator Joseph Lieberman of Connecticut. In this country, religious belief has become abhorrently coercive politically as a hallmark of good citizenship. This attitude, familiar from the religious Right, is epitomized by the late Richard John Neuhaus, a convert to Roman Catholicism who became a priest and edited a journal called *First Things*.

In a 1991 article there titled "Can Atheists Be Good Citizens?"[7] Neuhaus not only answers this question negatively but gives *rashly insolent* bad reasons for it. First, Neuhaus acknowledges grudgingly that Sidney Hook, a lifelong ardent secular humanist and atheist, was a dedicated, fearless critic of totalitarianism for decades and received the Medal of Freedom from the then president of the United States, Ronald Reagan. Wasn't Hook therefore a good citizen? Neuhaus fell into a fairly common confusion between the *semantic content* of a doctrine and the degree of *epistemological confidence* that a given supporter of the doctrine may have in it, believing incorrectly that Hook could not have been a full-fledged atheist because he did not claim *certitude* for his belief. The *content* of theism is the assertion that there is a personal God with specified attributes, while the *content* of atheism is the denial of that claim. But neither content tells us with what *degree of confidence* a given proponent avows the given tenet epistemologically.

The Roman Catholic Church claims absolute dogmatic, irrevocable certitude for its theism, while the late, appallingly crude Madelyn Murray O'Hair had proclaimed her atheism just as dogmatically with certainty. Alternatively, both theism and atheism alike can be espoused with varying *lesser* degrees of cognitive confidence.

Theoretical beliefs, however well supported by known evidence, are still fallible or revocable because of potentially adverse evidence. It is therefore the better part of wisdom to stop short of espousing one's hypotheses as irrevocably established. Thus, Sidney Hook, Sigmund Freud, Albert Einstein, and Bertrand Russell, among others, adopted this less-than-dogmatic attitude toward their belief in atheism (i.e., the denial of the existence of the personal God of theism), but without tampering with its semantic content. Notably, their lack of dogmatism therefore did *not* constitute a watering down of their atheism into the quite different doctrine called "agnosticism." Some theists obtusely, if not deliberately, misdepicted Einstein as a theist because he spoke of God *metaphorically*, as in his espousal of causal determinism, when he declared that "God does not play dice" with the universe and "the Lord is shrewd, but not mischievous."

In its contemporary technical meaning, as distinct from Thomas Huxley's original meaning, agnosticism *does not logically rule out* either theism or atheism: it pointedly makes no claim as to the existence of God one way or the other, even tentatively, because it regards the question as unanswerable in principle. Obviously, neither theists nor atheists are agnostics. And atheists clearly disavow agnosticism no less than theists do. Epistemologically, agnosticism deems theism and atheism to be *alike* unwarranted.

This state of affairs was untutoredly overlooked by Judge Robert Bork during his unsuccessful 1987 confirmation hearings to become a US Supreme Court justice. Eyes flashing, Bork irrelevantly told the senators fawningly that he was not an agnostic, presumably to convey that he was not irreligious, let alone an atheist. He did so in order to pander to the pro-religious prejudices of many senators, even though he knew very well that the US Constitution, to its permanent credit, states explicitly: "No religious test shall ever be required as a qualification to any office or public trust under the United States" (Article VI, Clause 3), which is the only passage in it in which religion is mentioned at all. But ironically and unintendedly (*malgré lui*), Bork's avowed *rejection* of agnosticism does not logically rule out his being an atheist.

Having wrongly assumed that atheism must be held to be true with certainty by its champions, Neuhaus concluded that, since Hook was a *fallibalist,* his rejection of theism must be tantamount to agnosticism after all.[8] But this conclusion is false: Hook's commitment, though fallibalist, was to atheism, not to agnosticism.

Deplorably, the ideologically coercive public religiosity in this country, which features much moral *self*-congratulation, has intimidated many atheists into mislabeling themselves as "agnostics." Theistic moral and civic arrogance are even built into the secondary dictionary meaning of the term "atheist." Thus, the 1958 unabridged second edition of *Webster's New Dictionary of the English Language* defines the noun "atheist" as follows: "A godless person, one who lives immorally as if disbelieving in God." Note the semantic theistic moral self-congratulation in the claim that "disbelieving in God" makes for immorality! No wonder that some pusillanimous atheists, who shun the label atheist as opprobrious, misleadingly call themselves "agnostic."

Returning to Neuhaus, the principal thesis of his 1991 article was that atheists cannot be good citizens. Therefore, Hook's actual atheism commits Neuhaus willy-nilly to the further conclusion that Sidney Hook was *philosophically unfit* to be a good citizen. But Neuhaus's central argument runs afoul of the *moral sterility* of theism. This ethical infertility, in turn, undermines his attack on the separation of church and state, as well as his irate indictment of those *religious* people who support that separation. I applaud such religious people and their frequent rejection of moral self-congratulation as a form of the "sin of pride." Indeed, I emphatically regard them as comrades-in-arms, as it were, in the quest for a civilized society.

But let me return to my boyhood. Although I was already inclined to embrace atheism, I am indeed indebted to a rabbi at the conservative synagogue I attended with my parents (in Cologne on its street Glockengasse) for having initiated my positive interest in philosophy.

In his sermons, the rabbi often talked about such philosophers as Kant and Hegel. Thereby, I became greatly interested in what they thought and in a series of German books called *Kultur der Gegenwart*, or Contemporary Culture. Thus, I picked up the volume on philosophy, which dealt with speculations about the origin of the world, Aristotle's four sorts of causes, and an array of related topics. I found it highly fascinating and decided that this was the field I wanted to study. At the time, I was about twelve years old.

It seemed to me that Arthur Schopenhauer was an excellent stylist and a very lucid writer, in striking contrast to Kant and Hegel. What I could understand of his writings completed my disenchantment with theism by the time of my bar mitzvah in 1936 at age thirteen. But I went through with that Jewish religious rite of passage at that stage of Nazi tyranny, when the Nazis had been in power for three years. My refusal to have a bar mitzvah would have betokened a lack of solidarity with the beleaguered Jewish community in Cologne, even though one Zionist youth group there (*Hashomer Hatzair*, meaning "The Young Guardians" in Hebrew) was openly atheist.

Schopenhauer's atheistic Buddhism with its emphasis on compassion for sentient beings appealed to me despite his general pessimism. As a pubescent boy, I eagerly picked up his essay "On Women." But I thought that his wholesale disparagement of women was nearly deranged. Soon I learned that, for whatever reason, his mother, a socialite hostess and wife of his well-to-do father, had completely rejected him. Hearsay also had it that he lost out in competing with the poet Lord Byron for the same woman, a defeat that, if not apocryphal, may also have been a factor. When my own mother saw me reading Schopenhauer's diatribe against women, she sternly informed me: "*That* essay does not belong in *this* house."

It was clear to me, even in my nascent dabbling in philosophy, that one could not get credible knowledge about such issues as the origin of the universe if one did not have a good grounding in the pertinent natural sciences. The tradition in philosophy of cleverly spinning ideas *on such topics* out of one's imagination struck me as clearly inadequate, if not retrograde. As we know, there have been a number of celebrated philosophers who have also been great natural scientists or mathematicians.

Yet a good many other philosophers were only very poorly or marginally educated in the natural sciences and thus wrote rather primitively about them. Present-day examples of such philosophers, who in my view have made abysmally ignorant and wrongheaded claims about the methods and content of the natural sciences, are Hans-Georg Gadamer and Jürgen Habermas in Germany. Thus, Gadamer wrote that "[i]t is the aim of science to so objectify experience that it no longer contains any historical element." But this claim is belied even by the elastic fatigue of ordinary rubber bands, and by the "hysteresis" behavior of ferromagnetic substances, not to speak of physical *cosmogony*, the grandest history of all. I have taken these two philosophers to task on this score in my 1984 book *The Foundations of Psychoanalysis: A Philosophical Critique* and in a journal article published in 2000.[9]

By age twelve, I realized that, in Germany at that time, very few of the people who obtained doctorates at universities then went on to attain professorships at them, because there were not enough jobs: so, in hoping to acquire a doctorate, I had no expectations of becoming a professor and actually making my living as an academic philosopher. But I did crave a university education at the level of the doctorate.

By the time I arrived in the United States at age fourteen, I had had four years of French in my Cologne high school, but only a few months of English. Upon our arrival in New York City, I went to DeWitt Clinton High School, in the remote Bronx, for the simple reason that there was one student in it, Josef Helmreich (now Helmrich), whom I had known from school in Cologne. It took me an hour and a half by subway each way daily to get there and back from where I lived in South Brooklyn. Yet although my knowledge of English was less than rudimentary, I was immediately put into a fourth-year English class in which we were reading Shakespeare, presumably because I was fifteen years old by then. Linguistically, it was a desperate situation at the beginning. But I was rescued by an edition of Shakespeare that had the English original on one page and a superb German translation on its opposite page. Indeed, I knew that some Germans had oddly claimed him as their own: "*Unser* Shakespeare," *our* Shakespeare.

In an honors class on trigonometry, there was a fellow student, Robert S. Cohen, who turned out to have a decisive influence on the shaping of my undergraduate and graduate education and who, I believed, became my very good friend for many years. Unfortunately, I had to end this friendship irrevocably in 2007. Yet I shall acknowledge his various good deeds below. And I shall refer to him by his formal name, since there are enough other Bob Cohens with whom he might be misidentified.

Having been held back by my need to learn English, I spent a little over a year at DeWitt Clinton High School. In the meantime, Cohen graduated in 1939 and had gone on to become a freshman at Wesleyan University, a liberal arts college in Middletown, Connecticut. But on a return visit he made to DeWitt Clinton, he and I happened to run into each other and struck up a conversation about our respective academic interests. We then discovered that our intellectual interests coincided in the philosophy of science, as informed by physics and mathematics.

Upon returning from his brief visit, Robert S. Cohen, though still only a freshman, took the initiative to secure a scholarship for me from the then Wesleyan director of admissions, Victor L. Butterfield. I started college there in September 1940, and graduated in October 1943, with a double major in mathematics and philosophy, receiving high honors and high distinction in both after writing an honors thesis in the former and taking a comprehensive examination in the latter.

Throughout my Wesleyan days, Robert Cohen and I were roommates and indeed soul mates. At Wesleyan, I had some teachers whom I cherished immensely (e.g., the philosopher Cornelius Krusé and the mathematician Burton Camp) and with whom I maintained warm relations for the rest of their lives. Curiously, in some respects, my life on the Wesleyan campus was very gratifying, while in others, it was quite hurtful and alienating. Intellectually and emotionally, my Wesleyan experience—both wonderful and often quite painful—was profoundly formative for the kid from Cologne whose transition to the culture of the United States and to the then Wesleyan University was not smooth.

But, memorably, even as an undergraduate, I was disappointed that, in the valuation of liberal education, the ethos at Wesleyan, and also at some other liberal arts colleges, tended to subordinate the scientific, rationalistic way of thought to nonrational modes, which were prevalent in the humanities. Thus, one highly respected and influential professor of English there wrote misguidedly that the greatest threat to the humanities comes from science. The mantra about the centrality of the humanities in liberal education became a dreary cult or fetish from which I sought deliverance.

Victor Butterfield, who later became the president of the college, was a Harvard PhD in philosophy, who had had the opportunity to study there with Alfred North Whitehead—the author of *Science and the Modern World*—and was, unfortunately, one of the principal purveyors of the cult of downgrading science ideologically in the face of the enormous role of science in our culture. No wonder that a deep temperamental and philosophical chasm separated me from him, although I always remained grateful to him for having granted me a freshman scholarship.

Butterfield was an apologist for compulsory chapel, which I opposed but which did give me the opportunity of hearing noted theologians like Reinhold Niebuhr and Paul Tillich. Yet I was unable to distinguish Tillich's theology of "ultimate concerns" from sheer atheism, except for Tillich's theological vocabulary. Nor could I understand Niebuhr's metaphoric proclamation that "Christ stands at the edge of history," which he said with a flourish. Like most students, believers and nonbelievers alike, I resented compulsory chapel, which Butterfield justified as valuable educationally.

But I did definitely disapprove of those students who displayed their protest rudely by pointedly reading newspapers in the pews. And, in lieu of compulsory chapel, I surely would have welcomed a required course in comparative religion.

One of the chapel speakers, who was unfortunately invited repeatedly, was a professor of sociology at Duke University. His name escapes me now, but he is otherwise very memorable to me qua purveyor of obscurantism. As he claimed, exact mathematical solutions of the hydrodynamic equations governing the motions of oceangoing vessels are not available. He then triumphantly offered the non sequitur that God exists, a God whose twilight existence is very precarious, as the speaker overlooked, because his God dwells in the *shifting* gap of human ignorance.

I was also alienated by Nathan Pusey, a classicist who later became president of Harvard, who taught a humanities course at Wesleyan and was of the same bent of mind as Victor Butterfield. Not surprisingly, Pusey later clashed with the Harvard faculty on the issue of religious observance on campus.

As a Wesleyan student, I was very poor financially. Soon after my parents, my two younger siblings, and I came to the United States, my father had to have major neurosurgery for spinal arthritis. The operation was quite unsuccessful, and he became partly paralyzed as well as unable to work for a living. My mother's very modest salary in her job was hardly adequate. The less so, since my sister, Sue, and brother, Norbert, also aspired to receiving a higher education, which I was determined they should get no less than I. Fortunately, despite terrible odds, my aspiration for them was later fulfilled: Sue became the chairperson of a department of modern languages in the high school system in upscale Great Neck, NY, and I was able to raise money for Norbert from the Dazian Foundation to go to medical school so as to become a psychiatrist.

The Wesleyan campus was visited rather regularly by a Mr. Lewis Fox, a member of a family that owned a successful department store in nearby Hartford, Connecticut. It was known that Mr. Fox bestowed scholarships and other largesse on students who met his criteria for being of good character: being religious, not smoking, and—of course—foreswearing sex. Although I never smoked, I was persona non grata with him on the other counts. Despite money being dangled in front of me and my being badly in need of it, I made no bones about where I stood, and indeed was somewhat vocal about it.

I was simply not going to compromise my integrity for a golden calf. Besides, I was well aware that a number of the students who were in Mr. Fox's good graces tailored their behavior opportunistically to his expectations whenever he was visiting the campus. Apparently, he was comfortable with being coercive. After a considerable time, he apparently decided that I was not totally undeserving after all and granted me a check for one hundred dollars with words to that effect.

During my Wesleyan studies as a mathematics major, the navy sent V12 cadets to the university for training in mathematics and science. Some of my fellow undergraduate majors in that subject were given the remunerative job of teaching these cadets. Although my grades in mathematics were clearly superior to theirs—indeed, I received high distinction for an honors thesis on Fourier series in the field—my application for a like teaching appointment was denied. Since I desperately needed that income, I went to see Victor Butterfield to ask why I was not also appointed.

His answer, which is indelibly etched in my memory, was "you are terribly philosophical," as if my pedagogic judgment was too unsophisticated to be aware of the obviously practical curricular needs of the V12 cadets. It was and remains clear to me that Butterfield's avowed reason for having rejected me was a euphemism, if not a code word, for the attitude that the cultural profile I would project is not one that he wanted the V12 cadets to carry away from their Wesleyan experience.

I appreciate all too well that the Wesleyan of today is light-years away from anything like that, having already greatly evolved culturally even by the time my daughter Barbara was there in the class of 1979. But, for me, that cannot undo the past, which still rankles, I must admit.

I had taken enough physics by 1943 to work then, for a short time, in a war research unit on vacuum tube development and radar at Columbia University in New York (housed at the Empire State

Building) after my graduation from college. But, soon thereafter, I entered the US Army as a draftee and received full infantry training at Camp Wheeler, Georgia, as well as my US citizenship in 1944. I was then sent for US Army military combat intelligence training at Camp Ritchie, Maryland. There, other German-speaking refugees, such as Thomas Mann's son Klaus Mann and Sigmund Freud's nephew Harry Freud, were taught, as I was, how to interrogate senior military prisoners of war and highly placed civilians from Hitler's Germany. As a member of a mobile field interrogation unit (MFIU #2) staffed by "Ritchie Boys," I went to France and England.

But on the day after the Soviet Army had conquered the Nazi capital, my unit became part of the advance party of the US Army in Berlin, where we were stationed in Heinrich Himmler's dreaded former suburban Gestapo headquarters, at a beautiful lake called Wannsee. The "Wannsee-Haus" had been the site of the infamous January 1942 conference at which—in a session of only ninety minutes—fifteen top Nazi officials planned the genocide of millions of European Jews and other "undesirables" under the chairmanship of SS security chief Reinhard Heydrich. He was killed by Czech commandos, whereupon the Nazis ordered the retaliatory total destruction of the Czech village of Lidice. Adolf Eichmann also attended the Wannsee session to plan the logistics of the Holocaust, and his verbatim transcript of it was found after the war in the archives of the German Foreign Office. Thus, while stationed at the Wannsee House, my unit had no inkling that it was the locus of devising the "final solution" which Göring had decreed to Heinrich Himmler to implement.

In Berlin, at age twenty-two, I interrogated an array of German Army (*Wehrmacht*) and SS personnel, including generals, members of the Nazi German professoriate, as well as other very highly placed civilians, such as the director of Berlin's famed Charité Hospital, Dr. Ernst Ferdinand Sauerbruch, who had been a pioneer of lung surgery. Alas, he had also been a leading figure in the Nazi medical establishment, when he became the personal physician of Joseph Goebbels and authorized medical experiments on inmates of concentration camps (cf. the German magazine *Der Spiegel*, February 2, 2009, p. 60), although he claimed to have had second thoughts later on and even questioned Hitler's sanity.

As Sauerbruch told me, he, and Max Planck, the German Nobel physicist and pioneer of quantum physics, along with other luminaries, had belonged to a group that met weekly for discussion on Wednesdays (the so-called *Mittwochsgesellschaft* or Wednesday Club). Reportedly, four of its members had been executed by the Nazis for complicity in the failed attempt on Hitler's life in July 1944. Through Sauerbruch, I got to know Nelly Planck, the physician widow of Max Planck's son, Erwin, who had likewise been executed for that reason.

Max Planck was the first German physicist to be invited to England shortly after World War II, where Nelly accompanied him as his medical doctor. She gave me a copy of his German book *Wege zur physikalischen Erkenntnis* (*Pathways to Physical Knowledge*), which sets forth his philosophy of science. I maintained a cordial correspondence with her after returning home to the United States until her death in the mid-1970s.

One of my experiences as an interrogator was formative for me as an appalling case of *trahison des clercs* (the betrayal of intellectual standards by academics). I was questioning the renowned mathematics professor Ludwig Bieberbach, from whom the eminent philosopher of science Carl G. Hempel, later my PhD dissertation supervisor at Yale, had had a course at the University of Berlin. As I learned later from Hempel, Bieberbach had lectured to his students at the University of Berlin in Nazi SA uniform even before Hitler's ascent to power in 1933.

I became aware that Bieberbach had delivered an outrageously ill-reasoned published lecture at Germany's University of Heidelberg on "The Racial Foundations of Science," or "*Die völkische Verwurzelung der Wissenschaft.*" In it, he argued, in absurdly contorted fashion, that in mathematics, the Nordic mind operates wholesomely in a spatio-geometric mode, whereas the Jewish mind functions only abstractly, in a reality-alien fashion. Apparently, such glaring counterexamples as the *geometrization of gravitation* by the Jewish physicist Albert Einstein did not give him any pause. This brazen piece of sloppy, dishonest, and tendentious argument left me permanently with a low threshold of indignation in the face of any such argument.

Even more contemptibly, during my interrogation of him, Bieberbach lamely tried to convince me that he really did not support the Nazis, since he took exception to the cutback in funding for *basic* mathematical research at the height of World War II. He was not even embarrassed when I remarked that, during a war, any government whatsoever has to have priorities in resource allocation.

As a final example of the intellectual dishonesty I encountered, let me mention German Army general Fritz von Bayerlein, who commanded the tank forces that achieved a breakthrough in the Ardennes Forest during the final offensive of General von Rundstedt in 1944–1945. In my interrogation of him, von Bayerlein told me that he fought for Hitler to save *Christian* civilization from the atheistic Soviets, as if genuine Christianity would have flourished after a Nazi victory!

One quasi-comic episode during the military interrogation phase of my experience deserves mention. Our mobile unit had the services of German prisoners of war of diverse backgrounds. Some of them worked as our drivers, while others had more valuable qualifications. I well remember one of them, a very able genealogist whose last name was an eponym of the famous eighteenth-century Swiss mathematician Leonhard Euler, and hence especially memorable to me. Our man Euler had a stunning command of the genealogical trees of prominent German families in various walks of life, including the ancestry of senior members of the traditional officer class.

Euler's treasure trove of knowledge comprised scandalous episodes, and even some details of dubious liaisons and escapades. Before interrogating officers from such families, my fellow interrogators and I would have the great benefit of a briefing session with Euler, making a special point of rehearsing compromising information to the point of being able to rattle off selected tidbits, as if in casual conversation. Even shrewd interrogatees, who were prepared to tell us no more than their name, rank, and serial number, as required by the

Geneva Convention, were simply bowled over by the cornucopia of our knowledge so as to loosen their tongues.

In one deliciously unforgettable case, I had to keep myself from breaking out in laughter, because the interrogatee declared: "We are supposed to act like enemies, but I cannot refrain from expressing my immense admiration for the gigantic achievement of the American Military Intelligence Service, which must be unprecedented in the annals of warfare," or words to that effect.

After that service in Berlin, I was honorably discharged from the United States Army in 1946.

Upon my discharge, I embarked upon graduate study at Yale University that I had only very briefly begun there before my military service. For my first two years, I studied philosophy and physics there in parallel, receiving first a master of science degree in physics (1948) en route to completing a PhD in philosophy, with a focus on the philosophy of science.

In June 1949, just before the start of my PhD dissertation year at Yale, I married the very lovely and intelligent Thelma Braverman, having met her in Brooklyn while I was still in my last year at DeWitt Clinton High School. For a couple of years before our marriage, I was one of several Yale graduate students who enjoyed the hospitality of Robert S. Cohen and his wife, Robin, in their home in Hamden, Connecticut, near New Haven, which we all ran as a kind of co-op, taking turns at cooking and other household chores under Robin's direction. At first, my apprehension of cooking successfully for this group exceeded my concern to do well on the five written comprehensive examinations for the PhD in philosophy.

During my first year of graduate study in 1946, I met the symbolic logician Ruth Barcan Marcus, who received her degree in 1947, soon became distinguished, joined the Yale philosophy faculty for good in 1973, and later became truly eminent internationally. Ruth has been a very dear friend of mine for decades.

Thelma and I have a beautiful and very intelligent daughter, Barbara, who was born in 1957 and graduated from Wesleyan in the class of 1979 with a major in studio arts. After Wesleyan, Barbara took a master's degree in photography at Rochester Institute of Technology, where she met a fellow graduate student, Ronald Gregory, a splendid man, whom she married in 1986. In 1988, they became parents of lovable, fine identical twin boys, Benjamin and Eli, and they make their home in Rockville, Maryland. Barbara has since been a producer of TV documentaries for Montgomery County, Maryland. In 2007, Ben enrolled as a freshman at the University of Pittsburgh, having lost a year from a nearly fatal car accident in August 2006. At that time, Eli began his freshman year at Indiana University in Bloomington, Indiana.

For many years, Thelma, now retired, was a first-rate pedagogue as a teacher of (high school) mathematics, while running our home in the Point Breeze section of Pittsburgh, after an initial twenty-five years in its eastern suburb of Churchill Borough. Throughout my long academic career, both past and present, she has sacrificed by generously accommodating in our lives the demands of my work on my time.

My doctoral thesis, which I completed by September of 1950 before formally receiving the degree in February 1951, was directed by Professor Carl G. Hempel and dealt with issues first posed, some 2,500 years earlier, by the ancient Greek, Zeno of Elea, in his celebrated paradoxes of motion and extension, which I treated, of course, in the context of modern mathematical physics. My PhD thesis was titled "The Philosophy of Continuity." One of its main inspirations was Bertrand Russell's 1926 declaration: "Zeno's arguments, in some form, have afforded grounds for almost all the theories of space and time and infinity which have been constructed from his day to our own."[10]

By the mid-1960s, this dissertation had evolved into my three Matchette Lectures on Zeno's Paradoxes of Motion and Extension in the context of modern mathematical physics. The venue for them was Wesleyan University in Connecticut, my undergraduate alma mater. I developed them into my book *Modern Science and Zeno's Paradoxes*, which was published in 1967 by the Wesleyan University Press and is dedicated to our daughter, Barbara.[11] But spurred on by H. Weyl's Zenonian question "Can an infinitude of operations be performed in a finite time?" I immediately prepared a second edition, which appeared in England in quick succession in 1968 under the imprint of George Allen & Unwin (London).[12]

I addressed Weyl's question very specifically in my two papers "Are 'Infinity Machines' Paradoxical?" and "Can an Infinitude of Operations Be Performed in a Finite Time?" having delivered the latter as a Monday Lecture at the University of Chicago in April 1968.[13]

Carl Hempel was called "Peter" by his friends for decades. So, as a mark of my enormous affection for him, I did likewise. When he was still at Queens College in New York City, the Yale Philosophy Department refused to give tenure to the ethicist Charles Stevenson, and it was no secret that Stevenson's *emotivist* conception of ethics had been the decisive factor. He had been the only representative of logical empiricism there when he had to leave. Therefore, when I became a graduate student at Yale in 1946, I promptly expressed strong disappointment that its large philosophy department did not include *any* representative of logical empiricism, a school of philosophy that was then quite influential in some quarters in the United States, although I was surely *not* an adherent of the phenomenalist ontology of the early logical empiricists.

Thereupon, the aesthetician Monroe Beardsley, then an assistant professor in the department, fibbed to the chairman, Brand Blanshard, that I was about to spearhead an *exodus* of graduate students to get a more balanced education elsewhere. Promptly, Peter Hempel, then a renowned logical empiricist, was invited to give a colloquium in the department, and he soon agreed to join it as a tenured associate professor. Professor Blanshard told me at the time that he deplored Peter's then emotivist stance in ethics, yet considered him the most acceptable of the logical empiricists. After Peter had become established in the department, it was rated the top philosophy department in the nation at the time, a rating that plummeted, however, not long thereafter.

Hempel's teaching soon became legendary. Naturally, he left an indelible mark on my further philosophic development. After his years at Yale, and then retirement from Princeton, I was overjoyed when he joined our Pittsburgh faculty as a university professor of philosophy in

1977, and graced it with his presence along with his very endearing wife, Diane, until 1985, when poor eyesight forced him to retire. Though Peter had begun as a logical empiricist, he became a vigorous and unsparing *internal* critic of his own philosophic patrimony and contributed decisively to its transformation.

Hempel distilled from the logical empiricist tradition what, I believe, was best in it. *I maintain that there is indeed a valuable residue from logical empiricism.* Alas, we are living in a period of historically untutored retroactive derogatory mythologizing of its tenets, mythmaking that has become fashionable in the service of ideological special pleading.

Thus, the arch-positivist Philipp Frank had soundly emphasized that sometimes a new theory *fails* to explain phenomena that *were* explained by its superseded predecessor. Yet Thomas Kuhn, an outspoken critic of logical empiricism, has been improperly credited with the discovery of just that phenomenon of explanatory forfeiture, a misattribution codified under the misnomer "Kuhn *loss*." Indeed, Frank had explicitly called attention to the phenomenon in a Boston lecture at the American Academy of Arts and Sciences attended by both Kuhn and myself very early in our careers. Many years later, I related this fact to Kuhn, who realized then that he had unwittingly internalized Frank's point only to "discover" it later.

As for the valuable residue from logical empiricism, I think that its pejorative notion of *pseudoproblem* is a precious legacy from it, because that notion can usefully derogate a question that rests on an ill-founded or demonstrably false presupposition. For example, Leibniz's question "Why is there anything at all, rather than just nothing contingent?" is a pseudoproblem, I claim, because it is predicated on an inveterate but *baseless* Christian ontological assumption, going back to the second century CE: the doctrine that, in the "natural" course of things, there should be and indeed would be just nothing contingent in the absence of an overriding (divine) cause (reason).

My reasons for this adverse judgment are set forth in careful detail in my lengthy essay "The Poverty of Theistic Cosmology," in the December

2004 *British Journal for the Philosophy of Science*, which is reprinted in this volume.[14] I extended its argument in my August 2007 presidential address on the topic "Why Is There a Universe AT ALL Rather Than Just Nothing?" in Beijing, China, at the quadrennial World Congress of the Division of Logic, Methodology, and Philosophy of Science, one of the two units of the International Union of History and Philosophy of Science, which will receive mention in the appendix below.

I should point out that, in advocating the use of the notion "pseudoproblem" when appropriate, I definitely do not intend to hark back to early logical positivist indictments of "meaninglessness" on the basis of the "verifiability theory" of factual meaning, made familiar to Anglo-Saxon readers by A. J. Ayer's book *Language, Truth and Logic*.[15]

The most decisive influence on the direction of my work during the first twenty-five years after receiving my doctorate came from Hans Reichenbach, Hempel's own graduate teacher at the University of Berlin. While I was still a graduate student, Robert S. Cohen, who was a serendipitous bibliophile, brought me an out-of-print copy of Hans Reichenbach's classic 1928 German work on space-time philosophy, titled *Philosophie der Raum-Zeit-Lehre*. This book did not become available in English translation until 1958 under the title *The Philosophy of Space and Time*. When I read it in German, its effect on me was truly electrifying, and I was swept into working on the sort of issues that Reichenbach had treated so magisterially in that book. Years later, I developed a cordial professional and personal friendship of long duration with his widow, Maria Reichenbach, who was an indefatigable translator of his German writings into English.

After I had been publishing on the topic of space-time philosophy, well-meaning friends pointed out to me that even among philosophy colleagues who were broadly interested in the philosophy of science, only a minority had the technical scientific competence, at that juncture, to understand my writings on space-time philosophy. They opined that I could achieve professional recognition much more expeditiously if I wrote on generally much more accessible topics such as scientific explanation. But I forsook that practical advice and have never regretted it. Indeed, for a quarter of a century after my doctorate, my writings focused primarily, though by no means entirely, on the philosophy of physics (space-time).

Thus, in 1963, I published my book *Philosophical Problems of Space and Time* (*PPST*), under the imprint of A. A. Knopf and dedicated it to my wife, Thelma.[16] This first edition was followed, a decade later in 1973, by a second, a treatise of about 885 pages, nearly twice the size of the first. This new edition appeared as volume 12 of the Boston Studies in the Philosophy of Science coedited by Robert S. Cohen and Marx Wartofsky.[17] In a review of this second edition in the *Journal of Philosophy*, Lawrence Sklar, a well-recognized author in the field, declared:

> ... The core of this book, the first edition of *Philosophical Problems of Space and Time*, has received as much attention in the journals as any work of a living philosopher. . . . Over-all, the new additions to the basic text reveal once more Grünbaum's intense dedication to the fundamental problems of the philosophy of space and time, his careful and thorough pursuit of each and every detail essential to his over-all program, and his astonishing breadth of learning in the literature both of philosophy and physics.[18]

In June 1974, in the wake of the publication of the second edition of my *PPST*, physicist John Stachel, a well-respected expert on the general theory of relativity and a recognized authority on the whole corpus of Albert Einstein's life work, organized a conference of twenty-six participants on "Absolute and Relational Theories of Space and Space-Time." This conference was held in Andover, Massachusetts, under the auspices of the Boston University Institute of Relativity Studies, a research group of which Stachel was the dynamo. The Andover meeting's Proceedings appeared in a 1977 book titled *Foundations of Space-Time Theories* (coedited by J. Earman, C. Glymour, and Stachel) as volume 8 of the Minnesota Studies in the Philosophy of Science, published by the University of Minnesota Press.

Therein, Stachel reported in the opening sentence of his "Notes on the Andover Conference" (p.vii): "The Boston University Conference [in Andover, MA] was held to mark the issuance of the second, considerably revised and expanded edition of Adolf Grünbaum's seminal book *Philosophical Problems of Space and Time* [as vol. 12] in the *Boston Studies* series." But beyond his very generous organizational effort as a tribute to my *PPST*, John Stachel's personal friendship was *semper fidelis* (ever loyal): Thus, I was very touched when he journeyed to Pittsburgh from Boston just to attend the 1978 ceremony at which Pitt Chancellor Posvar appointed me chairman of our Center for Philosophy of Science, after my eighteen-year service as its founding director. Again, John contributed a very illuminating essay on "Special Relativity from Measuring Rods" to the 1983 *Festschrift Physics, Philosophy and Psychoanalysis: Essays in Honor of Adolf Grünbaum* (ed. R. S. Cohen and L. Laudan), whose "Grünbaum Editorial Committee" included my very distinguished Pitt colleagues C. G. Hempel, N. Rescher and W. C. Salmon, besides the two very savvy editors.

One result of the substantial size of the second edition of my PPST was that in Daniel Dennett's good-natured spoof, *The Philosopher's Lexicon*, the entry under "grünbaum"—which is German for "greentree"—reads: "n. (in German folklore) A tree which, when one of its fruits is bruised, produces another of the same shape, taste and texture but five times as large." Lightheartedly, decades later, when we bought a summer home in 2004 at Lake Deep Creek in McHenry in western Maryland, our daughter, Barbara, and son-in-law, Ron, named it "Greentree Haven."

During the Christmas/New Year holiday of 1959–1960, I was a participating speaker at a conference on "The Axiomatic Method," which had been organized by Alfred Tarski and Patrick Suppes and was held on the campus of the University of California, Berkeley. At a New Year's Eve party at the home of Alfred Tarski during that conference, I had the opportunity, which I also had at other conferences, to engage in conversation with the 1946 Nobel laureate physicist, Percy W. Bridgman, who had received that prize for his work on the behavior of matter under extremely high pressures.

But he also achieved world renown by his then highly influential 1927 book, *The Logic of Modern Physics*, in which he credited Einstein's special theory of relativity (hereafter "STR") with being the inspiration for his espousal of his "operational" account of scientific concept-formation in the empirical sciences. In a nutshell, he held that we mean by any concept just a set of operations, so that, in effect, the given concept is the counterpart of a corresponding set of operations. As for such mathematical concepts as transfinite cardinal numbers, they were "operationalized" by paper and pencil. But, soberingly, so were nonsensical sentences and the word salad of psychotics.

Bridgman proclaimed that intellectual clarity would surely be served if the operational mode of thought were adopted in *all fields of inquiry.* Einstein himself did not endorse Bridgman's notion as having captured a philosophical lesson spelled by his STR. Instead he favored Henri Poincaré's notion that theoretical concepts in science are free creations of the soaring human mind. I myself have offered a critical discussion of the claim that the STR supports Bridgman's operational concept-empiricism in my "Operationism and Relativity," in the October 1954 issue of the *Scientific Monthly*.[19] It was reprinted in Philipp Frank's edited volume *The Validation of Scientific Theories* in 1957 and again in 1961.[20]

Psychologists and economists, among others in the human sciences, thought Einstein had endorsed Bridgman's operational concept-empiricism. And being eager to emulate the hard sciences in their own quest for scientific respectability, they were eager to "operationalize" the concepts ingredient in their theories. Giving lip service to doing so became a ritualistic mantra in many quarters, so that "operationalizing" simply became synonymous with providing any sort of empirical anchorage.

But it turned out in studies by Rudolf Carnap and other "logical empiricists," such as the aforementioned Carl ("Peter") Hempel, that the tying down of theories conceptually to their empirical base was much more complicated than giving an "operational" empiricist pedigree of their theoretical concepts à la Bridgman's formulation of concept empiricism.

In 1961, at age eighty, while afflicted by bone cancer, Bridgman managed to complete a book, *A Sophisticate's Primer of Relativity*. Very sadly indeed, when the pain of his hopeless affliction became intolerable, he took his own life on August 20, having said that a person has a right to ask his doctor to end it for him because "It isn't decent for society to make a man do this thing himself."

Bridgman had had a lifelong association with Harvard as undergraduate, graduate student, and for nearly half a century as a member of its physics department. These collegial bonds were very strong in some cases. As I was told, Professor Wendell Furry in that department felt a special gratitude to him. During the McCarthyite political hysteria in this country, Bridgman, a Republican, had helped in protecting Furry's job, when the latter was being harassed for allegedly pro-Communist prior political associations.

Despite Bridgman's eminence, it appears that two publishers had rejected his *Primer* for publication, as I learned from him in early February 1961. During a visit to Pittsburgh by then president Victor Butterfield of Wesleyan University, he told me that Bridgman had expressed the wish to have the Wesleyan University Press publish it. On November 25, 1961, the press asked me whether I would read his *Primer* and advise it in less than a month as to its merits for publication. The Wesleyan press then accepted my unreservedly favorable recommendation that it be published.

About a year and a half earlier, on August 15, 1959, Bridgman had written me from his country home in Randolph, New Hampshire, amiably requesting a reprint of my 1955 paper "Logical and Philosophical Foundations of the Special Theory of Relativity."[21] As he explained, he was engaged in clarifying the fundamentals of this theory in his own mind.

Then in an almost full-page letter of January 26, 1961, some nine months before the request of the Wesleyan press that I referee his book, Bridgman wrote me again, avowedly to get the benefit of what he called my "extensive knowledge of the literature of relativity theory." In a two-page letter of February 3, 1961, I happily replied to him, addressing his questions and comments.

Later that year, I learned from the Wesleyan University Press that they would publish his book. Both its then director, Willard A. Lockwood, and its Wesleyan faculty advisor, Professor Robert Rosenbaum, a national leader in the college teaching of mathematics and a very fine person, invited me enthusiastically in December 1961 to write two supplements to Bridgman's *Primer*, a "Foreword" and an "Afterword," later renamed "Prologue" and "Epilogue," respectively, at the advice of the press. In 1962, it published his book, and there was also a Routledge Kegan Paul edition in England in 1963, each containing my prologue and epilogue. So much for my interaction with Bridgman, whom I had first met in December 1959 at a conference on "The Axiomatic Method" in Berkeley, California.

In 1963, almost contemporaneously with the appearance of the first edition of my *PPST*, Hilary Putnam, whom I had known for at least a decade as a friend, was prompted to publish a fifty-page article titled "An Examination of Grünbaum's Philosophy of Geometry." The burden of Putnam's essay was to show that "Grünbaum has in my opinion failed to give a true picture of one of the greatest scientific advances of all time."[22] He was referring to the supersession of Newtonian space and gravitation by Einstein's. He balked at the role I had accorded to conventional elements, as distinct from factual ingredients, in Einstein's edifice. I had done so in the tradition of Hans Reichenbach, who—ironically—had been the supervisor of Putnam's doctoral dissertation. Putnam's charge provided a great deal of food for thought.

Thereupon, I published a very detailed and lengthy "Reply" to him in two places, explaining why I was definitely not persuaded by his arguments: (i) in my book *Geometry and Chronometry in Philosophical Perspective* and (ii) in my monograph-length "Reply," which also appeared, but in a somewhat different version, as the principal part of volume 5 in the Boston Studies in the Philosophy of Science.[23]

Commenting on this spirited exchange, I declared, "I am most grateful to Hilary Putnam for having written the essay to which the present paper is a response; his work has been a valuable stimulus to me to clarify my views both to others and to myself. Critical severity is linked here with friendly respect."[24]

The year 1968 was significant in my service to the Philosophy of Science Association (PSA). Before 1965, Ernest Nagel of Columbia University, the then doyen of American philosophy of science, had been president of the PSA. During his term of office, it continued to publish its journal, *Philosophy of Science*, which had been founded in the early 1930s, but it conducted professional meetings only *jointly* with Section L (History and Philosophy of Science) of the American Association for the Advancement of Science. In 1965, I had the honor of succeeding Nagel as president, and was dauntingly elected to serve for two consecutive two-year terms (1965–1967, and 1968–1970).

Early in my second term as president, I inaugurated the first separate biennial meeting of the PSA in the fall of 1968 at the Webster Hall Hotel in the Oakland section of Pittsburgh, near the city's main educational institutions, Pitt and Carnegie Institute of Technology, later Carnegie Mellon University (CMU) after its merger with the Mellon Institute. I organized a rich program of invited and contributed papers, lasting from Friday morning, October 11, until Sunday midday, October 13, 1968. In running that meeting, I had the very valuable assistance of Jerry Massey, then still at Michigan State University, who was both secretary-treasurer of the PSA and the managing editor of its journal.

Forty years later, in November 2008, the PSA conducted its biennial meeting in Pittsburgh again, this time concurrently with the History of Science Society (HSS). The PSA Program Committee of this 2008 meeting decided to mark the fortieth anniversary of my 1968 launching of autonomous PSA meetings by *reprinting* the full 1968 agenda within its 2008 Program. Besides, Jim Bogen, a resident fellow of our center, and chairman of the 2008 Local Arrangements Committee, very ably portrayed the saga of philosophy of science in Pittsburgh on the center Web site of the November 6–8, 2008, powwow.

Quite unexpectedly, but very gratifyingly, even the first edition of my *PPST* had an enormous impact in the Soviet Union not only among philosophers (of science) but also among physicists and mathematicians. When Progress Publishers in Moscow was about to publish a Russian translation of this first edition, I made revisions and additions for incorporation in it. The modified text appeared in 1969 under the title *Filosofski problemy prostransiva i vremeni*, and it sold an enormous number of copies, although I cannot vouch for the accuracy of the translation. But, in those days, the Soviet publishers paid no royalties. Furthermore, as I learned in Moscow in 1988, the second edition of my *PPST* (1973), though in English, also became well known in the Soviet Union.

Sometime in the late 1960s, Professor Vladislav Lektorski, the very genial and warmly collegial editor in chief of the leading Russian philosophy journal *Voprosy Filosofii* (*Problems of Philosophy*) (hereafter "*VF*"), initiated the publication of Russian translations of an ongoing series of seven selected articles of mine that had appeared in English on various topics in the philosophy of science. These translations began in 1970 and even featured two of my articles on psychoanalysis.

Indeed, as Lektorski wrote me more than once, my writings "are very interesting and very important for our [Russian] philosophers. I am proud that you are my friend and that my journal can publish your articles" (e-mail of October 3, 2004).[25]

My Russian publications had contributed to my being invited, under the auspices of the USSR Academy of Sciences, to visit Moscow in late June 1988 and deliver three or four lectures at the Institute of Philosophy, the Institute for Systems Studies, the Institute of the History of Natural Science and Technology, and, tentatively, the Institute of Atheism at the Academy of Social Sciences. The invitation was issued by Professor Vadim Sadovski of the Institute for Systems Studies. My wife accompanied me to Moscow, and Sadovski was our principal and congenial host there during our stay and kindly extended gracious hospitality to us, along with a driver as guide.

My first formal lecture, presented at the Institute for Systems Studies, was devoted to the philosophy of current physical cosmology under the title "Is There a Problem of the Creation of the Universe?" with intermittent translations into Russian by the able Soviet interpreter Helena Vyshinskaya. (Incidentally, when she admitted in a private conversation to some embarrassment that her namesake, the infamous Soviet state prosecutor Andrei Vyshinski, had been tainted by

Stalin's crimes during the notorious purges, I reminded her that one could not sink lower than Adolf Hitler, whose first name I share!)

The main thesis of my cosmology lecture was that none of the recent scientific theories of the origin of the universe—such as the big bang theory—pose a *well-conceived* problem of creation. *A fortiori*, I argued, none of these cosmologies lend credence to the claim that a divine creator is required to "solve" the pseudoproblem of creation. Sixteen years later, I developed my views on this issue at great length in my essay "The Poverty of Theistic Cosmology," which appeared in the December 2004 *British Journal for the Philosophy of Science* and is reprinted in this *Festschrift*.

Above, I concisely indicated à propos of pseudoproblems, my reason for claiming that Leibniz's primordial existential question "Why is there something [contingent] rather than nothing [contingent]?" is indeed a pseudoproblem. As I noted only briefly, this question is predicated on an originally Christian, both *a priori* and empirically groundless doctrine that I have designated as avowing "the spontaneity of nothingness" (SoN). As will be recalled, this tenet has perennially claimed the following: "*De jure*, there *should* be nothing contingent at all, and indeed there *would* be nothing contingent in the absence of an overriding external cause (or reason), because that state of affairs is the 'most natural,' or 'normal' one."[26]

But since this crucial presupposition of Leibniz's question is demonstrably unjustified, either *a priori* or empirically, I argued, his perennial ontological question is a *non*-starter pseudoquestion. For just that reason, it cannot serve as a springboard for the theistic answer of creation-out-of-nothing.

Some of the philosophers of physics in my Russian audience identified themselves in positive terms as atheists, and welcomed my own kindred stance, but they distanced themselves pointedly from shallow political atheists in their country, hacks whom they derisively dubbed "professional" atheists. I remarked that anyone who holds—as I do—that there are very cogent reasons for atheism should indeed reject *all* poor defenses of it. After all, fallacious arguments for it can only discredit it among those who are *unaware* of the telling reasons in its favor. And intellectual honesty demands not to win recruits by salesmanship in this way.

I was interested in sampling actual attitudes of the Soviet philosophers toward theism, if only because a highly placed Soviet academic administrator had told me on the previous day that an important renaming was probably in the offing: the Institute of Atheism at the Academy of the Social Sciences was to be denominated thereafter as the "Institute for the Study of Religious Questions," because it was thought that the atheist label "begs the question." I responded that *if* there were such merely semantic question begging, then the designation of theological seminaries at American universities as "divinity schools" would be committing a like *petitio principii*. And, moreover, by the same token, Greek orthodox seminaries in Russia should also be renamed innocuously "Institutes for the Study of Religious Questions" no less than the "Institute of Atheism," which could be *neutrally* relabeled "Institute for the *Study* of Atheism."

In due course, I met with the director of the Institute of Atheism, Professor Victor Garadzha, at the imposing building of the Academy of Social Sciences, which featured a huge marble foyer with a large statue of Lenin at the center. I had seen Garadzha's name on the list of people scheduled to participate in the Tenth (Secular) Humanist World Congress to be held in late July 1988 on the SUNY campus in Buffalo, New York. My guide, an avowed adherent of Russia's Greek Orthodox Church, informed me that Garadzha is a medievalist from whom he happened to have taken a solid course on Aquinas and Augustine. Thus Garadzha was hardly a "professional" atheist in the pejorative sense intended by my philosophy of science colleagues.

In my 1984 book, *The Foundations of Psychoanalysis: A Philosophical Critique*, and in various articles about which I'll report more later on, I had argued that the evidential support for psychoanalytic theory and therapy is quite weak. This book was known to some of the philosophers there, if only because one of them had contributed a commentary to the 1986 review symposium on it by forty-one commentators in the journal *Behavioral and Brain Sciences* (hereafter "*BBS*"). Thus I was asked to give a lecture on psychoanalysis at the Institute of Philosophy, quite near the Pushkin Museum of Fine Arts.

Though psychoanalysis had been denounced for decades in the Soviet Union by Pavlovians and others, I spoke to a standing-room-only audience. The tenor of the questions and comments addressed to me during the ensuing discussion revealed the same distribution of viewpoints on the topic that I had encountered again and again after similar lectures in the United States and elsewhere in Europe. I was much impressed that the Institute of Philosophy employed a large number of scholars who do full-time research in various areas of philosophy and that a fair number of them spoke excellent English.

In response to an invitation from Professor V. Lektorski, I met with his entire editorial staff at his *VF* offices and presented a report on my research in the philosophy of science since the time of the 1969 appearance of the Russian translation of my space-time book. The ensuing question-and-answer period was marked by an atmosphere of free, ideologically unconstrained discussion, no less than all of the others I had had, both formally and informally, throughout the week.

Thus, there was not even any malaise in the give-and-take, let alone tension. Though Karl Popper's social philosophy was taboo in Soviet Russia, one colleague in my audience roundly declared himself a Popperian in opposition to my detailed 1976 criticisms of Popper's falsificationist philosophy of science. Occasionally, there were touches of levity: a psychologist remarked that in the mornings, Soviet clinicians would toe the party line and attack Freud as a bourgeois ideologue, but in the afternoons, they would practice his type of psychotherapy.

Further details on my 1988 Moscow visit are given in my published report "Soviet Atheism and Psychoanalysis under Perestroika."[27]

But let me return to the aftermath of my graduate work. In the fall of 1950, having just completed my Yale PhD thesis at the end of that summer, I received my first faculty appointment in philosophy at Lehigh University in Bethlehem, Pennsylvania, where I began as an instructor, rather than as an assistant professor, because my doctorate was not formally conferred until February 1951 after I had missed the submission deadline of April 1, 1950. Yet, only five years later, I became a full professor of philosophy there (1955) and was then appointed to a named chair (William Wilson Selfridge Professor) a year later.

My academic life at Lehigh was made very agreeable by Howard Ziegler, an ex-clergyman and the chairman of the department, who was a highly conscientious and very kind man. For me, the attraction at Lehigh was mainly that I would teach a course in philosophy of science, enrolling, besides superior undergraduates, the cream of its graduate students in the sciences and engineering, fields in which Lehigh enjoyed a good reputation.

In 1954–1955, while teaching at Lehigh, I had a year's leave as a Faculty Fellow of the Ford Foundation. During that time, I received an invitation from Henry Kissinger to contribute an article to the magazine *Confluence*, which he edited at the time while being a member of the Harvard faculty. I was to get fifty dollars for my paper. As I learned from my good personal friend Joseph T. Clark, SJ, he too had been invited to write an article.

I had the impression that the aim of *Confluence* was to convey to Europeans that intellectual culture was valued in America, while encouraging those religio-political parties in Europe whom the State Department wished to favor, notably parties led by Christian democrats in West Germany and in Italy.

The article I submitted to Kissinger was titled "Science and Ideology." It was written from my explicitly secular humanist perspective. Among other things, it *challenged ideological constraints* on scientific cosmology such as the *a priori* frowning on a spatially *finite* universe of big bang cosmology as reverting to the Bible, disapprovals which were misguidedly featured in 1952–1953 French and East German Marxist philosophical journals.

But, in my article for Kissinger, I minced no words in likewise discrediting the theological miracle concept as ill conceived because it was predicated on the illusory notion that miracles can be identified as violations of the *true* laws of nature. After all, our knowledge of these laws is irremediably fallible, so that violations of the presumed laws are not supernatural but only evidence of their fallibility.

Having sent Kissinger my paper, I went to his Harvard office to get his reaction to it when I had occasion to be in Boston. He flatly turned it down as unsuitable for his journal without any further explanation. But, as he told me then, he had also rejected the paper he had invited from the renowned secular humanist philosopher Sidney Hook, claiming most unconvincingly, if not preposterously, that it was of the genre of the apocryphal question "How many angels can dance on the head of a pin?" Kissinger's disingenuous excuses disappointed me greatly.

Besides, he had apparently also been nervous about the Jesuit Clark's essay, and rejected it as well, presumably because it did not meet his theological expectations from a Jesuit author.

Clark, who taught then at the top Jesuit graduate school of Bellarmine College, was a powerful critic of Thomist philosophy, which was then in vogue as the ideological armamentarium of Catholicism. But he argued most tellingly that its neo-Aristotelian concept-empiricism was untenable in the face of modern hypothetico-deductive scientific theories. Quite generally, he was very engaging philosophically. Hence, when I ran a public lecture series at Lehigh University, I had repeatedly invited him to be a speaker, the more so because he was a magisterial performer and soon developed a following there.

Indeed, I had persuaded the Lehigh administration to let Clark teach my courses during my 1954–1955 leave, although the president asked me concernedly whether my Jesuit colleague would be teaching in his clerical garb, to which I replied reassuringly that he would not.

But, alas, just then, Clark's critique of Thomism came to haunt him ecclesiastically: he was not only denied permission to take my place for a year, but he was also severely disciplined by his order, being humbled to teach elementary Latin and even provide student services at the lesser Jesuit Canisius College in Buffalo, New York, where he died.

As soon as Kissinger had rejected my "Science and Ideology," it was accepted for publication in the *Scientific Monthly*, where it appeared promptly.[28] Thereupon, I returned to my teaching at Lehigh.

Among my undergraduate students, there was Brian Skyrms, a brilliant economics major. In 1956, during my service at Lehigh (1950–1960), I helped recruit Nicholas Rescher from the prestigious Rand Corporation for the Lehigh Department of Philosophy. Later, during my first year at the University of Pittsburgh (1960–1961), I was very happy to recruit him again, this time for the Department of Philosophy at the University of Pittsburgh (known colloquially as "Pitt"), where we have continued to be very good friends ever since his arrival in 1961. Brian Skyrms then followed us to do graduate work in philosophy at Pitt and has since become a distinguished philosopher of science at the University of California at Irvine.

After ten happy years at Lehigh, I decided that I wanted to be in a university where I would also teach graduate students *in philosophy*. I had attractive offers of tenured professorships from excellent institutions such as the University of Chicago and the University of Minnesota, as well as "feelers" from others. But Rudolf Carnap cautioned me that a power-hungry historian of philosophy at Chicago could make life uncomfortable for me, as had been the case for him.

The University of Minnesota, at whose pioneering Center for Philosophy of Science I had repeatedly been a visiting professor, including the fall term of 1956–1957, was located in Minneapolis, where the winter climate was quite harsh. Yet, not joining the Minnesota faculty deprived me of becoming a colleague there of its Center's founder Herbert Feigl, a much-cherished friend and esteemed mentor. His precious virtues included complete guilelessness, a rare commodity among academics. It was at the Minnesota Center where I first met Wilfrid Sellars and acquired a permanent cordial friend, an association that I shall chronicle further below.

Except for its medical school, whose faculty included Jonas Salk of polio vaccine fame for some years, the University of Pittsburgh had been a decidedly mediocre institution until the mid-1950s, when the powers that be decided that the city of Pittsburgh was to undergo a major renaissance. Integral to that renaissance was to be a radical qualitative overhaul of the University of Pittsburgh, which was to become a "state-related" institution of the Commonwealth of Pennsylvania, as well as an upgrading of some of the lower-tier schools at the then stratified Carnegie Institute of Technology (later Carnegie Mellon University).

The ensuing changes in their bearing on the pursuit of philosophy at the University of Pittsburgh are recounted by Karen Kovalchick, assistant director of its Center for Philosophy of Science, in her very ably written *History* of that center (cited hereafter as the "center *History*"). Her forty-year account was published to mark the Center's fortieth anniversary in 2000. In her "Overview of the First 40 Years," she wrote:

> By 1960, the University of Pittsburgh was undergoing its own transformation. Two years earlier, Chancellor Edward Litchfield, in a landmark speech delivered on December 16, 1958, formally announced the establishment of 10 A. W. Mellon Professorships and Fellowships in the academic disciplines. Academic Vice Chancellor [later Provost] Charles H. Peake was charged with the responsibility of filling those chairs. This was to be an augury of the University of Pittsburgh's renaissance.
>
> Peake had secured the advice of a board of outside eminent scholars to advise him on suitable occupants of the chairs in each of the 10 fields in which the professorships had been established. The advisory board had highly recommended Adolf Grünbaum, then at Lehigh University, for the chair in philosophy. As described by the editors [John Earman, Allen Janis, Gerald Massey, and Nicholas Rescher] in their preface to the 1993 *Festschrift* titled *Philosophical Problems of the Internal and External Worlds: Essays on the Philosophy of Adolf Grünbaum:*

In the fall of 1960, Adolf Grünbaum left Lehigh University to join the faculty of the University of Pittsburgh as Andrew Mellon Professor of Philosophy and as founding director of the Center for Philosophy of Science. Ten professorships at the University of Pittsburgh had been endowed by the A. W. Mellon Foundation during the 1950s, and for the initial period these chairs were filled on a visiting basis. When the time came to begin to fill these chairs on a permanent basis, the then Provost, Charles Peake, in what was to prove a brilliant administrative move, took the bold step of offering the Andrew Mellon chair in philosophy to an unusually promising young scholar, someone so young [thirty-six] that the age threshold of forty years for the Mellon Professorships had to be waived in order to secure Grünbaum for the chair. Perhaps no appointment at any university has returned greater dividends than this one.[29]

Thus, Vice Chancellor Peake, whose vision of the role of philosophy in a university was much like mine, recruited me for the fall of 1960 to become the first permanent Andrew Mellon Professor of Philosophy and indeed the first such Mellon Professor in any of the ten fields that had such a chair. Peake's aim was to transform completely the then philosophy department, which had, alas, been an academic backwater and wasteland, into a nationally and internationally visible major graduate one. He asked me to be the principal architect of that transformation starting from scratch, by being a magnet attracting first-rate philosophers and a new chairman.

Peake and I likewise saw eye to eye on the creation of a Center for Philosophy of Science (initially labeled a "Program"), modeled on the pioneering Minnesota Center for Philosophy of Science at which I had held visiting appointments. I was to be the founder and first director of Pitt's Center. Our fledgling Center would offer an Annual Lecture Series by prominent philosophers of science from universities in the United States and abroad, a series which I would organize each year. My hope then was that the papers in each series would merit publication in book form, an aspiration that was then realized.

I was to have the advice of a multidepartmental faculty committee, with which I met for the first time well before I came to the Pitt campus for good in August 1960. At the behest of that advisory committee, I delivered the first lecture in the inaugural 1960–61 series.

Had it not been for Charlie Peake's provostial magnetism, and the extraordinarily challenging opportunity to build a world-class Department of Philosophy and a leading Center for Philosophy of Science from the ground up, I would not have accepted his offer from Pitt, but would have chosen one of the other opportunities that I had in 1959–1960.

As chronicled further by Karen Kovalchick in her center *History*:

The inaugural edition of the Annual Lecture Series, which was supported by a grant from the United States Steel Educational Foundation, had an illustrious cast: Adolf Grünbaum, Carl G. Hempel (two lectures), Michael Scriven (two lectures), Wilfrid Sellars (who was still at Yale at the time), Ernest Nagel, Ernst Caspari (biologist), and Paul K. Feyerabend (two lectures).

Grünbaum presented the first lecture in the inaugural edition of the series: "The Nature of Time." The talk, which received coverage by *The Pittsburgh Press*, met with resounding success. There was not a vacant chair in the auditorium. In fact, one of Grünbaum's most distinct memories of that lecture is Jonas Salk sitting on the floor in front of him, having arrived too late to get a seat. Academic Vice Chancellor Charles H. Peake, in later correspondence, recalled the event:

The first lecture of the now famous lecture series in the philosophy of science was to be given by Adolf ("The Nature of Time"), and I was to introduce him. There was a serious question as to where the lecture was to be held— the Public Health School Auditorium (attractive but rather small) or Clapp Hall Auditorium (very large). Adolf insisted on Clapp Hall, raising the specter of a small elite gathering lost in cavernous space. When we arrived, we found that almost every seat was taken, and that people were beginning to sit in the aisles and stand in the rear!

Papers delivered in the inaugural version of the Annual Lecture
Series were published in 1962 by the University of Pittsburgh Press
as "Frontiers of Science and Philosophy," in the first volume of the
University of Pittsburgh Series in the Philosophy of Science, with a
publication subvention from the National Science Foundation.[30]

There were other sources of encouragement before I arrived. I met
several congenial Pitt colleagues whose expertise was very germane to
my own interests in the philosophy of physics, especially space-time.
One of them, the theoretical physicist Allen I. Janis, who specialized in
Einstein's general theory of relativity, soon turned out to be a stalwart of
the Center ever since, besides being my invaluable scientific helper and
occasional coauthor, as well as a much-prized personal friend.

I was also heartened, after my earlier 1959 lecture visit to the then
Carnegie Institute of Technology (later Carnegie Mellon University),
when Herbert Simon sent me a letter of welcome to the wider university community in Pittsburgh shortly after he learned that I would
become one of its members. We were honored that Herb then became
a valued and active member of the Center fellowship, and remained so
after receiving the Nobel Prize in economics until his death in 2001.

Before I arrived, the late Robert Colodny of the Department of
History at Pitt paid me a welcome visit at Lehigh University to express
his enthusiasm for the new plans and to volunteer his services as editor
of the Pittsburgh Series in the Philosophy of Science that I was to
launch. He then bestowed great care on seeing the first seven volumes
of the series through the press.

I found Peake's charming personality so agreeable, and his values
as an educational and administrative leader so very congenial, that I
felt inspired in all of my activities on campus. Happily, he, for his part,
was so very pleased with my performance as scholar, faculty recruiter,
and Center administrator that he made life at the university almost a
bed of roses for me until he retired in about 1972.

When I came to Pitt in 1960, a department of history and philosophy of science had not yet even been envisioned. But, by the mid-

1960s, it had become clear that the study of scientific rationality needed to be informed by historical models. Yet the philosophers of science who then were members of the Pitt Department of Philosophy had no professional training in the history of science. The imperative to take adequate cognizance of the historical dimension of science began to be translated into university appointments some years after I had met Larry Laudan at Princeton University during the academic year 1963–1964. I was there to give a colloquium lecture in the Department of Philosophy on Pierre Duhem, on whose philosophy of science I had been publishing articles, and Larry was then a graduate student there, charged with looking after the visiting speakers.

In conversation with him about Duhem—an avatar of scholarship in the history and philosophy of science—it struck me that Larry was developing just the sort of dual competence that could forge a genuine marriage between the history and the philosophy of science. In 1968, I was able to convince Vice Chancellor Peake that Pitt should try to recruit Larry from the University of London, where he had recently been appointed, which we then did. To signal the university's hospitality to the discipline of the *history* of science, Larry's primary appointment was initially in the Department of History, with a secondary appointment in philosophy. He arrived in 1969.

Soon after he arrived, Larry eloquently urged the creation of a separate Department of History and Philosophy of Science (HPS), whose primary appointees would genuinely integrate the history of science with its philosophy. It would offer both undergraduate and graduate instruction leading to degrees, unlike the Center, which does not grant any degrees and is fully devoted to research, although its lectures and seminars are entirely open to students as well. For financial and administrative reasons at a time of budgetary retrenchment, the then dean of Arts and Sciences, Jerome Rosenberg, was reluctant to approve the creation of a new department.

I then pleaded our case to Peake, who came to the rescue and swayed the dean. Thus, our HPS department came into existence in 1971, with Larry Laudan as its first chairman. Very soon he and I began the tortuous negotiations that issued in the acquisition of the Rudolf Carnap collection by our nascent Archives of Scientific Philosophy in the Twentieth Century at the Hillman Library. Its inventory includes his valuable and extensive correspondence.

That collection was our first, and we could not have persuaded Carnap's quirky daughter to sell it if not for Larry's tenacity and skill as a negotiator. Again in 1971, James McGuire, whom I had met in 1965 at the three hundredth celebration of Newton's *annus mirabilis* (miraculous year) in Austin, Texas, also joined the fledgling department as a primary appointee. With the creation of this new department, the Center fellowship now included a group of specialists devoted to integrating the historical and philosophical study of science.

In 1978, I gave up the directorship of the Center, after eighteen years, and—at Larry's suggestion to the provost—I was appointed to its newly created position of chairman, which I have held ever since. I was very glad that Larry then succeeded me as director, and I refer readers to Karen Kovalchick's splendid fortieth anniversary *History* for a very interesting account of the ensuing developments, including new programs.

From these beginnings, the Departments of Philosophy and History and Philosophy of Science, have since grown to be among the top-ranked in the country.

During my first year of teaching at Pitt (1960–1961), Thelma and I became very good friends of two medical colleagues and of their wives: Abraham Braude, head of the Division of Infectious Diseases, and Gita Braude, and Elliott Lasser, chairman of the Department of Radiology and his wife, Phyllis. We met them because Gita and Phyllis were auditing my senior undergraduate course in philosophy of science in my first term. Soon we saw them not only in Pittsburgh, but also on summer vacations elsewhere, such as when we rented Jonas Salk's cottage with the Braudes at Lake Deep Creek in western Maryland, where we acquired a summer home years later in 2004. When both couples moved to La Jolla, California, where Elliott and Abe became professors at the new medical school of the University of California (San Diego), we repeatedly spent our summer vacations with them and their children. Thus, our daughter, Barbara, also had company there.

In La Jolla, I also developed friendships with some of the colleagues in the Department of Philosophy of the University of California there, and I resumed contact with its members Paul and Patricia Churchland, whom I knew while they were still graduate students in philosophy at Pitt, having taught Paul when he enrolled in my graduate course on space-time philosophy.

Early on in Pittsburgh, Thelma and I had also become warm friends of Myron (Mike) and Ruth Garfunkel. Mike was professor (and later chairman) of physics, a department to which I built collegial bridges. Later on, Thelma and I became very close friends of another Pitt couple: Hugo Nutini, a distinguished cultural anthropologist, and his wife Jean Nutini, a researcher at the School of Public Health. Hugo had done graduate work in philosophy at the University of California at Berkeley before becoming an anthropologist.

I also developed some durable friendships in the course of attending professional meetings abroad. Thus, in October 1992, when I gave an invited paper at the international conference on "Time in Science and Philosophy" in Naples, Italy, I was captivated by the presentation of Massimo Pauri, an Italian theoretical physicist and splendid philosopher of science who taught at the venerable University of Parma. Our intellectual and interpersonal chemistry was auspicious from the start, and our cordial friendship has flourished ever since.

In 1993–1994, Massimo was a visiting fellow in our Center for Philosophy of Science and a participant in my graduate seminar, as were other visiting Center fellows before and after. Presumably at Massimo's initiative, his university awarded me a medal in 1998 in recognition of my "prestigious career." I also developed a warm personal and professional friendship with the Florentine philosopher Alessandro Pagnini, who publicized my work on psychoanalysis in Italy, and published his Italian translation of my essay "Psychoanalysis and Theism" (1991).

Happily, the original 1960 aspiration to transform Pitt's philosophy department was realized to such an extent that, by 1982, the national rating of philosophy departments by the National Research Council ranked Princeton, the University of Pittsburgh, and Harvard as the top three departments in the United States, in that order. But, very notably, to our pleasant surprise, as early as twenty years earlier, in 1962, only two years after my arrival, the department had already been rated nationally by the National Research Council as sixth in one category, and eighth in another, thus placing it well within the top ten departments of philosophy.

Very remarkably indeed, at least by the time of the aforementioned 1982 national rating, HPS, which was created only about a decade before that rating, tied for fifth place in the United States among departments of *philosophy*, even though it is much more specialized than they are and hence does not compete on a level playing field at all with such departments. In addition, the worldwide range and caliber of the programs of our Center for Philosophy of Science have since earned it international recognition as the world leader in the field, as can also be gleaned from the Center's forty-year *History*.

Early in the 1990s, the Center received a major and completely unexpected financial boost. One of my Lehigh students in the mid-1950s was an industrial engineering major named Harvey Wagner, who took every undergraduate course I gave there and always stayed after class to ask me very searching questions. Thirty-five years later in 1990, when I was a visiting Mellon professor ("associate") at Cal Tech in Pasadena, California, where my philosophy colleague James Woodward and I became warm friends, Harvey telephoned me there. I returned his call upon my return to Pittsburgh, whereupon he told me that he credited my teaching of critical thinking with his great subsequent success as a technology entrepreneur and that I had been "the principal intellectual influence on his life." Though he thought that I would not remember him, he declared that the time had come to pay me back for these contributions to his life.

I told him on the telephone that I shall not try to disabuse him of his attribution of his economic success to my teaching, and I thanked him for his generous offer. He then invited my wife and me to his home at Lake Tahoe, in Incline Village, Nevada, where we met his sweet, very affable wife, Leslie. It was she who had encouraged him to phone me, when he hesitated, because he thought I would not remember him. At their home, Harvey led us into his study and pulled out a drawer containing all the lecture notes he had taken in my philosophy courses. He said that he was still referring to them.

Soon I told him that I had founded a Center for Philosophy of Science at Pitt upon my arrival there from Lehigh in 1960. I explained that, ever since, I had hoped for an endowment for the Center to assure its financial viability in the face of the uncertainties inherent in seeking periodic grants that are then spent as operating money. Fortunately, the Wagners found my case persuasive. In 1992, they gave the Center an initial endowment of one million dollars in my honor.

In conveying the gift, Mr. Wagner added, "One encounters a teacher like Professor Grünbaum only once in a lifetime, if one is lucky." The substantial constructive impact of that major gift is recounted in the Center *History*, which identifies the various Center programs that are supported or assisted by the income from the Wagner endowment, as is my budget for research expenses. Thus I have warmly dedicated this *Festschrift* to both of the Wagners.

Maureen Macaleer was the Major Gifts Officer of Development for the School of Arts and Sciences at Pitt. As she reported to me, Harvey Wagner's warm appreciation of my teaching was shared by an array of other former students. Thus, touching the soft underbelly of my pedagogical vanity, she wrote: "It's rare that a student singles out one professor of influence, but you seem to do that often" (letter of June 29, 2006). And earlier (e-mail of May 4, 2005), she had reported:

> When I met the alumnus Dr. William Mackey [BS physics '62] last month, Dr. Mackey mentioned that the best course he had in all of his years of education was with you. . . .

Dr. Mackey considers you to be the best professor he was taught by. Given the number of fine institutions that he has been educated at, you have had a singular influence.

. . . You are and have been distinctive and instrumental in transforming the minds of countless students throughout your career. What a wonderful legacy you have!

Ever since the resumption of my contact with Harvey, the Wagners have invited us for a visit repeatedly as their guests in a very comfortable condominium that they own near their home in Incline Village. Indeed, not very long after their endowment gift, Harvey and Leslie Wagner, along with Thelma and me, came to feel that we were part of one family. In late July 2007, they hosted us very generously at the splendid Four Seasons Hotel in San Francisco on our way to Beijing, where they gave us a bon voyage party in style.

Going back to the early 1960s, I was fortunate to bring off a major recruitment coup in 1962, after Kingman Brewster, then the provost at Yale, and later Yale's famous anti-Vietnam War president and American ambassador to England, had come to my office in Pitt's Cathedral of Learning (2510 CL) and spent the afternoon with me on October 4, 1962, to offer me a full professorship of philosophy at Yale. At the time, the renowned philosopher Wilfrid Sellars was there, having left the University of Minnesota. As mentioned above, Wilfrid and I had become very good intellectual and personal friends during my visits to the Minnesota Center, and we then hoped to work together in the same department, just as Wesley C. Salmon and I wished to do so later on.

I was captivated by Sellars's advocacy of the cognitive superiority of science to the presumptions of common sense and by his naturalistic philosophy of mind. Wilfrid was not the only attraction for me at Yale. The logician Alan Ross Anderson as well was a drawing card there, not only professionally but as a congenial person of sterling character whom I had met there when we were fellow students.

I thought, however, that the distinctive promise held out by Pitt, and the commitments I had made to Charlie Peake, were very substantial indeed. Much as I was tempted to join Sellars at Yale and to accede to Provost Brewster's enticement, I regretfully declined the Yale offer in November 1962.

Just about a month thereafter, Wilfrid met me at the December 1962 meeting of the Eastern Division of the American Philosophical Association. Unforgettably, his opening words were: "If Muhammad won't come to the mountain, then the mountain will come to Muhammad." I promptly reported this beckoning episode to Provost Charles Peake, who had had no prior inkling at all of Sellars's availability. But when I then urged Sellars's appointment at Pitt as a "university professor," I was pushing against an open door. Within two weeks, Wilfrid had accepted that appointment.

When he notified the then Yale philosophy department chairman Frederick Fitch of his decision to leave for Pitt and declined to negotiate with him, Fitch drew the far-fetched, odd inference that I had somehow placed Sellars under an obligation *not* to do so. And when Fitch phoned me urgently and asked me to release Sellars from that presumed obligation, I replied that I had never dreamed of obligating him in this way. It seemed that Fitch's tacit moral maxim was that it had been quite proper for Ivy League Yale to try to lure me away from Pitt, but illicit for Pitt to raid Yale! I felt uneasy in having this exchange with Fitch because he had been my agreeable graduate teacher of symbolic logic.

Indeed, Pitt compounded the supposed felony; in the wake of Sellars's move, a whole contingent of philosophers from Yale came to Pitt: the logicians Alan Anderson and Nuel Belnap, as well as Jerome Schneewind, the ethicist, and the philosopher of linguistics Richmond Thomason. This exodus earned *me* the sobriquet "The Pittsburgh Pirate." But I felt no disloyalty to my alma mater Yale as its graduate alumnus, and no compunction, because the dissensions among cliques in the philosophy department there had already planted the seeds of the department's major and precipitous decline, which then materialized.

Indeed, the Yale administration did not hold Sellars's departure and the ensuing exodus to Pittsburgh against me: although bad memories often linger, in 1990, Yale generously awarded me its Wilbur Lucius Cross Medal "for outstanding achievement." Alan Anderson's tenure at Pitt included the departmental chairmanship. Yet, tragically in 1973, he died of cancer at the age of only forty-eight.

My warm friendship with Wilfrid continued to flourish after he came to Pitt in 1963, where he spent the rest of his career until he died in July 1989. He had not told me that, in his will, he had designated me to be the executor of his estate. Thus, I learned of it only when the will was probated. But I carried out my task with the valuable advice of his very devoted secretary, the late Mary Connor, who was also the assistant to the director of the Center.

Very early in the 1960s, I met Richard Gale, a recently minted philosophy PhD from New York University who shared my keen interest in the philosophy of time. Soon thereafter, he came to Pitt as a National Science Foundation postdoctoral fellow, whereupon he joined the faculty of the Department of Philosophy and taught there until 2004.

Beyond establishing a strong reputation for his work on philosophical questions pertaining to temporality as encountered in ordinary human experience, Gale became recognized nationally and internationally among philosophers of religion, and as a scholar of American philosophy, notably of the work of William James and John Dewey. Besides auditing some of my graduate seminars, ranging from spacetime to psychoanalysis, Richard has been a dedicated, magnanimous, and indefatigable critical commentator on drafts of my writings dealing with topics of shared overlapping interest. Moreover, Thelma and I regard him and his wife, Maya, as close, warm friends. Indeed, he very generously dedicated to me his collected works, which are being published by Prometheus Books, like the present *Festschrift*.

My sketch so far of some of the very early history of the University of Pittsburgh Department of Philosophy in its new era was contextualized and amplified by the following excerpts from Robert C. Alberts's valuable 1986 book, *PITT: The Story of the University of Pittsburgh 1787–1987*, published to mark the bicentennial.

Speaking of a report by the Andrew Mellon Educational and Charitable Trust, Alberts quotes from it as follows:

It is evident that the Mellon Professorships have been a central influence in the development of scholarly programs in the Arts and Sciences at the University of Pittsburgh. For example, the first two appointments in Anthropology and Philosophy made possible the development of those departments into leaders in their fields throughout the United States. . . .

The Mellon Professorship that started the Department of Philosophy on its path to preeminence was awarded in 1960 to Adolf Grünbaum, a brilliant young (thirty-seven) pioneer in a new field—the philosophy of science. . . .

Grünbaum brought with him a Lehigh colleague as the associate director of the new Center for the Philosophy of Science: Nicholas Rescher, an internationally recognized authority on the theory of knowledge and a prolific author (he now has over forty books to his credit).

Late in 1962, Yale University attempted a raid: It invited Adolf Grünbaum to set up a Center for the Philosophy of Science at New Haven, making what [then Provost] Charles Peake called a most appealing offer. Grünbaum's strongest advocate at Yale had been Wilfred [sic. Wilfrid] Stalker Sellars, a star of the department. When Grünbaum declined the Yale offer, writing a letter "painstaking in its presentation of the many considerations" that swayed him in his decision, the Yale department head bowed out gracefully. ("We have lost a great opportunity and I am sorry, but I wish you well. . . . We tried hard to get the best man we knew.") Grünbaum thereupon proposed [after Sellars had explicitly volunteered that he wants to join him at Pitt] that since he could not join Sellars at Yale, perhaps Sellars would like to join him at Pitt. Sellars agreed to come early in 1963. The Yale Philosophy Department held an emergency meeting on a Sunday morning and, after a unanimous vote on the matter, made a dramatic telephone call to Grünbaum. They appealed to him not to take Sellars on the ground that his departure would wreck their department. Grünbaum [knowing that Sellars no longer wished to stay at Yale] refused to withdraw the offer. Sellars said that the decision had been made and that the subject was closed. . . .

The department's claim to preeminence was fortified in 1962 when Kurt E. Baier, forty-five, head of philosophy at the Australian National University in Melbourne [actually, Canberra] became chairman. Norwood R. Hanson, Professor of Philosophy at Yale, wrote on June 10, 1965, "You people are doing something which is invaluable to the future of my professional discipline. . . . All we professionals look to Pittsburgh now as the virtual heart of the serious metabolism in Philosophy of Science within this hemisphere.[31]

It would require a separate essay of its own to give an adequate account of the action-packed roles at Pitt of my very valuable colleague and cordial friend Gerald (Jerry) Massey, whom I recruited in 1970 to become chairman of Pitt's philosophy department. For no less than seven years (1970–1977), he chaired that department outstandingly during a period of its further expansion. And, to boot, for the nine years of 1988–1997, he was a highly innovative director of our Center for Philosophy of Science. His extraordinary stewardship was marked by entrepreneurial vision and yet imbued with meticulous attention to detail, as is his wont in whatever he touches.

His multiple scholarly, professional, and administrative services to the university and the profession were very deservedly acknowledged at Pitt by his elevation to the rank of Distinguished Service Professor. But the magnitude of his achievement in fostering German-American cooperation in philosophy of science via the Pitt and University of Konstanz Centers for Philosophy of Science has been signally recognized by the president of the German Federal Republic, who conferred on Jerry the highest civilian honor bestowed by that country: the *Bundesverdienstkreuz erster Klasse* (freely translated as the Officer's Cross of the Order of Merit).

Jerry generously also found time to give me the substantial benefit of his competence on logical issues in space-time philosophy. The intellectual debt I owe him in that area is apparent from the array of my references to him in the second, enlarged edition (1973) of my aforementioned treatise *Philosophical Problems of Space and Time*. But the benefits he solicitously bestowed upon me were hardly just academic: a number of times, he ran interference for me at Pitt through thick and thin.

His affectionate generosity toward me is illustrated by the great labor he bestowed on conceiving, initiating, and organizing a special three-day international "Colloquium in Honor of Adolf Grünbaum" on October 5–7, 1990, "in celebration of the 30th Anniversary of his appointment as Andrew Mellon Professor of Philosophy and his founding of the Center for Philosophy of Science at the University of Pittsburgh." Mary Connor of the Center's staff vigorously assisted Jerry in his endeavor, but she also became the Mistress of Shenanigans in targeting me for a good-natured "roast" at the Saturday night conference banquet.

The Colloquium occasioned a gracious editorial titled "Pitt's Pre-eminent Philosopher" in the *Pittsburgh Post-Gazette* of Saturday, October 6, 1990, which read in part:

> Among the University of Pittsburgh's many eminent scholars, Adolf Grünbaum may be the most celebrated. Brought to Pitt 30 years ago as part of then-Chancellor Edward Litchfield's "spires of excellence" program, Professor Grünbaum is widely credited with having led the university's philosophy department to its status as one of the nation's best.
>
> It is fitting that the Center for Philosophy of Science is hosting a colloquium this weekend in his honor. . . . During the three-day conference, scholars from around the world will gather to pay tribute to Professor Grünbaum. For all his honors, he is treasured for his humility.
>
> The conferees will also discuss three of Professor Grünbaum's major concerns—the philosophy of physics, the general philosophy of science, or scientific rationality, and the philosophical problems of psychoanalysis. . . .
>
> It is fitting that a man both highly regarded and humble should be honored by his peers.

For over half a century, I was also the beneficiary of a very affectionate, personal friendship with Wesley C. Salmon, ever since he first visited me while I was still at Lehigh. Besides being a prized friend, he has always been a philosophic comrade-in-arms, as it were. Our shared intellectual interests ranged over space-time philosophy, the theory of scientific rationality, the critique of religious belief, and Freudian theory. Thus, his *Space, Time and Motion*[32] is dedicated to me. And, speaking of my contribution to the topic in the preface of his 1970 anthology *Zeno's Paradoxes*, Wes wrote: "I cannot overestimate the amount I have learned from him [Grünbaum] regarding the basic philosophical issues that are involved in the analysis, clarification, and resolution of Zeno's marvelous paradoxes."[33]

Furthermore, in 1999, referring to Hans Reichenbach, who directed his UCLA doctoral dissertation, Salmon said: "Adolf Grünbaum, who, though he never studied with Reichenbach, was deeply influenced by his work on space, time, and relativity."[34] In an appended footnote, Wes added a remark concerning our shared interest in Freudian theory:

After completing monumental work in this area [space-time philosophy], Grünbaum has become the world's leading philosophical authority on psychoanalysis. It is interesting to note that Reichenbach [a cofounder of the Los Angeles Psychoanalytic Society] encouraged me to work in that area, but I found the literature in that field insurmountable. I greatly admire Grünbaum's determination and stamina in mastering that corpus.[35]

Yet, despite his disclaimer here, Salmon developed some helpful ideas on the empirical testability of psychoanalytic theory. In 1958, sixteen years before I began to examine the Freudian corpus systematically, he contributed a paper on "Psychoanalytic Theory and Evidence" to Sidney Hook's conference volume *Psychoanalysis, Scientific Method, and Philosophy*.[36] An important feature of this paper is his lucid account of the sort of empirical evidence that would *disconfirm* Freud's motivational explanation of so-called counterwish dreams. Such dreams feature the frustration of a wish or the occurrence of something clearly unwished-for.

For example, a onetime fellow student of Freud's, who had envied his primacy at school, became a trial attorney and then dreamed that, very disappointingly, he had lost all of his court cases. Freud claimed that the *non*-fulfillment of the attorney's wish to win in court allowed the cunning oneiric fulfillment of his supposedly unconscious wish to *disprove* Freud's thesis that dreams are universally wish-fulfilling.

But, as I have argued in my 1993 book *Validation in the Clinical Theory of Psychoanalysis: A Study in the Philosophy of Psychoanalysis*, Freud's purported explanation of counterwish dreams is, unfortunately, riddled with fallacies. Indeed, I was driven to say there about Freud's particular explanatory reasoning in this case: "Alas, I would be hard put to find any other few sentences in the writings of a comparably influential thinker that contain so high a density of fallacies as his ensuing passage."[37]

For years after I had come to Pitt, I endeavored doggedly to bring Wes Salmon and his academic wife, Merrilee Salmon, to the university. Naturally, I exulted when they decided to join me in 1981, he as university professor of philosophy and successor to Carl Hempel and she, in due course, as professor of anthropology and as professor and chair of HPS. The HPS Department was attractive to Merrilee, because we had succeeded in enlisting as primary appointees in it such outstanding colleagues as Clark Glymour, James Lennox, John Earman, Kenneth Schaffner, and James ("Ted") McGuire. Wes's sudden death from presumably natural causes while driving an automobile in April 2001 has left a large gap in my life. I commemorated him by the dedication of my aforementioned presidential address to the 2007 World Congress of the Division of Logic, Methodology, and Philosophy of Science in Beijing.

James Lennox's presence at Pitt became especially beneficial to me in my job as chairman of the Center when he became its director in 1997. He is a very fine Aristotle scholar and well-known philosopher of modern biology. It has been a sheer delight to work with him. His professional performance has been splendid and creative, while he and his wife, Pat, have developed into highly congenial, heartwarming friends.

The results of my decades-long endeavor to build the triadic philosophy complex at Pitt have been assessed most generously by Robert S. Cohen in the biographical memoir he contributed to a 1983 surprise *Festschrift* for me on my sixtieth birthday, coedited by him and Larry Laudan. As mentioned above à propos of John Stachel's role, this *Festschrift* was titled *Physics, Philosophy and Psychoanalysis: Essays in Honor of Adolf Grünbaum*. It was published by D. Reidel (Dordrecht, Holland) and reprinted in 1992, as volume 76 in Robert S. Cohen's series, Boston Studies in the Philosophy of Science. It contains a bibliography of my writings up to 1983, which was compiled by Larry Laudan. In his memoir, Robert S. Cohen wrote:

> Indeed Adolf was an astounding organizer, the master recruiter of the most extraordinary faculty of philosophy since Whitehead collaborated with the young Russell. . . . Dozens of the stalwart creative new philosophers of the past quarter century have come from the Pittsburgh philosophers. . . . The assessment of the Center [for Philosophy of Science], and the Department of Philosophy, and its younger sibling Department of History and Philosophy of Science, will someday receive the attention of historians of philosophy and sociologists of institutional development; for with all their differences of scale and goals, the story of Grünbaum's philosophical Pittsburgh—a little jewel of imaginative organization—belongs with the stories of such ambitious places as Flexner's Princeton Institute [for Advanced Study] and Hutchins's [University of] Chicago.[38]

Let me resume the retrospect of my philosophic odyssey. In 1952, I wrote my first paper on the problem of free will versus determinism. During the next twenty years, I published a series of further, successively revised articles on that topic, which culminated in my 1972 essay "Free Will and Laws of Human Behavior."[39] It appeared in the volume *New Readings in Philosophical Analysis*, edited by Herbert Feigl, Wilfrid Sellars, and Keith Lehrer.

This essay was discussed critically by Philip L. Quinn, my quondam PhD dissertation student, in the aforementioned first of three Festschrift volumes (of 1983, 1993, and 2009), which my friends and colleagues very kindly dedicated to my work. In the later Festschrift of 1993, Arthur Fine dealt approvingly yet very penetratingly with my views on the free will issue.

Only about four or five years after I had come to Pitt, Bas van Fraassen, now McCosh Professor of Philosophy at Princeton after a meteoric career dotted with honors, came to Pitt in 1964 on a Woodrow Wilson fellowship to study philosophy of physics with me. He had turned down an offer from Harvard to go there with that transportable fellowship. Then I happily directed his PhD dissertation on "Foundations of the Causal Theory of Time."

Also in the early or mid-1960s, an Argentinean named José Alberto Coffa wrote to me from Buenos Aires. He said that he had earned a master of science degree in engineering from the university there and that he was earning his living as an engineer. In remarkably good English, he told me of his strong interest in the philosophy of science. When explaining what particular issues excited him, he unfolded an astonishing mastery of the literature that would have done any fully trained professional in the field proud.

Yet he had acquired it autodidactically. The point of his letter was to explore possibilities of coming to this country to work with me and earn a doctorate. The letter was not only impressive; it was positively moving for me. The more so, perhaps, because—having been an immigrant myself—I strongly identified empathically with his aspiration.

After a considerable delay, he did come to Pitt, where his intellectual talents soon became patent. But I was also struck by his wonderful joie de vivre and a touch of mischievousness, which often broke through his genuine shyness in the form of infectiously hearty, even roaring laughter.

In his 1973 doctoral dissertation on "Foundations of Inductive Explanations," which I supervised, he presented a highly interesting alternative to the widely renowned models of scientific explanation developed by Carl Hempel and Wesley Salmon. Hempel's inductive-statistical model had relied on the notion of logical probability, and Salmon's statistical relevance model was grounded in a frequency interpretation of probability. Instead, Coffa proposed an account of inductive explanation based on a dispositional version of the propensity interpretation of probability, which had been proposed by Karl Popper.

After making significant contributions in other areas as well, Coffa was soon recognized by being given a senior appointment in the excellent Department of HPS at Indiana University in Bloomington, where Wesley Salmon had been appointed the Norwood Russell Hanson Professor of History and Philosophy of Science in 1967. I had originally been instrumental in getting Hanson appointed at Indiana, where he then founded its HPS Department. Ever the promoter, he had even misleadingly listed me as one of its initial members in a flyer announcing its creation. Yet I did become an advisor to it very gladly.

When Coffa coined the fetching phrase "deductive chauvinism" as a caveat, he recognized—and focused sharp attention upon—a pervasive and often unwitting tendency to force philosophical concepts and theories into a deductive mold. Whether the tendency is legitimate or illegitimate in one context or another, he opined, we need to be aware that we are doing so when we indulge in this practice.

Thus, Coffa applied the label "deductive chauvinism" to the view that the higher the probability assigned to an explanandum event by a proposed explanation of it, the better the explanation yielding it. His further admonition was this: Unless we can assume an underlying, hidden determinism, we must conclude that the actual irreducible probabilities exhibited by the stochastic facts of the world are sometimes low. Consequently, they do not accommodate the deductivist's explanatory goals.

A cognate but much more elaborate formulation became a major thesis of Wesley Salmon's, who maintained, contra Hempel, that explanations are *not arguments* showing that their explananda were to be *expected*; instead, they are articulations of factors *causally relevant* to the occurrence of the explanandum events. Thus, on Salmon's view, untreated syphilis helps to explain the affliction of general paresis by being causally relevant to its occurrence, although *most* untreated syphilitics do *not* ever become paretics.

Tragically, in 1984, at the age of only forty-nine, José Alberto Coffa succumbed to a streptococcal infection of his kidneys after having waited too long to seek medical help, because he wanted to be home with his family during the Christmas holidays. I felt his loss as if he were a son of mine. Wesley Salmon and I then coedited a 1988 volume titled *The Limitations of Deductivism*,[40] which we dedicated "In affectionate memory of J. Alberto Coffa (1935–1984)." It contains appreciations of Alberto by Wes Salmon and myself, and appeared in our Pittsburgh Series in Philosophy and History of Science, then published by the University of California Press.

In the mid-1960s, Philip L. Quinn, later a distinguished chaired professor at Notre Dame University until his recent untimely death, who had received a master's degree in physics after his baccalaureate in philosophy, came to work with me in the philosophy of science, and I supervised his doctoral dissertation on Pierre Duhem. Thereafter, his interests shifted mainly to the philosophy of religion. But the gap between his Roman Catholicism and my atheism did not interfere in the least with his being a steadfast, invaluable, generous, and indeed delightful interlocutor in recent decades on the spectrum of issues I discussed in my series of critical writings on theism and on belief in it.

I have been told that I am a very exacting director of PhD theses. Thus, in reminiscences delivered by Bas van Fraassen—at the aforementioned 1990 banquet marking my thirtieth year of service to Pitt— Bas reported that when he gave me the first draft of his proposed doctoral thesis on "The Causal Theory of Time," I returned it with critical comments on *every* line, in a display of prismatic colors. For him, he said, this experience was at once the most traumatic and the most exhilarating in his life as a student. He soldiered on with enormous success, as reported above. I was delighted to accept the invitation, in virtue of being his quondam "*Doktorvater*," to introduce him at Carnegie Mellon University to deliver the 2002 Ernest Nagel Lectures.

Perhaps I acquired the reputation of a severe taskmaster as a PhD dissertation director, which then drove away other students. Indeed, I gather that even such very renowned philosophers as Willard V. O. Quine, Alonzo Church, and Paul Feyerabend each supervised very few PhD dissertations, if any. And, as I recall, when the great nineteenth-century mathematician Karl Friedrich Gauss appraised the paucity of his own publications, he aptly coined the Latin epigram *Pauca sed Matura*, few but seasoned.

Nancy Cartwright, who was a senior undergraduate physics major at the University of Pittsburgh, claims to have learned her philosophy of science from me. She has been a recipient of a MacArthur Fellowship, informally known as "the genius award," and, after a professorship at Stanford, has directed the University of London Centre for the Philosophy of the Natural and Social Sciences, while also teaching intermittently in the Department of Philosophy at the University of California at San Diego (La Jolla, California).

At this point, it behooves me to deal with another topic, and report on an episode in the institutional history of the American Philosophical Association, Eastern Division, featuring events in 1981–1983 that involved my election to its presidency.

The association has been divided into three divisions—the Eastern, Central, and Pacific Divisions—each of which elects its own officers (president, vice president, etc.), although there is also a national board of officers with its own overall chairperson. Of the three geographic divisions, the membership of the Eastern Division was and remains by far the very largest, because of the relatively larger number of colleges and universities in that part of the country.

After World War II, when considerable numbers of returning veterans enrolled in colleges and universities, there was a scramble among these institutions to recruit newly minted PhD graduates as faculty to staff their departments. In a significant number of colleges, which previously had not been offering advanced degrees in philosophy, the new faculty staff was being attracted by the promises to create graduate programs at the master's or even PhD level. Thus very large numbers of students flocked to these nascent graduate programs, especially on the eastern seaboard of the United States, which is relatively studded with institutions of higher learning.

In due course, by about five or ten years later, the great many graduates of these programs, which were sometimes marginal, entered the job market but found that it had contracted considerably in the interim. Unable to find a job, these untenured academic "proletarians" very understandably became disgruntled, thus turning into ready supporters of disaffected more senior members of the Eastern Division, who were smarting under their perception of insufficient recognition in the American Philosophical Association.

Thus, a mythic old-boy-network of narrowly sectarian but supposedly privileged "analytic philosophers" was invented and demonized. Allegedly, these doctrinaire sectarians eschewed dealing with the grand questions of perennial philosophy, thus supposedly impoverishing the field by a misguided focus on petty logical technicalities. The malcontents sailed under the banner of being philosophically eclectic, styling themselves as cosmopolitan "pluralists." And they chose as their immediate agenda getting one of their own elected to the presidency of the Eastern Division.

For years, it had been the practice of the division *to elect a Nominating Committee by a mail ballot of its entire large membership*, whereupon the elected nominating committee would choose the next vice president of the division without a further confirmation by the membership. The vice president, in turn, would serve for a year, but would then *automatically* become the president of the division for the following year.

But during the early years of the twentieth century, the divisional membership had been only a very small fraction of its size in the 1970s and 1980s. At that very early stage, business meetings were quite naturally well attended and were therefore suitable occasions for representative decision-making by the membership. Thus, the divisional bylaws written at that time provided scope for nominating and also selecting divisional officers at the business meeting itself.

But as the Eastern Division grew by leaps and bounds after World War II, attendance at its marginalized business meetings declined very drastically and became so unrepresentative that the early bylaws fell into chronic disuse, and indeed were very largely forgotten by most of the active members of the division. Unfortunately, however, the old bylaws were left on the books without any updating to take account of the enormous growth of the division and of such resulting changes as the marginalization of the business meetings.

Thus, any invocation of the anachronistic bylaws by as late as the last quarter of the twentieth century became, in effect, a case of taking advantage of a *loophole*, in contravention of the then existing established divisional practices, such as the de facto selection of the vice president and then president of the division by the elected nominating committee, without pro forma approval by the membership.

In mid-November 1979, I was very glad to receive the following letter from the chairman of the then nominating committee, Professor Monroe Beardsley, himself a humanist specializing in ethics, aesthetics, and philosophy of literature:

> The Nominating Committee of the Eastern Division, American Philosophical Association, met today, and it gives me enormous satisfaction to be the one to inform you that the committee has nominated you to be Vice-president in 1980 and President in 1981. You were our unanimous choice; and I hope very much that there will be no bar to your acceptance of this nomination.
>
> Let me know, when convenient, of your acceptance. I am already making firm plans to be present at your [presidential] address two years from now. Meanwhile, my best regards to you and your wife. (Quoted with permission of David E. Schrader, executive director of the APA, July 13, 2007)

I was, of course, greatly touched by Beardsley's very warm letter, and I wrote him a grateful response of acceptance. But I had no inkling at the time that the "pluralists" were preparing a kind of coup d'état by packing the December 1979 business meeting with their supporters, including students who had no credentials as members. Indeed, no arrangements had been made to check the voting credentials of the atypically much larger number of attendees. Invoking the bylaws of old in this way, the self-styled aggrieved pluralists nominated John Smith of Yale from the floor of that business meeting for the 1980 vice presidency, and they promptly elected him to that office at that same business meeting as the incongruous Ivy League standard-bearer of those who saw themselves as wrongfully marginalized.

Thus, my nomination to that 1980 office became defunct. But, as I was told emphatically by all and sundry, including some of the pluralists themselves, the brewing pluralist discontent in the division came to a head just when I happened to have been nominated, and furthermore, the pluralist revolt was not directed at all against me personally: I just happened to be caught in the cross fire. These sentiments were expressed in warm and sometimes affectionate personal letters of concern for my feelings, letters that I have prized ever since.

Professor J. B. Schneewind, a onetime Pitt colleague, then vice president of Hunter College in New York City, and thereafter professor and chairman of philosophy at Johns Hopkins University in Baltimore, Maryland, wrote me very cordially on January 7, 1980, as follows:

> I was simply appalled by the events at the recent APA meetings. Not the least of my concerns was the possible effect of this little *coup d'état* on your feelings. You surely deserve to be the vice president and the president of the Eastern Division. . . . The only consolation is that they would have behaved in the same way no matter who had been the nominating committee's candidate.
>
> I do hope you won't take it personally to heart. You have a great many friends and admirers, and all of us are concerned about the matter. (Quoted by permission from J. B. Schneewind, August 5, 2005)

Indeed, quite unexpectedly, I soon learned that as many as nine former presidents of the Eastern Division of the APA openly denounced the coup d'état, debunked the pluralist sophistry, and called for the defeat of their candidates in the impending 1980 election. Thus, in December 1980, they sent a major open letter to five hundred selected members of the division.

That letter is, I believe, a milestone document in the history of professional American philosophy. Hence I reproduce most of it below, having added italics to the sentences in it that, in my view, state the gravamen of their case against the pluralists.

Dear Colleague:

At last year's Eastern Division A.P.A. business meeting the Committee for Pluralism in Philosophy succeeded in electing its candidates for the Vice Presidency and the Executive Committee. This was the result of a determined campaign on its part to get out the vote [for *bloc* voting!], together with insufficient determination or concern on the part of those favoring the candidates recommended by the elected Nominating Committee. This year the elections will again be held at the business meeting (perhaps for the last time, since the By-Laws may be changed so that future elections will be by mail ballot [of the entire membership]). The Committee for Pluralism is again waging a well-organized campaign for its Vice-presidential candidate, William Barrett. We believe that Barrett's election would not represent the judgment of the Eastern Division as a whole. We are therefore writing to urge you to attend the meeting, and to persuade your colleagues and friends in the profession to do so too. If you agree with the views expressed in this letter, we hope you will send copies of it to others.

This year the Nominating Committee has submitted four names for the office of Vice President [as against only one candidate]: William Barrett, Arthur Danto, J. N. Findlay and Adolf Grünbaum. . . . We believe that the Vice-presidential election is particularly important, and that all three of the other candidates are superior to Barrett in philosophical distinction. But we particularly urge support for the candidacy of Grünbaum, the defeated choice of the Nominating Committee last year: he deserves the benefit of the kind of serious and well-organized effort that was not made at that time, to maintain high professional standards in the selection of officers and to resist factional pressures. We expect the election to be held by single transferable vote, and we urge you to put Grünbaum first in your ranking. . . .

More is at stake here than an A.P.A. election. *No reasonable person can be opposed to the exploration and development of a variety of philosophical approaches and methods. True philosophical pluralism is not at issue.* It is our belief that the A.P.A. has fairly represented philosophy in America, giving appropriate emphasis to those contributions, which have been of highest quality. *The Committee on Pluralism seeks to obtain, through political means, a position of influence, which its members have not been able to obtain through their philosophical work.* We believe that the Committee favors the suppression of serious scholarly and intellectual standards under the false banner of open-mindedness. (As an example we cite John Smith's approach to the NEH asking them to stop support for the Council of Philosophical Studies, alleging that it did not represent the philosophical views of the A.P.A.) If the Committee [on Pluralism] succeeds, then the A.P.A. and the philosophical profession will suffer, and will suffer because the rest of us have not bothered to resist with sufficient energy, and in sufficient numbers. (italics added)

Signed by former Eastern Division Presidents: Monroe Beardsley, Nelson Goodman, Carl G. Hempel, Donald Davidson, W.V.O. Quine, Alice Lazerowitz, Hilary Putnam, Kurt Baier, Wilfrid Sellars

In the ensuing 1980 election at a hugely attended and dramatic business meeting, I was handily elected to the vice presidency for 1981, and the presidency in 1982, after defeating the other three candidates for vice president, perhaps with an additional boost from a backlash against the pluralists and their manipulative tactics. And, at the December 1982 meeting of the Eastern Division in Baltimore, Maryland, I took special pleasure in delivering my presidential address to a packed house, because the dear Professor Monroe Beardsley, who had chaired the 1979 nominating committee, was in my audience, as he had said he hoped to be.

Although I had been working on psychoanalysis systematically for only about five years, I devoted my address of December 28, 1982, to the subject of "Freud's Theory: The Perspective of a Philosopher of Science."[41] At the suggestion of Professor Richard T. Hull, editor of a whole series of volumes of the presidential addresses of the American Philosophical Association, 1901–2000, I updated it as of about the year 2008, and expect to update it again before publication in one of Professor Hull's forthcoming volumes, which are being published by Prometheus Books.

Unfortunately, the pluralist revolt ushered in an unwholesome politicization of the division, despite my defeat of their candidate. For example, some pluralist activists skewed the subsequent elections by organizing *bloc voting*, such that the pluralists shamelessly horse-traded their support of potential candidates for high divisional office by exacting promises from them to heed their agenda. Yet, after the "Committee for Pluralism in Philosophy" had shot its bolt, some of its members met with like-minded others in Cambridge, Massachusetts, in November 1987 to keep the pot boiling by forming a new organization pretentiously titled "the Society of Philosophers in America."[42] But as far as I know, nothing ever came of it, and it deservedly just faded away.

Ten years after the first appearance of my first *Festschrift* under the title *Physics, Philosophy and Psychoanalysis: Essays in Honor of Adolf Grünbaum*, my four Pitt colleagues and friends John Earman, Allen Janis, Gerald Massey, and Nicholas Rescher graciously coedited a second *Festschrift* dealing largely with my work.

The papers in this 1993 volume grew out of the aforementioned three-day 1990 "Colloquium in Honor of Adolf Grünbaum" that occasioned an editorial tribute in the *Pittsburgh Post-Gazette* (October 6, 1990). Incidentally, at the banquet for that colloquium, Robert Turnbull, then chairman of the national board of officers of the American Philosophical Association, represented that body and hailed my major role in the steep ascent of philosophy and philosophy of science at Pitt to national and international prominence, extolling my recruitment of Wilfrid Sellars.

The new 1993 *Festschrift* is a volume of nearly 625 pages, and is titled *Philosophical Problems of the Internal and External Worlds: Essays on the Philosophy of Adolf Grünbaum*.[43] Its topical range was concisely outlined in the opening paragraph of a five-page review essay on it, written by Chris Daly of Brasenose College, Oxford, and was published in 1995 in the *Canadian Philosophical Review*. There, Daly wrote:

> This is a collection of 24 papers organised into five main sections: space, time, and cosmology; scientific rationality and methodology; philosophy of psychiatry; freedom and determinism, science and religion; and moral problems. The wealth of issues covered by these sections indicates the range of Grünbaum's thought, and the high standard of the volume's papers is a fitting testimony to the eminence and quality of Grünbaum's own writings.[44]

I learned a very great deal from the sixteen essays in the original 1983 *Festschrift*, as well as from the twenty-four in its successor of ten years later. And I am most grateful to the forty scholars who contributed them.

Both of these *Festschriften* contain some essays dealing with psychoanalysis. Yet the first impetus for my subsequent extensive inquiry into the intellectual merits of the psychoanalytic enterprise came only in the early 1970s as just one part of my systematic critical scrutiny of Karl Popper's very influential philosophy of science qua theory of scientific rationality. Hence I am greatly indebted to Popper for that vital stimulus.

Soon I published the results of my appraisal of Popper in four papers, three of which appeared in the 1976 volume of the *British Journal for the Philosophy of Science*, while the fourth was my contribution to the Imre Lakatos memorial volume of that year. These four papers were very critical of all the major facets of Popper's falsificationist philosophy of science.

My interest in psychoanalysis was kindled by Popper's emphatic claim that it provides the prime illustration of his pivotal thesis that the received "inductivist" criterion for distinguishing science from pseudoscience (and nonscience) by reference to available evidential support is unacceptably permissive epistemologically, and wrongly accords bona fide scientific status to Freudian theory. Wherefore, he avowed that inductivism as a "criterion of demarcation" needs to be *superseded* by his supposedly much more stringent requirement of falsifiability (refutability, or at least disconfirmability), by potential adverse empirical evidence. Thus the susceptibility of a hypothesis to refutation by logically possible adverse evidence became Popper's linchpin for the *scientific entertainability* of a hypothesis.

Three centuries before Popper came upon the scene, Francis Bacon had pioneered *eliminative inductivism* as a scientific epistemology. True, Bacon had erred in supposing that, for any given set of observational data, there is only a *finite set* of alternative hypotheses, each of which might explain them. But he emphasized long before Popper that negative instances have greater probative force than positive ones; in fact, he scoffed at simple enumerative induction from positive instances as "puerile." Thus, Bacon envisioned the *inductive elimination* of all but one of the supposedly finite number of alternative hypotheses, which would thereby be shown to be true.

Yet, as we know, he erred, since, in principle, there is always a potentially *infinite* set of alternative explanatory hypotheses, some of which working scientists may eliminate by means of refuting (disconfirming) instances, while other such hypotheses survive, at least temporarily, as theoretical candidates for inductive acceptance. Therefore, in my 1984 *Foundations* book, I advisedly distanced my self somewhat from Bacon, and I spoke of the latter epistemic process as *"neo-Baconian* eliminative induction."

My use of this designation should not have occasioned any carping, although it did at the hands of Joseph Agassi, who untutoredly chose to label me "the elusive neo-Baconian."

Popper averred that, as against Francis Bacon's seventeenth-century eliminative "inductivism," his own rival account of the rationality of the scientific enterprise is far more nuanced for several reasons: (i) it uses falsifiability (or at least disconfirmability) *rather than confirmability* as the touchstone of the scientific entertainability and legitimacy of a theory, and moreover this criterion is refined by allowing for *degrees* of falsifiability; (ii) it purportedly yields a solution to the inveterate problem of induction; (iii) it provides an epistemological theory of corroboration; (iv) its falsificationist methodology opens the door to scientific theories of increasing verisimilitude; and (v) his degrees of falsifiability even give science a criterion for the admissibility of auxiliary hypotheses in the modification of prior theories.

I found Popper's views immensely and admirably thought provoking, if only because his brimming ideas invited detailed critical scrutiny. But as early as 1976, I fully realized that his knowledge of Freudian theory was demonstrably very inadequate. Thus, the Canadian philosopher and veteran practicing psychoanalyst Charles Hanly wrote in his 1988 review of my *Foundations* book:

> Grünbaum's scholarship on the Freud texts, given his purpose and interests, is *light years ahead* of that of philosophers such as *Popper*, Habermas, and Ricoeur, not to mention the scholarship of some psychoanalysts. (italics added)[45]

The cardinal role played by the status of Freud's theory in Popper's advocacy of a falsifiability criterion of demarcation can be gauged more fully than above from the following considerations:

(*a*) In Popper's view, Freud's theory simply does not have any potential empirical falsifiers, that is, imaginable observations that would contradict it, and is therefore nonscientific.[46] And, in virtue of Freud's own claim of scientific status for it, his edifice is a *pseudo-science* rather than just a nonscience by Popper's criterion of scientificity. But, in his judgment, the time-honored inductivist conception of scientific rationality was unable to detect this supposed fundamental flaw, because inductivism allegedly always countenanced psychoanalytic hypotheses as *confirmed, come what may!*

Thereby inductivism purportedly failed to reveal the scientific bankruptcy of the psychoanalytic enterprise. Wherefore, Popper concluded: "Thus there clearly was a need for a different criterion of demarcation" between science and pseudoscience, *other than* the inductivist one. In this way, psychoanalysis served as the gravamen and benchmark of his case for the indispensability of his own rival falsifiability criterion of demarcation. But Popper had very tendentiously and ahistorically glossed over the major probative discrepancy between sophisticated eliminative inductivism and its enumerative ancestor.

Thus, in the concluding chapter 11 of *Foundations*, I rejected his treatment of the demarcation problem as follows:

> It is ironic that Popper should have pointed to psychoanalytic theory as a prime illustration of his thesis that inductively countenanced confirmations can easily be found for nearly every theory, if we look for them. Being replete with a host of etiological and other causal hypotheses, Freud's theory is challenged by neo-Baconian inductivism to furnish a collation of positive instances from *both* experimental and control groups, if there are to be inductively *supportive* instances . . . to this day, analysts have not furnished the kinds of instances from controlled inquiries that are *inductively required* [for example] to lend genuine support to Freud's specific etiologies of the neuroses. Hence, *it is precisely Freud's theory that furnishes poignant evidence that Popper has caricatured the inductivist tradition by his thesis of easy inductive confirmability of nearly every theory*! (italics in original)

(*b*) Throughout his career, Popper has repeatedly made two claims: (1) logically, psychoanalytic theory is empirically irrefutable by any human behavior; and (2) in the face of seemingly adverse evidence, Freud and his followers always dodged refutation by resorting to immunizing maneuvers via ad hoc auxiliary hypotheses. But clearly, his charge (1) of unfalsifiability against psychoanalytic theory *itself* does not follow from the *sociological* objection that Freudians are not responsive to criticism of their hypotheses. After all, a theory may well be invalidated by known evidence, even as its true believers refuse to acknowledge this refutation.

Besides, ironically, if Popper were right in (1) that Freud's theory simply does not have potential falsifiers, which could count against it, why would it ever have been necessary at all for Freudians to *dodge* refutations by means of immunizing gambits, as charged by Popper's item (2)? Apparently, his (1) and (2) are *incoherent*.

Moreover, it emerges clearly from some of Popper's other doctrines that the recalcitrance of Freudians in the face of falsifying evidence, however scandalous, is not at all tantamount to the irrefutability of their theory. As he tells us, theories, on the one hand, and the intellectual conduct of their protagonists, on the other, "belong to *two entirely different 'worlds.'*"

As Popper reported later on in his 1983 book, *Realism and the Aim of Science*, it was not until 1983 that he "published his detailed analysis of Freud's method of dealing with falsifying instances."[47] Yet, again ironically, no such instances should exist at all if Popper's categorical charge of empirical irrefutability against psychoanalysis is to be believed. My 1984 book was already in press when I obtained access to his 1983 volume. Therefore, I deferred my most *comprehensive* critical appraisal of Popper's case against Freud to my 1993 second book on psychoanalysis,[48] which is mentioned further below, although I had already taken issue with his earlier writings on that topic in my 1984 book.[49]

Clearly, the psychoanalytic enterprise played a central historical and logical role in Popper's falsificationist philosophy of science. Hence in my critical concern with the warrant for his complaint of pseudoscience against Freud, it behooved me to study the foundations of psychoanalysis, which I began from scratch in the mid-1970s.

But what of Popper's claim (1) above that *none* of the deductive consequences of Freud's hypotheses are refutable (disconfirmable) by potentially contrary empirical evidence? As I have shown in detail in my writings, his (1) is clearly incorrect.[50] But, for here, suffice it to mention three further critical points against Popper's objections to Freud:

i) In a fine contribution to Sidney Hook's well-known aforecited 1958 symposium, *Psychoanalysis, Scientific Method and Philosophy*,[51] whose proceedings were published in 1959, the Australian philosopher Michael Scriven thoroughly articulated key features of a cogent design for empirically testing the treatment-outcome claim that psychoanalysis effects substantial improvements in most of its patients.

As Scriven's paper showed clearly, the claim of favorable therapeutic *outcome* from psychoanalytic treatment is indeed testable, and hence potentially disconfirmable. Thus, four years before Popper categorically reiterated the intrinsic empirical untestability of the whole of the Freudian corpus in 1962, Scriven had already implicitly *undermined* that Popperian thesis, though *only* with respect to psychoanalytic *therapy*.

ii) Unlike induction-*by-enumeration*, which Francis Bacon dismissed as "puerile," Bacon's own *eliminative* inductivism emphatically does *not* license mere positive instances of a causal hypothesis *H* as probative support for *H*, contrary to the blithe, unsound claims of ubiquitous confirmation made perennially by a bevy of psychoanalysts, whose behavior Popper misinvoked in support of the unfalsifiability of psychoanalysis.

In 1985, Philip Holzman, a noted psychoanalytic psychologist and schizophrenia researcher, commented on my pre-1993 critique of Popper as follows: "Grünbaum, in a series of superbly reasoned appraisals [three references omitted] has joined this issue [of Popper's case against Freud] with a powerful counterargument. Psychoanalysis is indeed falsifiable."[52] Relatedly, my exoneration of psychoanalysis from Popper's complaint of pseudoscience was greeted with much satisfaction by another well-known psychoanalytic psychologist, Robert Holt, who wrote in 1989:

"Fortunately, a leading philosopher of science, Adolf Grünbaum, became interested in the problem of the scientific status of psychoanalysis in 1976 and began studying its literature. . . . Here at last is a philosopher who has done his homework before criticizing Freud."[53]

Furthermore,

Grünbaum [three references omitted] has earned the gratitude of all of us [psychoanalysts] by taking Popper down a peg, showing that his arguments not only are based on ignorance of what Freud actu-

ally said, but have logical flaws as well.... Grünbaum has unan-
swerably established the claim of psychoanalysis to a place among
the sciences by Popper's own criteria.[54]

And speaking of my 1984 *Foundations*, Holt declared: "The power
and subtlety of the analysis and arguments Adolf Grünbaum presents
in this book far surpass those of any previous philosophical evaluation
of psychoanalysis."[55]

In his reply to his critics, Popper sums up the centrality of psychoanalysis to his plaidoyer for his theory of demarcation and states the
upshot in a clumsily worded sentence:

> ... my criterion of demarcation ... is more than sharp enough to make
> a distinction between many physical theories on the one hand, and
> metaphysical theories, such as psychoanalysis ... on the other. *This is,
> of course, one of my main theses; and nobody who has not understood
> it can be said to have understood my theory.* (italics added)[56]

In my 1993 *Validation* (chapter 2, pp. 65–68), I dealt with the
deplorable 1983 denouement of Popper's Freud-critique and of his
treatment of the demarcation between science and nonscience in his
postscript of that year. Alas, as I have documented there, his 1983
notion of demarcation has the earmarks of a *degenerative* philosophical research program in Imre Lakatos's sense of the term. In mid-
September 2007, at a conference on "Rethinking Popper" held at the
Czech Academy of Sciences in Prague, I gave an invited paper on
"Popper's Fundamental Misdiagnosis of the Scientific Defects of
Freudian Psychoanalysis, and of Their Bearing on the Theory of
Demarcation." It appeared in the journal *Psychoanalytic Psychology*,
volume 25, number 4, October 2008, and also in the conference pro-
ceedings titled *Rethinking Popper*, published in 2009 by Springer
(Dordrecht, Holland).

In the aforementioned 1988 memorial book, *The Limitations of
Deductivism*, which commemorates my onetime Pittsburgh PhD dissertation student J. A. Coffa (1935–1984), my coeditor Wesley Salmon

called attention to my critique of Popper's deductivist construal of scientific rationality as follows: "In a series of four articles published in 1976, Grünbaum provides an extended and detailed critical *tour de force* against Popperian deductivism."[57] Then he lists the specific references in the literature to which I alluded before.

I might mention that, after Popper had retired from his professorship of philosophy at the London School of Economics, I felt greatly honored when Ralph Dahrendorf, the then director of that school, offered me Popper's chair there. But I decided to remain at the University of Pittsburgh, where my roots had become very deep.

Unfortunately, in the early 1970s, my knowledge of psychoanalysis was not better than that of a mere undergraduate. But even then, I had a strong hunch that Popper's wholesale, undemonstrated indictment of the whole Freudian corpus as categorically immune to empirical refutation or disconfirmation, and *therefore* as pseudoscientific, was a fundamental and indeed rash misdiagnosis. Presumably, his clearly insufficient psychoanalytic literacy, though a fellow resident of Freud's in Vienna, contributed to his basic error.

Thus, as I acknowledged in 1984, a decade later, in my first book on psychoanalysis: "The first impetus for my inquiry into the intellectual [and evidential] merits of the psychoanalytic enterprise came from my doubts concerning Karl Popper's philosophy of science, [doubts] which I published in a series of essays in 1976."[58] Therefore, it became essential that I scrutinize the pillars of Freud's edifice not only vis-à-vis Popper, but *quite generally*.

It is incumbent on me now to provide the background and context of my odyssey into psychoanalysis much beyond the initial motivation I derived from Popper's linkage of it to the demarcation between science and pseudoscience.

This assessment of psychoanalysis on my part became the more imperative when I turned to the writings of philosophers on it. Soon, I was deeply disappointed to discover that, alas, most of the *philosophical literature* on psychoanalysis, notably also in the English-speaking world, did not come to grips with the key issues of its credibility and was distressingly unilluminating concerning its logical structure and evidential rationale.

Thus, I looked in vain for a *systematic appraisal of Freud's major specific arguments* for his cornerstone repression theory of psychopathology, with its associated hypotheses on the dynamics of therapy, his dream theory, his theory of psychosexual development, and his theory of slips.

Indeed, I was dumbfounded that Breuer and Freud's crucial, epoch-making 1893 "Preliminary Communication" for their *Studies on Hysteria*, a lucid foundational document of only fifteen pages, was almost entirely ignored by philosophers writing on psychoanalysis, or that its cardinal epistemic significance was simply lost on them when they did mention it! Yet, they typically purported to elucidate the logical architecture of Freud's reasoning.

The prolific English philosopher Richard Wollheim, who has written a great deal about psychoanalysis over many years, illustrates my incredulous disappointment. Thus, in Wollheim's 1971 book, *Sigmund Freud*, under the historical rubric "The First Phase," he tells us that the upshot of the 1893 "Preliminary Communication" was: "Hysterics suffer mainly from reminiscences." And to this diagnostic conclusion, Wollheim adds synoptically that this communication of Breuer and Freud's "also sets these findings inside a somewhat broader conception of mental functioning."[59] Well enough.

Yet in his mere two additional pages (pp. 13–14) on this pivotal paper, Wollheim says nary a word to point out a key *epistemological* fact: Therein—in only two of their pages in J. Strachey's monumental English *Standard Edition of the Complete Psychological Works of Sigmund Freud* (vol. 2, pp. 6–7)—the founders of psychoanalysis lucidly encapsulated the *pioneering evidential crux* for their cornerstone *repression etiology* of hysteria (neurosis).

But I felt entitled to expect from a philosopher like Wollheim, especially since he was a true believer in psychoanalysis, both an exposition and an epistemic appraisal of the highly specific *therapeutic* argument given there by the founding fathers for their seminal repression etiology of the psychoneuroses. Alas, my hopes were entirely dashed.

The historic epistemological importance of Breuer and Freud's paper on the significance of their cathartic method—whose use of hypnosis antedates, of course, Freud's own introduction of free association as both his paragon method of clinical investigation and his vehicle for the therapeutic lifting of repressions—can be gauged from Freud's 1924 retrospect on their 1893 "Preliminary Communication" in his "Short Account of Psycho-Analysis":

> The cathartic method was the immediate precursor of psychoanalysis; and, in spite of every extension of experience and of every modification of theory, is still contained within it as its nucleus.[60]

Furthermore, in this 1924 paper, Freud also emphasized the dual investigative and therapeutic role of the cathartic method as "the immediate precursor of psychoanalysis":

> Thus one and the same procedure served simultaneously the purposes of investigating and of getting rid of the ailment; and *this unusual conjunction was later retained in psycho-analysis.* (italics added)[61]

The "extension of experience" and "modification of theory" mentioned by Freud feature importantly the *enlargement* of the range of begetters of compromise-formations and pathogens *beyond* repressed traumata to repressed forbidden *wishes*, as in his theory of dream-generation.

Nor is there any of the required cogent philosophical illumination in the writings of such philosophers as Sartre, Wittgenstein, Ernest Nagel (a dear friend of mine), Sidney Hook, and Alasdair MacIntyre. Nor yet is there an appropriate scrutiny of Freud's epistemically fundamental method of clinical investigation by supposedly "free" association, which he claimed to be *causally probative*, a further egregious omission.

Some of the authors merely considered snippets from here and there in the Freudian corpus, morsels that were often selected to implement prior agendas. Regrettably, the writings of the philosopher Stephen Toulmin in particular even tried my philosophical patience beyond endurance, if only because he irrelevantly injected so-called ordinary language philosophy epistemically into characterizing causality in psychoanalysis, a style of philosophy that, in my view, then deservedly died an overdue death on both sides of the Atlantic.

Reading Breuer and Freud's 1893 "Preliminary Communication" in James Strachey's English translation from an *epistemological* point of view, with an occasional glance at its German original, was an electrifying eye-opener for me early on.[62] Alas, as shown by the writings of legions of psychoanalysts, the core reasoning in that historic paper seems to have been lost on them as well.

One deplorable result of this impoverishing neglect was the conceptual murkiness beclouding much of the expository literature on the foundational ideas of psychoanalysis. The "Preliminary Communication" convinced me that there was no excuse for this nebulousness. It inspired me to fill this void within my 1984 book, *The Foundations of Psychoanalysis: A Philosophical Critique*.

My unfavorable verdict on the quality of the philosophical literature on psychoanalysis, which I encountered at the outset in my study of it in the mid-1970s, was corroborated by the Canadian philosophy professor and practicing psychoanalyst Charles Hanly, in his review of my 1984 book in the *Journal of the American Psychoanalytic Association*. As will be recalled, therein Hanly went beyond Popper and averred:

> Grünbaum's scholarship on the Freud texts, given his purpose and interests, is light years ahead of that of philosophers such as Popper, Habermas, and Ricoeur, not to mention the scholarship of some psychoanalysts.[63]

But, to a limited extent, the Australian philosopher Michael Scriven, whose experimental design for testing the efficacy of psychoanalytic treatment I have adduced above against Popper, is *eo ipso*, an honorable exception to my complaint about the inadequacy of the philosophical literature on psychoanalysis. That serious shortcoming was not remedied until the appearance of the work of such fine authors as the philosopher Edward Erwin.

When I set out to examine the credentials of the major pillars of the psychoanalytic theoretical edifice, I surely harbored no expectations, one way or the other, as to my ensuing stance on its evidential merits. If anything, the great cultural prestige still enjoyed by psychoanalysis at the time imparted a positive élan to my endeavors. In this vein, the well-known aforecited psychoanalytic psychologist Robert Holt declared in the aforementioned 1986 review symposium on my *Foundations* in *BBS*: "I don't believe that he [Grünbaum] set out to do a hatchet job; it is quite evident that he is not driven by hostility or spite."[64]

To my good fortune, in my endeavor to master the literature as I began my daunting task, I was the beneficiary of substantive help with it and encouragement from some practicing psychoanalysts and psychodynamically oriented clinicians, as well as from some opponents of Freud's legacy. Hence I was most happy to dedicate my 1984 book gratefully to four such people and to mention a number of others in my acknowledgments there (pp. xi–xiv).

The four dedicatees were, in alphabetical order: Morris Eagle, noted professor of psychology and clinician, as well as philosopher of psychology, who later became president of the Division of Psychoanalysis (Division 39) of the American Psychological Association; Stanley Rachman, creative behaviorist-theoretician and professor in the Department of Psychology of the Institute of Psychiatry at the Maudsley Hospital in London; Benjamin Rubinstein, a privately practicing psychoanalyst and prolific, distinguished psychoanalytic theoretician, who had, moreover, a very impressive command of relevant general philosophical literature;[65] and Rosemarie Sand of the Institute for Psychoanalytic Training and Research in New York, who was a practicing lay psychoanalyst in New York City for decades; having also done graduate work in philosophy at Columbia University, she became a resourceful expert on the *philosophical* ancestry of Freudian ideas, about which she has written an important original (forthcoming) book. All four of these very congenial mentors became my very good friends. But I became especially close to both Rosemarie Sand and Morris Eagle.

Moreover, Rosemarie soon gave me some invaluable advice as to how I should express my critical ideas on psychoanalysis in writing, if I aim, as I do, to reach not only philosophical readers but also those psychoanalysts who might be hospitable to criticism. As she pointed out, psychoanalysts and other mental health professionals are typically quite unaccustomed to the sort of hard-hitting polemical style encountered in some of the philosophical literature. Therefore, she counseled that I temper my argumentative fervor, if I am to make my potential psychoanalytic readers at all receptive to my views. I have tried hard to take her wise counsel to heart, but with only mixed results, I believe.

While working on what became the core chapters of my *Foundations*, I met and developed good intellectual rapport with Dr. Bahman Fozouni, a genial Iranian who was in Pittsburgh to work at Pitt's Western Psychiatric Institute and Clinic, and whose very Americanized nickname was "Buzz." It was he who first made me aware of the presumptuous yet influential contention of the European philosophers Jürgen Habermas (Germany) and Paul Ricoeur (France, later US) that, paradoxically, Freud himself had basically misunderstood the very nature of his own psychoanalytical enterprise, when claiming the status of a natural science for it.

As they would have it, his project ought to be construed instead as a reconstructed "hermeneutic" endeavor to which natural science modes of validation are alien. Yet Fozouni's motive for his admonition to me was not partisan, but rather his correct belief that it behooved me to address the exegetical and substantive challenge from the hermeneuts, which had made some headway among defensive psychoanalysts, and in scientophobic humanities departments in universities.

In this vein, I noted the timeliness of Fozouni's advice in the acknowledgements of my *Foundations* (p. xiii): "Dr. Fozouni gave me the benefit of his extensive knowledge of the hermeneutic literature by steering me to what is most germane to my concerns." But once I had heeded his recommendation, I became an acerbic critic of both of the following: (i) Habermas's regrettably muddled, primitive thesis that psychoanalytic therapy "overcomes" or "dissolves" rather than utilizes the presumed *causal connections* between the pathogens of the patient's affliction and the ensuing neurotic mentation, and (ii) Ricoeur's gross misassimilation of the Freudian corpus to the semiotics of psychoanalytic language in a so-called "*semantics of desire*" (*Foundations*, introduction, sections 2 and 3, respectively).

My initially academic friendship with Bahman Fozouni blossomed into a very warm, long-term personal friendship that soon included both his lovely wife, the dental specialist Dr. Mahnaz Fozouni, and my wife, Thelma. This cordial relationship occasioned happy get-togethers of the four of us when Thelma and I visited them in their attractive home in Eldorado Hills, California, or met them at the Wagners in Incline Village, Nevada (at Lake Tahoe), or most recently, when Buzz and Mahnaz were among the invited guests at the ever memorable bon voyage party that Harvey and Leslie Wagner graciously and affectionately gave for us at the splendid Four Seasons Hotel in San Francisco in late July 2007, when we were en route to Beijing.

My initial writings on psychoanalysis in journals in the mid-1970s soon attracted lively attention, both pro and con. But I was altogether unprepared for the *éclat* of the reception of my 1984 *Foundations* book, not only in the professional literature but also in the mass media. After all, it is a closely reasoned piece of philosophy of science and not readily accessible to the general reader. Thus, it has generated an enormous and still burgeoning literature, both critical and supportive, from writers of very diverse persuasions as to the merits of psychoanalytic theory and therapy.

But in some of the mainstream literature produced by psychoanalysts or their partisans, I have been pilloried with monotonous regularity—often obtusely and, alas, sometimes even quite dishonestly.

The volume of these critiques is such that I have not managed to read more than a fraction of it, and I have previously replied to only such a modest fraction of it. But my two-part essay "The Reception of My Freud-Critique in the Psychoanalytic Literature," whose first installment appeared in the July 2007 issue of *Psychoanalytic Psychology*, will, I trust, contribute to closing that gap.[66] The second installment is scheduled for later.

Foundations had appeared in late November of 1984. Yet, in less than two months, on January 15, 1985, the Science Section of the *New York Times* featured Daniel Goleman's front-page article on it, carrying a photo of me, highlighting my dispute with Popper on psychoanalysis and extending to a later page. Barely a week thereafter, the *New Republic* also carried an article on it (January 21, 1985), titled "The Future of an Illusion," by the renowned literary critic Frederick Crews at the University of California (Berkeley), who became an acerbic and widely read critic of Freud's intellectual integrity, and besides a very good personal friend. Very thoughtfully, he dedicated his 1986 book, *Skeptical Engagements*, to me "in friendship and gratitude."[67]

In his 1985 article, Crews had written: ". . . with the publication of Adolf Grünbaum's monumental new book, people will now begin to comprehend that the entire Freudian tradition—not just a dubious hypothesis here or an ambiguous concept there—rests on indefensible grounds."[68] Moreover, as I mentioned above only in passing, less than two years after the appearance of *Foundations*, the journal *BBS* devoted a review-symposium to it in which some forty-one contributors, drawn from the whole spectrum of diverse views on psychoanalysis, offered their commentaries, preceded by my "Précis," and followed by my "Author's Response" to all of them.[69]

This review-symposium appeared in Italian translation *Psicoanalisi: Obiezioni e risposte*, published by Armando Editore, Rome, 1988, under the editorship of Marcello Pera, who wrote a twelve-page introduction to it titled "*Le Sfide di Grünbaum*" (The Challenges of Grünbaum). Later, Pera became president of Italy's Senate.

Relatedly, in the year before, the Italian magazine *L'Espresso*, the counterpart of the American *Time* magazine, built a seven-page article around my *Foundations* (October 18, 1987, pp. 118–25), which it described in its opening paragraph as having presented the challenge of the century to psychoanalysis, though from an (unimpressively) small inner office, on the twenty-fifth floor of the University of Pittsburgh's Cathedral of Learning (which, let me hasten to say, is *not* a Divinity School!). Indeed, the magazine's photographer found that office unsuitable for a photo in the article, and I had to pose in front of the huge forty-two-story Cathedral building for the published picture.

Giovanni Forti, the journalist who interviewed me for this article, reported that, at the 1987 summer sessions of the World Congress of Psychoanalysis in Montreal (Quebec, Canada), my critique in *Foundations* "dominated" a number of them, presumably because of worries that I had provided fuel against third-party reimbursement for psychoanalytic treatment. Yet, in regard to that treatment outcome, such anxiety was quite disproportionate to what I had written there:

> In recent decades, comparative studies of treatment outcome from rival therapies have failed to reveal any sort of superiority of psychoanalysis within the class of therapeutic modalities that exceed the spontaneous remission rate gleaned from the (quasi-)untreated controls. (references deleted)[70]

Of course psychoanalysts carried the burden of having to justify the far greater expense and duration of their modality.

As of two decades later, in the summer of 2006, the Thompson Scientific *ISI Web of Knowledge* lists over three hundred articles citing my *Foundations*.

In 1985, almost immediately after its appearance, Barbara von Eckhardt, a philosopher of psychology and a granddaughter of famed, originally German, dissident and feminist psychoanalyst Karen Horney, published a very valuable essay, "Adolf Grünbaum: Psychoanalytic Epistemology," in *Beyond Freud: A Study of Modern Psycho-*

analytic Theorists, edited by Joseph Reppen, who has been, until
2007, editor of the journal *Psychoanalytic Psychology*, the official
journal of the Division of Psychoanalysis of the American Psychological Association. The very distinctive merit of von Eckhardt's essay
lies in its careful, lucid, and thorough fifty-page digest of some ten of
my articles that charted my course toward my 1984 *Foundations* book,
although that book did not, by any means, just recapitulate the bulk of
them. She expressed the view that these ten papers "have succeeded in
completely changing the state of the art." And her paper espoused the
following thesis:

> Grünbaum's contribution in . . . psychoanalytic epistemology . . . is
> unparalleled on . . . [two] counts. Not only does he bring to bear a
> very great sophistication in the philosophy of science, but, in addi-
> tion, he has done his psychoanalytic homework.[71]

Von Eckhardt's article also appeared in German translation in a 1991
volume titled *Kritische Betrachtungen zur Psychoanalyse* (*Critical
Reflections on Psychoanalysis*), which I edited and was published by
Springer-Verlag in Berlin-Heidelberg. That volume also contains a
German translation of the *BBS* review-symposium on my *Foundations*.

In *Foundations*, I deliberately, though incongruously, labeled its
first ninety-four pages "Introduction," which is a "Critique of the
Hermeneutic Conception of Psychoanalytic Theory and Therapy," and
whose substance is heralded above. But as I explained on its opening
first page, my rationale for the odd designation "Introduction" is based
on my main aim in the book: to examine *the principal clinical arguments* for the avowed cornerstone of his theoretical edifice put forward
by Freud, after they were first launched by Breuer and himself in
1893. Yet, as I pointed out there, I first needed "to expose a widespread
exegetical myth." Accordingly, I wrote:

> It is precisely that myth, the contrived reading, which has served as the
> point of departure for convicting Freud of "scientistic self-misunder-
> standing." This demonstrably ill-founded charge was leveled by the

philosophers Jürgen Habermas and Paul Ricoeur, champions of the so-called "hermeneutic" version of psychoanalytic theory and therapy. Indeed, their rendition has gained widespread acceptance in various quarters as now being at the cutting edge of the field, if not *de rigueur*. But besides resting on a mythic exegesis of Freud's writings, the theses of these hermeneuticians are based on profound misunderstandings of the very content and methods of the natural sciences.

Hence, it will be useful that I address, at the outset, not only the fabrication of the textual legend but also the multiple ontological and epistemic blunders inherent in the currently fashionable hermeneutic construal of psychoanalysis. The more so since [in 1971] Habermas has [very wrongheadedly] deemed precisely this reading of the Freudian corpus to be potentially prototypic for the other sciences of man.[72]

I developed these ideas considerably further in a 1999 paper titled "The Hermeneutic versus the Scientific Conception of Psychoanalysis: An Unsuccessful Effort to Chart a *Via Media* for the Human Sciences."[73] It was enlarged somewhat for reprinting in 2003 under the title "The Poverty of the Semiotic Turn in Psychoanalytic Theory and Therapy" after being invited very open-mindedly to appear in a ninetieth birthday *Festschrift* for Paul Ricoeur, although the paper is very critical of Ricoeur's stance on Freud.[74]

In a promotional quotation for the dust jacket of *Foundations*, the noted American anthropologist and practicing psychoanalyst Melford Spiro opined:

> With this book Adolf Grünbaum has established himself as the most important philosophical critic of the hermeneutic conception not only of psychoanalysis but also of the social sciences, most especially sociology and anthropology. His criticism of the logical and methodological foundations of this style of research and theorizing is masterful and, in my view, definitive.

Relatedly, in 1986, Morris Eagle, one of the aforementioned eminent psychoanalytically oriented dedicatees of the book, provided a masterful twenty-four page in-depth review article on it in the journal *Philosophy of Science*, where he wrote in his concluding paragraph: "Grünbaum's devastating critique of hermeneutic conceptions and formulations . . . alone represents a major contribution of Grünbaum's book."[75]

Sidney Hook, whose well-known 1958 New York University conference on the merits of psychoanalysis I had occasion to mention above, writing in another promotional quotation for *Foundations*, described that book as "critically sympathetic to psychoanalysis" and yet as a "formidable critique of the probative relevance of the clinical data" offered in support of the theory. And he is quite specific on the usefulness of my critique of hermeneutics:

> Especially noteworthy is Professor Grünbaum's repudiation of the obfuscatory efforts of Habermas and Ricoeur to save Freud from his alleged "scientism" by reading into his text dubious metaphysical notions.

Furthermore, writing from a perspective very unfavorable to psychoanalysis, the English behaviorist theoretician Hans Eysenck of the Department of Psychiatry at the Maudsley Hospital, London, reviewed the book in the British journal *Behaviour Research and Therapy*, where he declared:

> [It's] an absolutely essential critique of psychoanalysis. . . . It goes to the roots of psychoanalysis in a manner which other philosophers . . . have tried to do, but have conspicuously failed to achieve. . . . A brilliant book . . . [by] a razor-sharp intellect . . . the most important discussion of the topic to be found in the literature . . . only a philosopher could have written it.[76]

Psychoanalysis was the topic of my ten (actually twelve, as it turned out) Gifford Lectures, which I felt highly honored to deliver in March 1985 at the University of St. Andrews in Scotland. In November of that year, I was privileged to give the Werner Heisenberg Lecture on another facet of Freudian theory at the Bavarian Academy of Sciences in Munich. While I was in St. Andrews, the English analytical psychologist Ann Casement wrote me from London to obtain more information about my lectures than she found in the London *Times*. As a sound reviewer for the London *Economist*, and as the editor of the anthology *Who Owns Psychoanalysis?* she disseminated my ideas on Freud in the UK. And, happily, we have become cordial friends ever since.

In his 1993 book, *Freud and His Critics*, the clearly pro-Freudian intellectual historian Paul Robinson devotes one chapter each to Frank Sulloway and Jeffrey Masson, and a third to my *Foundations* book under the title "Adolf Grünbaum: The Philosophical Critique of Freud."

At the start of this chapter, he provides a survey of some of the reactions, both pro and con, to my book by some authors who champion rival perspectives on psychoanalysis. Robinson introduced his chapter by supplying some context, saying:

> By that time [1984], Grünbaum, who was born in 1923, already had behind him a distinguished career as a philosopher of science. He was especially admired for his "magisterial" studies in the philosophy of time and space [footnote reference to Robert S. Cohen's aforecited 1983 *Festschrift* article "Adolf Grünbaum: A Memoir" partly omitted]. In contrast to Sulloway and Masson, then, Grünbaum came to his engagement with Freud relatively late in life with an impressive record of accomplishment in another field of inquiry.[77]

Thus, unlike some of my psychoanalytic critics, Robinson treats my earlier work in the philosophy of science as an asset, rather than as a liability. These Freudian critics have relied on a red herring as follows: Purportedly, I have *misextrapolated* the epistemology of *physics* to the evidential appraisal of psychoanalysis. This supposed methodological malfeasance on my part allegedly undercuts my Freud critique. But Robinson wrote most appreciatively about my 1984 book:

> The *Foundations of Psychoanalysis* has been widely hailed as the most substantial philosophical critique of Freud ever written. One

might suspect the enthusiasm of so resolute an anti-Freudian as Fred-erick Crews, who greeted Grünbaum's book, in *The New Republic*, as "monumental" and "epoch-making." "After Grünbaum," wrote Crews, "the wholesale debunking of Freudian claims, both thera-peutic and theoretic, will be not just thinkable but inescapable [refer-ence omitted]." Yet even a critic of the book like the philosopher David Sachs acknowledged it as "an 'event' in philosophical criticism of psychoanalysis" [reference omitted]. When, in 1986, the book was made the subject of "Open Peer Commentary" in the journal *The Behavioral and Brain Sciences*, forty-one of Grünbaum's colleagues paid tribute to his achievement. Robert R. Holt, Professor of Psy-chology at New York University, wrote that "the power and subtlety of the analysis and arguments Adolf Grünbaum presents in this book far surpass those of any previous philosophical evaluation of psycho-analysis," and Irwin Savodnik of UCLA called it "the most exhaus-tive and powerful critique of psychoanalysis to date" [reference omitted]. Psychoanalysts were hardly less admiring—in sharp con-trast to their response to Sulloway and especially Masson. The ana-lysts Robert Wallerstein and Judd Marmor praised Grünbaum's inci-siveness and his mastery of Freudian theory, while Marshall Edelson paid him the high compliment of writing an entire book to answer his criticisms [reference omitted]. . . .

Although philosophical critiques of psychoanalysis have existed almost since the doctrine first made its appearance at the turn of the century, Grünbaum's analysis is distinguished from these earlier efforts by several features. Most notable are his extraordinary rigor and precision. Grünbaum is manifestly both very smart and very sophisticated, and his critique maintains an unprecedented level of dialectical intensity. At least for the philosophically untutored (to borrow one of Grünbaum's favorite words), virtually every sentence must be carefully unpacked, so thick and unforgiving (although never obscure) is his habit of thought. At the same time, Grünbaum surpasses all previous philosophical critics of psychoanalysis in the breadth and suppleness of his knowledge of Freud's writings.[78]

Yet, alas, as in the case of some other intellectual historians, Robinson's skill in historical reportage and knowledge of sources is not matched at all by the dialectical acumen required to deal with the specifics of my arguments or other pertinent philosophical issues. In my aforementioned 2007 essay, "The Reception of My Freud Critique in the Psychoanalytic Literature," the interested reader can find my documentation that Robinson's *own exegesis* of my critique of Freud is, alas, a mere figment of his philosophically untrained imagination. And there I show that his resulting misreading spells an important lesson of broad relevance: *The failure of an intellectual historian to understand the logical architecture of a theory, and its philosophy, begets false or at best misleading intellectual history of it.* In my 1993 book, *Validation in the Clinical Theory of Psychoanalysis,* I have given a further telling illustration of this vital moral.[79] And I had good reason to issue the same caveat apropos of the history of Einstein's special theory of relativity.[80]

In Philip Holzman's general "Introduction" to *Validation,* which he had sponsored for publication in the well-known Psychological Issues Series issued by the International Universities Press, he evaluated its content in terms that bear quotation, if only because they are antithetical to radically unsound dismissals of my Freud critique by some, though commendably *only some,* other psychoanalysts. Thus, Holzman wrote:

> During the past decade [1983–1993], I have found it especially welcome and instructive to read Adolf Grünbaum's brilliant and incisive critical analysis of the nature of the scientific enterprise within psychoanalysis. . . .
> . . . Beginning in 1975, he [Grünbaum] began to read Freud's works systematically. Soon, in a series of superbly reasoned and vigorously argued papers, he established a new level of criticism that was both appreciative of the brilliance of Freud's writings and yet uncompromisingly clear about their defects.

And speaking of six of the chapters in my *Validation* volume, Holzman pointed out that they:

address further crucial questions in psychoanalysis as well as offer a
new systematization of the theory of placebogenic phenomena not
only in psychiatry but in medicine in general. . . . In pursuing these
issues, Grünbaum demonstrates his superior sophistication about
psychoanalysis compared with those philosophers who participated
in the celebrated debates of 1959 [1958].[81]

It behooves me to return now to my earlier comment on Breuer and
Freud's pioneering 1893 "Preliminary Communication" because of an
important *spin-off* from it. As I have reported earlier, this paper of theirs
was a pathfinder for me early on, because it encapsulated—on its crucial
pages 6 and 7—the fundamental logical architecture of their epoch-making
argument for postulating that unsuccessful repression is pathogenic.

The incipient founders of psychoanalysis told us that their patients
experienced symptom relief, when they enabled them hypnotically to
recall forgotten presumably (repressed) traumatic memories, along
with the verbalized ("cathartic") release of the accompanying negative
affect. The two authors then argued (p. 7) that these remissions of
symptoms were *caused* by the insightful lifting of the patient's previously ongoing repression of the pertinent traumatic memory, emphatically *not* by his/her hopeful expectation of symptomatic improvement.

Incidental treatment factors, such as the arousal of the patient's
hope of improvement, are often called *placebo* factors, as distinct from
those treatment components that are mandated by the therapist's
theory of the dynamics of effecting improvement. But, importantly,
Breuer and Freud's reason (p. 7) for *denying* the rival hypothesis that
the improvements shown by their patients were *placebogenic* was fallacious, as I explained in my article "Critique of Psychoanalysis" in
Edward Erwin's *2002 Freud Encyclopedia*.[82]

The important point is this: To discredit the hypothesis of placebo
effect, it would have been essential to have a comparison with treatment outcome from a suitable control group whose repressions were
not lifted. If that control group were to fare equally well, treatment
gains from psychoanalysis would then be placebo effects after all. Nor

have other analysts after Freud discredited the ominous hypothesis of placebo effect by the required comparison of treatment outcome.

Yet, having concluded unwarrantedly that the patient's improvements were wrought by the mediation of psychoanalytic insight rather than placebogenically, Breuer and Freud used this conclusion, in turn, to infer (again fallaciously) their *repression-etiology of the neuroses*: It asserted that unsuccessful—and hence only *partial*—repressions, accompanied by affective suppression, are themselves *causally necessary* for the very existence of a neurosis.[83] On this view, neurotic symptoms are the products of unsuccessful repressions by being "compromises between the demands of a repressed impulse and the resistance of a censoring force in the ego."[84]

To draw this inference, Breuer and Freud reasoned (incorrectly) that their supposed insight-dynamics of therapy supports their repression-etiology, simply because the latter *entails* the former.[85] But, as I have shown in my aforementioned contribution to the *2002 Freud Encyclopedia*, their inductive argument here is vitiated by what I have dubbed *"The fallacy of crude hypothetico-deductive pseudo-confirmation."*[86] (All but one section on "The Future Prospects of Psychoanalysis" from this *Freud Encyclopedia* article are reprinted, but coupled with a special new preface, in my paper "Is Sigmund Freud's Psychoanalytic Edifice Relevant to the 21st Century?" which appeared in the Special 150th Freud Anniversary Issue of the journal *Psychoanalytic Psychology* 23, no. 2 (Spring 2006): 257–84.) The reader will find *my most articulated statement and critique* of Breuer and Freud's 1893 foundational reasoning in my aforecited 2007 paper, "The Reception of My Freud-Critique in the Psychoanalytic Literature."

In any case, the still unrefuted rival hypothesis of placebo effect gravely jeopardizes the historic clinical argument for the cornerstone theory of repression as well as what Freud built on it: The repression-etiology is aborted by not being able to get off the ground, as it were, because the challenge of the unrefuted placebo hypothesis undermines the initially inferred psychoanalytic dynamics of therapeutic gains on which the repression-etiology had then been predicated.

My appreciation of the import of the placebo challenge for the clinical foundations of psychoanalysis generated a fruitful spin-off for me: I was prompted to give a new rigorous analysis of the placebo concept in both general medicine and psychiatry after I found that the extant literature on placebos was an utter *conceptual* miasma as well as a *terminological* Tower of Babel.

Soon I submitted an article for publication in which I rebuilt the placebo concept from scratch, along with revamping its inherent gobbledygook vocabulary. But the referees of the journal to which I originally sent it were steeped in the received ill-conceived notions of the field. Hence their unfavorable reports were exasperatingly obtuse. Yet, in due course, after a merely embryonic 1981 article, I published my full-fledged paper, "The Placebo Concept in Medicine and Psychiatry," in 1986 in the highly reputable British journal *Psychological Medicine*.[87]

This article has since been reprinted four times, notably in a 1989 volume of the World Health Organization titled *Non-Specific Aspects of Treatment*,[88] a title which, alas, stubbornly uses the term "non-specific," one of the several terms in the prior literature that I deplored as very ambiguous or misleading. In a review of a 1994 volume, *Philosophical Psychopathology*, edited by George Graham and G. Lynn Stephens, which contains one of these reprintings, the physician-author Philip R. Sullivan wrote: "Grünbaum's paper, 'The Placebo Concept in Medicine and Psychiatry,' should be read by any and every medical experimenter," adding generously that the paper is a "top-flight reprint."[89]

Soon after 1970, I had the great luck to meet Thomas Detre, an enormously and multiply talented academic psychiatrist. He had come to Pitt from Yale as chairman of the Department of Psychiatry, which had theretofore been a kind of satrap of the psychoanalysts in the department. It did not take him long to transform the then pedestrian department into one of the foremost psychiatry departments in the country. Tom's creative and administrative talents then earned him appointment as senior vice chancellor of the health sciences until his fairly recent retirement.

Our interpersonal chemistry was splendid. Despite the burdens of greatly overhauling the Department of Psychiatry, he soon found time to talk to me about my ideas on placebos as well as about my incubating, nascent critical views concerning psychoanalysis. To my delighted surprise, Tom then took the initiative of proposing my secondary appointment as research professor of psychiatry, which I gladly received in 1979, while, of course, retaining my Andrew Mellon chair of philosophy. I have always felt very much at home with my colleagues in psychiatry.

Tom Detre was married to an eminent, internationally renowned medical epidemiologist, Katherine Detre, who directed major, far-flung, and well-funded inter-university research projects. To my and Thelma's great joy, the Detres became our very dear, close friends with whom we also shared the ups and downs of life. But, tragically, to our deep sadness, Katherine died in 2006 of cancer. We miss her very much.

By the time of my psychiatry appointment, I was offering a graduate course in the Philosophy of Psychoanalysis, which—for some time—enrolled as many as ten physician-residents from the Department of Psychiatry, along with a contingent of graduate students in philosophy, history and philosophy of science, and some other departments of the faculty of arts and sciences, although the literary people tended to stay away. The sometimes quixotic graduate students and the down-to-earth psychiatrists made for a fermenting mix, and I found it edifying to orchestrate the seminar discussions.

Alas, the sharp cutback in government funding for the training of residents in psychiatry across the country soon forced the Pitt residents to devote all of their time and energy to the clinical care of patients at the expense of their own postdoctoral studies. Thereafter, most of them understandably dropped out of my elective course on psychoanalysis.

What of the reception of my Freud-critique by psychoanalytic partisans? Here I can refer the reader to my aforementioned 2007 essay "The Reception of my Freud-Critique in the Psychoanalytic Literature."

A number of philosophic partisans of Freud, no less than some of the psychoanalytic contributors to the "Symposium on the Grünbaum Debate," have misread and misportrayed my views, setting up straw men for easy demolition. Thus, in a chapter on "Desire, Belief, and Professor Grünbaum's Freud," in his book *The Mind and Its Depths*,[90] Richard Wollheim—a true believer in psychoanalysis, and especially an aficionado of Melanie Klein's egregiously far-fetched version of it—engaged in just such misstatement of my views, as I documented carefully in a lengthy letter to the editor of the *New York Review of Books*.[91] But Edward Erwin has published a much more detailed defense of my views against Wollheim in Erwin's searching 1996 book, *A Final Accounting*.[92]

Now let me comment on the disappointing objections raised by Jonathan Lear, a professor of philosophy and lay analyst, who is more or less the philosophical apologist of the American psychoanalytic establishment.

In a 1995 article "The Shrink Is In," "a counterblast in the war on Freud," Lear wrote very begrudgingly against the wide attention accorded to my views in mainstream publications:

> To see nuance disappear, one has only to look at the supposed debate over the scientific standing of psychoanalysis. In a series of books and articles, Professor Adolf Grünbaum of the University of Pittsburgh has argued that psychoanalysis cannot *prove* the cause-and-effect connections it claims between unconscious motivation and its visible manifestations in ordinary life and in a clinical setting. Grünbaum argues correctly that Freud made genuine causal claims for psychoanalysis; notably, that it cures neurosis. But Grünbaum goes on to argue, much less plausibly, that in a clinical setting psychoanalysis cannot substantiate its claims. It is remarkable how many mainstream publications—*Time, The New York Times, The Economist* to name a few—have fallen all over themselves to give respectful mention to such abstruse work as Grünbaum's. Mere

mention of the work lends a cloak of scientific legitimacy to the attack on Freud, while the excellent critiques of Grünbaum's work are ignored.

There is no doubt that the causal claims of psychoanalysis cannot be established in the same way as a causal claim in a hard-core empirical science like experimental physics. But neither can any causal claim of any form of psychology which interprets people's actions on the basis of their motives—including the ordinary psychology of everyday life.... We cannot *prove* that our ordinary interpretation [of motivated behavior in ordinary life] is correct....

What are we to do, abandon our ordinary practice of interpreting people ...? No historical causal claims can be verified in the same way as a causal claim in physics.... Yet as soon as one enters the realm of meaningful explanation one has to employ different methods of validating causal claims than one finds in experimental physics. And it is simply a mistake to think that therefore the methods of validation in ordinary psychology or in psychoanalysis must be less precise or fall short of the methods in experimental physics.[93]

It is embarrassing that a philosophy professor like Lear ran afoul of the following considerations:

1. He rehearsed the dreary red herring that I have made inappropriate and unreasonable probative demands on psychoanalysis by misextrapolating the epistemology of (particle) physics to the evidential appraisal of the Freudian corpus. It is deplorable that Lear very naively speaks of "proof" in physics as categorical, while miscontrasting it with interpretation in psychology in regard to fallible epistemological status, as I shall show.

As for the canard of epistemological misextrapolation, I had occasion to deal with it in rebutting the animadversion by the philosopher Peter Caws that I displayed "physics envy" in my Freud critique.[94] As I put it:

As against Caws phantom, note that ever since Francis Bacon four centuries ago, it has been clear that *there is nothing at all endemic to physics* in the requirements for the validation of *causal* hypotheses, with which psychoanalytic theory and therapy are replete. And I demanded the fulfillment of just these entirely legitimate epistemological and methodological requirements [as I would in any other field of empirical knowledge] when I argued that the principal tenets of the Freudian corpus, which are causal, have not been validated even a century after their enunciation. (emphasis in original)[95]

2. Lear speaks evasively in mere generalities without providing chapter and verse from a single one of my epistemic indictments of psychoanalysis to illustrate his philosophical complaint against me.

For specificity, consider my reasons for claiming that, in Breuer and Freud's "Preliminary Communication," these founding fathers of psychoanalysis built their historic repression-etiology of the neuroses clinically on inferential quicksand. Let me recall that the reader will find my careful statement of these reasons conveniently (pp. 124–26) in my aforementioned article "Critique of Psychoanalysis" in the *2002 Freud Encyclopedia*, and even more explicitly in my aforecited July 2007 paper, "The Reception of My Freud-Critique in the Psychoanalytic Literature" (pp. 552–53).

Hence, I put the question to Lear: Where, oh where, in this reasoning did I (tacitly) assume that psychoanalysis has failed to substantiate its causal (etiologic) claims, *because* it cannot—in Lear's wording—*"prove"* them in the epistemic manner allegedly *peculiar to physics*? Or, to take another major specific example, where did I engage in such philosophic misextrapolation, when I indicted the cardinal psychoanalytic method of clinical investigation by free association as causally nonprobative, if only because it is guilty of fallacious *causal inversion*?[96]

The hollowness of Lear's charge of epistemic misextrapolation against me is of-a-piece with a series of further errors as follows:

(a) Repeatedly, he trades on the serious ambiguity in his tricky use of the phrase "establish [verify] . . . in the same way" as used, for example, in his sentence: "There is no doubt that the causal claims of psychoanalysis cannot be established *in the same way* as a causal claim in a hard-core empirical science like experimental physics." And again, "No historical account is immune to skeptical challenges; no historical-causal claim can be verified *in the same way* as a causal claim in physics" (italics added).[97]

But here, Lear is fallaciously relying on a mere platitude to infer an intrinsic asymmetry of epistemic fallibility. For example, obviously, an oral lesion in a person is not identified (or verified) histologically in quite the same way as the gunshot wound suffered by President Kennedy in Dallas. But that patent difference hardly makes for a generic *epistemological asymmetry* between the credibility of their respective diagnoses.

Similarly, *within physics*, the relative hardness of minerals is detected by the "scratch test," whereas radiations given off by the radioactive release of atomic energy are detected by Geiger counters or scintillation counters. Thus, causal claims in mineralogy are not established "in the same way" as those in nuclear physics. Yet again, this difference in method does not generically militate against there being epistemic parity in regard to fallibility between the evidential warrants of some claims in the one field and some in the other.

Again, the *subject matter* of physics differs, at least phenomenally, from that of psychology, but the respective interpretations (hypotheses) are alike fallible or revocable. Indeed, even within physics, theories can differ greatly in regard to credibility, because they differ in the extent of the evidential support they command. Thus, for a good many years after Einstein's 1915–1916 enunciation of his general theory of relativity, only three observational tests had provided support for it, by contrast to, say, the physics of metallurgy.

(b) Lear complains that my Freud-critique exemplifies the disappearance of "nuance." But methinks the shoe is on the other foot, as illustrated by his failure to allow for the following sorts of very familiar cognitive states of affairs:

When I take a sip of a liquid and say that it is water (H_2O), or say of a white powder that it is table salt (sodium chloride), or yet say of a nocturnal gleaming silver globe in the sky that it is the earth's monthly satellite, I am putting forward familiar but fallible, revocable interpretations (hypotheses). They are fallible, no less than if I say that it is Joseph Lieberman's ambition to become president of the United States by running as an independent. As for the revocability of hypotheses in physical science, one need only read a newspaper to learn about the recent demotion of far-out Pluto from its erstwhile hypothesized status of ninth planet to a minor dwarf among others orbiting in a distant ring of icy debris.

3. In regard to historical-causal claims concerning the outbreaks of wars, Lear rightly deems none of them "immune to skeptical challenge." But he adds anew that no such claims "can be verified in the same way as a causal claim in physics," again intoning the slippery mantra of "in the same way." Hence I must ask: Has Lear never heard about the historical Kant-Laplace nebular hypothesis, which offered to explain *causally* how our solar system was formed from a nebula? Even the popular *World Book Encyclopedia* stressed the *epistemic fallibility* of that hypothesis, pointing out that this historical-causal "theory has been changed by new discoveries *and different analyses of known facts*" (italics added).[98] Thus, it was replaced by the Chamberlin-Moulton "Planetesimal Hypothesis."

More generally, the grandest historical-causal claims of all are featured *within physics* by large-scale physical cosmogony and cosmology, which inform us concerning the galactic and stellar evolution of the universe from the big bang. On a spatiotemporally much smaller scale, there are the historical-causal branches of geology and geophysics, such as paleontology (fossils), volcanology, glaciology, sedimentology, and more broadly, earth history divided into "eras," "ages," and "epochs."[99]

Somatic and psychosomatic medicine likewise spell the same moral of epistemic revocability in asserting (long-term) historical causation in carcinogenesis and other kinds of pathogenesis.

In short, all of the empirically based theories, however diverse their specific subject-matter, are alike *interpretive* and more-or-less fallible by transcending the evidence for them conjecturally.

It emerges that Lear's complaint against me is just a red herring, vitiated by his unawareness that, in important respects, the history of physical and biological science is the chronicle of the *supercession* of ineluctably fallible, conjectural theories by epistemically more preferable ones.

By the time Lear wrote in 1995 that "the excellent critiques of Grünbaum's work are ignored," he ought to have done his homework in the literature to find out whether the supposedly "excellent critiques" of my work were indeed being ignored. Had he done so, he would have found the opposite. As I see it, powerful, in-depth rebuttals to the allegedly "excellent critiques" appeared as follows:

(i) My cherished friend Edward Erwin, the editor of the *2002 Freud Encyclopedia* and a leading, prolific author in the field whose writings Lear has ignored, issued a major (52 page) essay titled "Philosophers on Freudianism: An Examination of Replies to Grünbaum's *Foundations.*" It appeared in a 1993 *Festschrift* for me. As Erwin explains:

> In the first round of replies to Grünbaum's *Foundation of Psychoanalysis: A Philosophical Critique* (1984), most of the critics were non-philosophers. Philosophers now have had their turn. In what follows, I discuss some of the issues they have raised.[100]

There Erwin presented cogent and well-honed counterarguments to some of my pro-Freudian *philosophical* detractors (pp. 409–60) and in a subsequent powerful assessment of 1997 titled "Psychoanalysis: Past, Present, and Future," Erwin extended his polemic against further recent philosophic defenders of Freud.[101] This important article supplements the arguments given by him in his 1996 book, *A Final Accounting.*

(ii) Moreover, the philosophically sophisticated clinical psychologist Morris Eagle, one of the dedicatees of *Foundations,* took up the cudgels for me against the charge of anachronism. It had been contended in criticism of me that post-Freudian versions of psychoanalysis—either so-called self-psychology or object-relations theory—had remedied the epistemological and methodological difficulties that I had urged against classical Freudian theory. But in his contributions to both of my *Festschriften* (1983; 1993), Eagle showed incisively that this apologia is without merit. Besides, in a highly instructive and very valuable twenty-four-page in-depth review of my *Foundations* in the journal *Philosophy of Science*, Eagle vigorously articulated and defended the various theses in that book and offered keen apercus concerning psychoanalysis as a "movement."

(iii) In my "Author's Response" to the forty-one commentators in the aforecited *BBS* review symposium on *Foundations*, I replied to all of my critics therein, although the commentators represented a spectrum of attitudes toward psychoanalysis.[102]

Besides, in a 1996 letter to the editor of the *New Republic*, I responded to Lear's December 1995 article in part as follows:

> Jonathan Lear (December 25) could have spared your readers his contention that I wrongly assimilated the validation of cause-and-effect connections in psychoanalysis to those in particle physics, when I found psychoanalysis wanting. He needed only to have read my "'Meaning' Connections and Causal Connections in the Human Sciences" in the *Journal of the American Psychoanalytic Association* (vol. 38, no. 3, 1990). There I explained why "meaning"-connections between mental states do not, by themselves, vouch for *causal* connections between them and how psychoanalysts have failed to validate their major etiologic hypotheses concerning psychopathology, transference, or dream production. My alleged emulation in psychoanalysis of the standards for "proofs" in particle physics is just a red herring, if only because Lear's belief that the latter "proofs" achieve certainty is primitive, [a belief which is resoundingly belied by the history of physics].[103]

I should recall for the reader that, apropos of the late Pope John Paul II's inference of a *causal* connection between two events from *a mere thematic kinship* between them, I pointed out that this inference committed what I have dubbed the "thematic affinity fallacy." In my 1984 *Foundations* (pp. 55, 62, 198 and 227–28) and in my 1993 *Validation* (pp. 121–38) book I had called attention to its repeated commission in Freud's 1909 case history of the "Rat Man."[104] Then I was enormously heartened when the eminent Minnesota psychologist, philosopher of psychology, *and veteran psychoanalyst* Paul E. Meehl hailed my challenge to causal inferences from mere thematic connections as "the biggest single methodological problem that we [psychoanalysts] face." I hope I shall be forgiven for proudly feeling vindicated by *Meehl's watershed acknowledgment of my critique*. Fittingly, I think, he published it in the *Journal of the American Psychoanalytic Association* in his 1995 "Commentary: Psychoanalysis as Science." Indeed, Meehl speaks very soberingly concerning the import of the failure to meet the major epistemological difficulty I posed. As he put it in context:

> His [Grünbaum's] core objective, the epistemological difficulty of inferring a causal influence from the existence of a theme (assuming the latter can be statistically demonstrated), is the biggest single methodological problem that we [psychoanalysts] face. If that problem cannot be solved, we will have another century in which psychoanalysis can be accepted or rejected, mostly as a matter of personal taste. Should that happen, I predict it will be slowly but surely abandoned, both as a mode of helping and as a theory of the mind.[105]

Meehl's paper and Lear's "The Shrink Is In" both appeared in December 1995. Meehl was a very senior psychoanalyst and psychologist, as well as a renowned philosopher of psychology, having also been a strong supporter, along with Wilfrid Sellars, when Herbert Feigl founded the pioneering Minnesota Center for Philosophy of Science at which many budding whippersnappers (like myself in the 1950s) cut his/her teeth. Hence, Meehl's paper should have been very sobering to Lear, if he had seen it then, and it should still be so now.

The principal organ of the German psychoanalysts, their counterpart of the *Journal of the American Psychoanalytic Association*, is the

journal *Psyche*. Drawing the wagons in a circle, it published a German
translation of Lear's 1995 article from the *New Republic* (*Psyche* vol.
50, July 1996) on the heels of an article by the German analyst Werner
Bohleber, a member of *Psyche's* editorial board. Its English *Summary*
(ibid., p. 577) afforded me much amusement, because much like the
analyst Elisabeth Roudinesco in France, it characterized my attitude
toward psychoanalysis very pejoratively as giving me the status of
"The main representative [of] . . . a species of scientific Puritanism
born of blind belief in the 'exact sciences.'" Thus my demand for adequate evidence is held to bespeak just that belief. Though Bohleber
never had me on the couch for analysis, he declared in context:

> Closer inspection of the motives and roots of the attacks on Freud
> lead to the hardly surprising conclusion that the popular parlour
> game "Freud Bashing" has created the wave in a social climate
> marked by fundamentalist conservatism, the new prudishness, and a
> species of scientific Puritanism born of blind belief in the "exact sci-
> ences." The main representative of this attitude in connection with
> psychoanalysis is Adolf Grünbaum, while revisionists like Jeffrey
> Masson and Frederick Crews are to be regarded rather as exponents
> of the sexual counter-revolution.

Bohleber's article is intended as an introductory commentary for the
unavailing translated essay by Jonathan Lear that follows it.

For her part, Roudinesco considerably surpasses Bohleber in con-
victing me of fanaticism in her 2001 polemic *Why Psychoanalysis?*,
an English translation of her 1999 French original *Pourquoi La Psychanalyse?*[106] So ingrained is the ignorant idea that a philosophy of science appraisal of Freud's edifice must embody a methodological misextrapolation from physics that, revealingly, Roudinesco's fantasy
even turns me into a physicist, and, to boot, fabricates out of whole
cloth a totally undocumented misattribution to me:

> The most representative attitude in today's scientistic crusade is that
> of Adolf Grünbaum. A well-known physicist [*sic!*], a philosopher,

and then a professor of psychiatry, he became a specialist in anti-Freudianism around 1970. In his 1984 book *The Foundations of Psychoanalysis*, which caused a huge stir in the United States, he picked up again the classical argument of advocates of the brain mythology, reproaching Freud for having abandoned his *Project* and given up on making psychoanalysis a natural science.[107]

Not only am I, of course, not a physicist, but Roudinesco seems unaware that the charge of "scientism" was leveled by Jürgen Habermas against none other than Freud himself, whom he accused of having incurred a "scientistic self-misunderstanding." Besides, I never "reproached Freud for having abandoned his *Project*," let alone for having given up his aspiration to construct psychoanalysis as a natural science. Like all the rest of her fatuous, hostile rant against me (pp. 71–74, 76, 84–85, 87–88, 96–97, and 111), this attribution is completely erroneous. One will look in vain for a reference by her to where I am supposed to have said anything like what she claims. But, without any indication of where the reader could look for her documentation, she refers (p. 74) to endnote 16 (p. 162), which is hollow indeed, since she merely lists *Foundations* without any chapter, let alone page numbers.

Her various attacks on my supposed "fundamentalism" (p. 76)—whatever that is—are beneath rebuttal, if only because of her primitive incomprehension of the philosophy of science, and her personally abusive style of argument.

But just one further item will illustrate the extent of her enmity. In chapter 7 of my 1993 *Validation* book, I gave a quite sympathetic, though still somewhat critical discussion of Freud's three-pronged psychology of belief in theism and of his ensuing argument for atheism, while making no bones about favoring atheism myself. Roudinesco asserts that, as a partisan of scientism, I share "an absolute rejection of religion" with the exponents of "the reduction of the psyche to the neural" (p. 88). But now comes her ludicrous, demagogic clincher: "This atheism, it must obviously [*sic!*] be pointed out, bears no resemblance to that of Freud or the heirs of the Enlightenment" (pp. 88–89). A brief anecdote will put this further crude fabrication into perspective.

In a recent biography of Freud's wife, Martha (née Bernays), it was reported that she came from a religiously observant Jewish family (with some ancestral rabbinic background). Hence, upon her marriage, she expected to follow the ritual tradition of lighting the Friday night candles. But she discovered promptly that her husband, Sigmund, would autocratically forbid it very firmly. Yet after his death, she immediately resumed the practice.

In sum, Roudinesco's comments on my views are entirely without merit. Yet the Columbia University Press saw fit to publish her book in its European Perspectives Series. This series, one learns, "presents outstanding books by leading European thinkers."

Going further afield in France, I have had no interest in dealing with Jacques Lacan's version of psychoanalysis, if only because of his obscurity and his indifference to familiar canons of evidence.[108]

My *Foundations* had appeared in French translation (Presses Universitaires de France, Paris, 1996) after a shorter book *La Psychanalyse À l Épreuve* (*Psychoanalysis Under Scrutiny*) (Paris, Editions de l Éclat, 1993), as well as a few journal articles in French translation. And the former, the 1996 *Les Fondements de la Psychanalyse*, was reviewed very favorably in the leading Parisian newspaper *Le Monde*.[109] Indeed, a truly glowing review of it, calling it "a model for any philosophical analysis of any theory that lays claim to be scientific" (my English translation from the French), appeared in *Revue de L'Association Henri Poincaré*.[110]

My very good friend and warmly esteemed French academic colleague Joëlle Proust at CREA, a Center for Applied Epistemology at the École Polytechnique in Paris, and then of the Centre National de Recherche Scientifique (France's National Science Foundation), was very instrumental in bringing my American psychoanalytic writings to the attention of the French public. First she translated my three essays in the 1993 book, and then she initiated the 1996 French translation of my *Foundations* book and generously used her expertise to correct it.

The French journal *Sciences et Avenir* (The Sciences and the Future) devoted its July/August 2001 issue to the hypothesis of the unconscious (*L' hypothèse de l' Inconscient*). It contained my article *"L' inconscient à l' épreuve"* (translated from my English "The Unconscious Under Scrutiny"), pp. 42–49. In an editorial preface to the issue, the editor, Laurent Mayet, introduced my paper (p. 3) as a contribution from *"le plus éminent des sceptiques"* (the most eminent of the skeptics; italics in original). In 2004, the journal *Le Nouvel Observateur* (October/November, p. 11) carried my *"Quelques Objections Fondamentales"* (Some Fundamental Objections).

Since I had never written anything about Lacan, I was very surprised by the remarkable invitations from two Parisian Lacanien professional associations to deliver lectures to them. Indeed, the larger of the two, the leading *École Lacanienne de Psychanalyse* (The Lacanian School of Psychoanalysis), held an all-day *"Rencontre-Débat Avec Adolf Grünbaum"* in Paris on June 28, 1997. And the smaller one, *Fondation Européenne pour la Psychanalyse*, held its own *"Soirée-débat avec Adolf Grünbaum"* in Paris on June 26, 1997.

But there is much doctrinal discord between these two rival groupings such that, normally, they hold their respective meetings at conflicting times so that nobody can attend both of them. But since neither of the two rivals was willing to be outdone by the other in having me as a speaker, they relented and met at different times, which enabled me to be a guest speaker at each of them. All the discussions at both events were very cordial.

On our visits to Paris, our daughter, Barbara, introduced Thelma and me to a brilliant retired physician and polymath, Dr. Itzchak Torchin, and his two highly educated daughters, Dahlia and Danielle. A truly wonderful intellectual and very warm permanent friendship blossomed between us. Alas, he died in 2007 and we now cherish his memory.

Another cherished friendship developed with Patrizia Lombardo, a native of Italy and a widely recognized scholar of French literature and culture. We first met years ago as colleagues at the University of Pittsburgh, where she was professor of French. Later, her academic renown earned her a coveted professorship at the University of Geneva in Switzerland. Happily, ever since, her precious friendship with both Thelma and me has continued to flourish across the Atlantic Ocean.

It will be recalled that in 1993, nearly a decade after the appearance of my *Foundations* book, I published *Validation in the Clinical Theory of Psychoanalysis: A Study in the Philosophy of Psychoanalysis*. Its ten chapters include one (chapter. 7) on "Psychoanalysis and Theism" appraising Freud's three-pronged psychogenetic portrait of belief in theism as engendered by three, significantly different sorts of powerful wishes. In a lead review of this *Validation* book in the *London Times Literary Supplement*, the reviewer Owen Flanagan declared most flatteringly at the outset: "Adolf Grünbaum . . . is one of the major philosophers of science of our time."[111]

But very oddly, Flanagan goes off on a tangent without even telling the reader what I covered in that 1993 book, let alone evaluating it! However, just such useful information about my *Validation* book is given in a review in the *American Journal of Psychiatry* by the psychoanalyst Aaron Esman, a professor of psychiatry at Cornell University (New York City) and a onetime public relations officer of the American Psychoanalytic Association. He also provides relevant historical context:

> The scientific status of psychoanalysis has been a matter of controversy—often heated—throughout its century of existence. Freud was unwavering in his conviction that the "science" he had created could stand on all fours with the physical sciences; others have challenged its theoretical concepts as unverifiable, even fanciful. The British philosopher Karl Popper has dismissed them as "unfalsifiable," and some Continental philosophers (Habermas, Gadamer, Ricoeur) and some American "anti-metapsychologists" (Klein, Schafer, Gill, Spence) have sought to evade the question by opting for a "hermeneutic" stance that would place psychoanalysis, like history, among the humanities.

Into this fray has stepped Adolf Grünbaum, Andrew Mellon Professor of Philosophy at the University of Pittsburgh, who for the past decade has devoted his formidable talents as logician, philosopher of science, and polemicist to the critical assessment of just this issue. In *Validation in the Clinical Theory of Psychoanalysis*, his second book on the subject, he continues to explore themes advanced in the first [reference omitted]—primarily the lack of probative value of clinical data—and to answer challenges to his earlier formulations. It would be fair to say that no other critic has rivaled him for depth of scholarship, sharpness of focus, or intensity of argumentation, and his criticisms have struck to the heart of those psychoanalysts whose concerns go beyond quotidian clinical practice.

In essence, Grünbaum contends that data drawn from the clinical psychoanalytic encounter are impossibly contaminated by suggestion to the degree that they are useless for validation of "Freudian" hypotheses, and that even if the data were reliable, the specific pathogenicity of repressed events, affects, fantasies, and "traumata" cannot be (or at least has not been) demonstrated clinically. Such validation, he contends can derive only from extra-clinical investigations, experimental or epidemiologic. His arguments are powerful, and, to his credit, no psychoanalytic scholar, as Grünbaum repeatedly points out, has succeeded in refuting them. Edelson [reference omitted], for one, has attempted to defend the "case study" method as a basis for clinical validation, but Grünbaum cogently undermines his efforts to do so. . . .

Appearances to the contrary notwithstanding, Grünbaum abjures hostility to the psychoanalytic enterprise per se. He acknowledges, even defends, the heuristic value of clinical data and forcibly, even devastatingly, rejects Popper's accusation of unfalsifiability, pointing out a number of occasions when psychoanalytic propositions have in fact been falsified, by Freud as well as by his successors. Notably, too, Grünbaum is vehement in his critique of the "hermeneuticists," who, he contends, would trivialize psychoanalysis by eliminating causality from the understanding of human behavior and replacing it with "meanings." Here he fails, I think, to do justice to the subtlety of thought of Schafer, who, concerned about many of the same issues, arrives at widely divergent conclusions.

In a remarkable chapter, "Psychoanalysis and Theism," Grünbaum demonstrates an evenhandedness toward Freud's thinking. He
offers strong support for Freud's views on the nature of religious
belief and explores the ultimately fideistic nature of arguments that
purport to justify it on logical, even "scientific," grounds. He is especially persuasive in his challenges to Meissner [reference omitted],
who, as a Jesuit priest and a psychoanalyst, has sought (in Grünbaum's view and mine, unsuccessfully) to reconcile what Freud
insisted were incompatible points of view.

Without question, Grünbaum presents a major challenge to
psychoanalysis, both as a theory of human behavior and as a therapeutic practice. He writes as a philosopher, and his diction is often
dense and demanding. Unfortunately, it is at times marred by repet-
itiveness, both of ideas and of actual paragraphs, and equally by
Grünbaum's rather caustic, even contemptuous, way of dismissing
the views of those who disagree with, much less criticize, his own.
Finally, preoccupied as he is with the specific problem of scientific
validation, Grünbaum seems inadequately to appreciate the profound diffusion of Freudian thought into the broad matrix of
modern culture. Still, no one who is concerned about the future of
psychoanalysis and its place in the world of ideas can afford to
ignore his incisive, penetrating efforts to compel analysts to engage
the canon of modern scientific thought.[112]

The psychoanalyst Stephen Mitchell, the late editor of the journal
Psychoanalytic Dialogues, described the adverse effect of my Freud-
critique on the *serenity* of psychoanalysts.

> . . . we [psychoanalysts] are particularly vulnerable to a clinical state
> I have observed in psychoanalysts that I have come to think of as the
> "Grünbaum Syndrome." This may afflict psychologist-analysts more
> than others. I don't know. I have come down with it several times
> myself. It begins with some exposure to the contemporary philoso-
> pher Adolf Grünbaum's (1984) attack on psychoanalysis.
>   . . . What follows for an analyst afflicted with the Grünbaum
> Syndrome is several days of guilty anguish for not having involved
> oneself in analytic research.

... There may be a sleep disturbance and distractions from work. However, it invariably passes in a day or so, and the patient is able to return to a fully productive life.[113]

Since the "Grünbaum Syndrome" is not an affliction of mine, let me mention that my doctors are much impressed by my general good health and the considerable volume of work, publication, and lecture travel (both national and international) that I continue to perform at my current age of eighty-five. I no longer do scheduled classroom teaching, but the pleasure I take in teaching is afforded by lectures I deliver by speaking elsewhere and by giving talks in our Pitt Center for Philosophy of Science.

Thus, in March 2006, I gave invited lectures at All Souls College, Oxford University, and at the Royal Institute of Philosophy in London. They have been published under the title "Is Simplicity Evidence of Truth?" and an enlarged version was invited by Nicholas Rescher for publication in the April 2008 issue of the *American Philosophical Quarterly*, where it has appeared.[114] But I gave a categorically negative answer to its titular question. A month later, at the University of Vermont in Burlington, I delivered an all-campus "President's Distinguished Lecture" and conducted the "Grand Rounds" in its Department of Psychiatry. Locally, at Pitt, I also give an occasional lecture to medical students, most recently on placebos.

Recently and currently, "the power behind the throne" in my Pitt office is my truly exemplary, very talented, greatly dedicated, and highly agreeable administrative assistant, Leanne Longwill. Almost routinely, she is one step ahead of me: More often than not, when I ask her to do something in my office, she has not only already thought of doing it, but has even done it by then. I am indeed fortunate to have her running my shop.

Formally, I work only part-time at the university, while still (i) occupying my tenured lifetime chair as Andrew Mellon Professor of Philosophy of Science, (ii) maintaining my departmental affiliation with the Department of History and Philosophy of Science as "Primary Research Professor," (iii) keeping my status of research professor of psychiatry, and, not least, (iv) continuing to serve as chairman of the Center for Philosophy of Science ever since 1978.

Very recently I was the president of the International Union for History and Philosophy of Science for 2006–2007 (see appendix). Thus, in early August 2007, I presided over the quadrennial World Congress of its Division of Logic, Methodology and Philosophy of Science in Beijing, China, where I delivered my presidential address on August 9 on the topic "Why Is There a Universe AT ALL, Rather Than Just Nothing?" I formally dedicated it to the memory of my long-term colleague and cherished friend Wesley C. Salmon. Substantively, my lecture offered a complete deflation of its titular question as an ill-conceived pseudo-issue, generated by a baseless presupposition that has haunted Occidental philosophy since the second century. This address is being published in 2009 in a Proceedings volume by King's College Publications, London, UK, titled *Logic, Methodology and Philosophy of Science*.

Sometime early in 1995, if not before, I became aware that the Library of Congress (LC) in Washington, DC, intended to mount a major Freud exhibit and that some influential psychoanalysts were expecting it to provide a shot in the arm for psychoanalytic therapy, which had fallen on hard times in the United States.

I was one of about fifty Freud-scholars representing a broad spectrum of views on the merits of psychoanalysis who was invited to sign a petition to the LC asking that the exhibit reflect the catholicity of the signatories rather than be confined to sectarian advocacy. But unexpectedly, the petition initiated a hue and cry, charging that the signees were intent on canceling the exhibit, if they could not censor or control its content.

In early January 1996, the *Pittsburgh Post-Gazette* asked me to write an account of the widely publicized controversy that ensued when a budgetary shortfall prompted the LC to announce a postponement of the exhibit until 1997, which actually turned out to be 1998.

My narrative appeared under the title "Freud, in Full View" in the

Sunday, January 7, 1996, issue (pp. B1 and B4). Since I wrote it when my impressions of the events were still fresh, I reproduce it here in full:

Last month, the Library of Congress postponed until 1997 a major exhibit, "Sigmund Freud: Conflict and Culture," one in the library's thematic series "Reviewing the Past Century." In news reports and editorials, national newspapers and magazines deplored this postponement as a feckless surrender to ideological pressures from Freud-bashers.

Emphasizing that "The (library's) decision followed protests by people who disapprove of Freud and his ideas," *New York Times* columnist Anthony Lewis opined that "underlying the attempt to block the Freud exhibit was a particularly dangerous notion: that an institution owned by the government should avoid doing anything controversial."

Unfortunately and remarkably, the text of the brief communication sent to the Library of Congress by a group of 50 independent scholars, including academics from 37 universities in 11 countries, was not quoted by the press, though it was made available to all interested reporters. When it was mentioned, its contents were not reported objectively.

In no way does the text warrant the report that its signers demanded the postponement or censorship, let alone cancellation, of the Freud exhibit. As one of the signatories, I can say with confidence that we simply requested that the framework of the exhibition be widened to include "the full spectrum of informed opinion about the status of Freud's contribution to modern intellectual history." The record shows clearly that the charges of censorship are irresponsible, if not willful misrepresentations.

Indisputably, Freud's immense influence on 20th-century civilization has been felt not only in psychology and psychiatry but also in literature and literary criticism, art and its history, the theater, religion, anthropology, political philosophy and psychohistory. As a theory of human nature in which sex and aggression are central, psychoanalysis has been seen as a liberating gospel in some quarters.

Even as, in recent decades, the status of Freudian therapy has been markedly declining in psychiatric theory and clinical practice, the sway of his legacy over American higher education has been phenomenal.

In a 1995 lecture to a conference on "The Flight from Science and Reason" at the New York Academy of Sciences, the well-known Freud critic Frederick Crews recounted a stunning statistic from a 1992 report in the *Chronicle of Higher Education*: 38 percent of all professors of literature in the United States were teaching psychoanalytic theory to undergraduates.

And at the beginning of this decade, when Stanford University notoriously enlarged its Western Civilization requirement to comprise eight alternative tracks of "Culture, Ideas, and Values," Freud was accorded a place in all those tracks, which put him into a tie for unsurpassed cultural authority with Shakespeare and the Bible, thus ranking him ahead of Plato, Aristotle, Homer, Sophocles, Virgil and Voltaire!

Yet the past 25 years have seen an efflorescence of sharp critical reappraisal of the merits of Freud's intellectual estate and of his personal integrity. As the Freud historian Peter Swales has stressed, this burgeoning revisionist scholarship originated almost entirely from outside the psychoanalytic establishment.

Understandably, practicing psychoanalysts now see themselves as beleaguered because of the seemingly relentless erosion of the pool of patients seeking their treatment; besides, rightly or wrongly, the litigious tragedies spawned by the "false memory syndrome" in child molestation cases are often laid at their door. Witness *Time*'s ominously titled 1993 cover story "Is Freud Dead?" which featured such tags as "Father Freud Is Under Siege" and "Repressed-Memory Therapy Is Harming Patients."

In the midst of this strife, the Library of Congress planned its exhibit on "Freud's Contested Legacies." Its timing was supposedly occasioned by the availability of substantial material that had been previously restricted, and by the imminent millennium of Austria, where Freud had lived in Vienna from the age of 4 until his rescue from the Nazis and emigration to London in mid-1938. The library's holdings of Freudiana are unmatched, having been massively enriched by donations from the Sigmund Freud Archives of New York and from Freud's daughter Anna, a psychoanalytic theoretician in her own right.

The library's announcement of the exhibit in its Information Bulletin (June 13, 1994) stated in part: "Freud's intellectual legacies have been under close critical re-examination since the 1970s. The exhibition will encompass modern reassessments of Freud's ideas and show how his insights have been useful in the humanities."

Yet when Peter Swales saw the list of the key planners and advisers as well as the name of the sole curator, he was dismayed to note the complete absence of any representation from those scholars who had been the prime movers, during the past quarter century, of just that "critical re-examination" and "modern reassessment" that the library was pledging itself to include in its exhibit.

Indeed, when Swales alerted scholars throughout the world, it was obvious that the planners chosen by the library were unlikely to countenance fundamental criticism of the Freudian enterprise. Thus the library's explicit promise of intellectual inclusiveness appeared hollow. After all, the exhibit was to feature not just documents, photographs, correspondence and memorabilia but also a catalogue of essays, public lectures, seminars and Internet outreach—evaluative and interpretive activities that would inevitably reflect the orthodox one-sidedness of the makeup of the planning committee.

In an effort to assure the fulfillment of the library's express commitment to doctrinal catholicity, the 50 scholars, almost all of whom had themselves published research on Freud, sent a petition on July 31, 1995, to the library's chief of the Manuscript Division.

The petition's operative statement read: "As (Freud) scholars working independently of one another, with no common doctrinal commitment nor shared institutional affiliation, we are all of us concerned that the Freud exhibition at the Library of Congress, scheduled for next year, should suitably portray the present state of knowledge and adequately reflect the full spectrum of informed opinion about the status of Freud's contribution to modern intellectual history."

For the sake of the library's convenience, it was proposed that

Henry Cohen, a legislative attorney in its own Congressional Research Service—and therefore on its premises—be appointed to the committee responsible for the exhibition in order to represent a viewpoint suitably wider than that of the original organizers, while keeping the signatories informed of its deliberations and evolving plans.

The letter added that "Mr. Cohen has himself consented to act on our behalf in just such a capacity." And it concluded with "so we ask please, that you direct to him for further dissemination any response that you may have." But the library official to whom the petition was addressed has yet to reply to it. Instead, the librarian of Congress announced that a financial shortfall to date compelled the postponement of the Freud exhibit.

Had Anthony Lewis and other critics of the petition bothered to look at the list of signatories, they would have noticed that it included, among other psychoanalytically oriented scholars, such luminaries as Morris Eagle, the president-elect of the division of psychoanalysis of the American Psychological Association, noted psychoanalyst and schizophrenia researcher Philip Holzman of Harvard, and even Sophie Freud, Sigmund Freud's granddaughter—hardly a group of Freud bashers. And had the petition's detractors read its text, they could not, in all honesty, have accused the petitioners of seeking to censor, let alone cancel the exhibit.

Plainly, the petition did not call for any diminution of the content envisioned by the psychoanalytic partisans whom the library had put in sole charge. Rather it proposed the augmentation of that content to include the fruits of recent, influential revisionist scholarship. Only by means of such inclusion, the signers believed, can the public be given an authentic picture of Freud's "contested legacies" for our times. And the public deserves no less.

Nonetheless I was penalized for having signed the petition. After I had accepted an invitation to deliver the Ortega Y Gasset Lectures in Buenos Aires, my prospective hosts apologetically withdrew their invitation because they feared my lecture would be picketed by Freudian partisans wishing to protest my having been a signatory of the petition after having heard, via Paris, that I had done so.

In 1998, more or less concurrently with the Library of Congress exhibit, A. A. Knopf in New York published a companion volume *Freud, Conflict and Culture: Essays on his Life, Work and Legacy*, edited by the curator Michael S. Roth, with an introduction by James H. Billington, the Librarian of Congress. Because of the pressure of the petition to the LC to acknowledge scholarship in its exhibit that takes issue with psychoanalysis, Roth quite belatedly and clearly grudgingly invited me (and a couple of other critics) to contribute a chapter to the book, which I did under the title "A Century of Psychoanalysis: Critical Retrospect and Prospect." His partisanship is pointedly criticized in reviews of Roth's companion volume.

Thus, in his editorial review, which can be found on the Amazon.com Web site, Richard Farr wrote: ". . . as a result of the 1996 controversy [regarding the professional integrity of the LC exhibit], top notch critics of Freud such as Adolf Grünbaum are now grudgingly represented."[115] And writing in the *Los Angeles Times* (February 28, 1999), Andrew Scull declared: *"Freud, Conflict and Culture* is a deeply schizophrenic production. . . . Certainly, the first three quarters of the book suggest that the original [petitioning] critics of the [LC] show were absolutely correct in their suspicions of what was in the works. In serried ranks, the faithful parade their loyalty to the sacred texts and the Founding Father. . . . [H]alf a dozen suitably sympathetic scholars (their devotion to the cause amply vouched for by prior publications) round out the remarkably unbalanced cast of contributors. . . . Grünbaum's analyses throw into stark relief how one-sided and willfully partisan much of the book is."

Besides, as I have explained in my 1993 *Validation* book (pp. 117–18), Roth's understanding of the logical architecture of Freud's theory of repression is deeply flawed, if only because he fails to see that, in psychoanalytic theory, pathological symptoms are no less *interpreted* motivationally than manifest dream content, *both* being alike "compromises between the demands of a repressed impulse and the resistances of a censoring force in the ego," as Freud had told us in 1925.

In 1999, the LC exhibit was brought to New York City, where it occasioned a discussion of Freud's legacy on May 12 on the televised *Charlie Rose Show*, which was aired on the TV station WNET of the PBS network. In addition to Roth and myself, the panel of invited discussants included the well-known Freud biographer, eminent cultural historian, *and* lay psychoanalyst Peter Gay, and the New York feminist psychoanalyst Lois Kaplan, with both of whom I clashed during the program, but emphatically more so with Kaplan than with Gay. In his 1988 biography *Freud* (p. 745), Gay had complained that my doubts about psychoanalysis are "obsessive": "The most formidable among the sceptics, who has made the credibility of Freudian science (or lack of it) into an obsessive concern for a decade, is the philosopher Adolf Grünbaum." Yet, for his part, Gay has been insouciantly uncritical concerning the evidential merits of psychoanalysis for much longer than a decade.

Having lived under the Nazis, it took years after World War II before I was prepared to have professional dealings with German colleagues, let alone develop personal friendships with them. But, to my great satisfaction, I found genuine camaraderie with Erhard Scheibe, then at the University of Göttingen, and with Jürgen Mittelstrass, Gereon Wolters, and Martin Carrier of the University of Konstanz, all of whom have become warm, valued friends, as have Joachim and Christa Pfarr of my native Cologne, Germany. Indeed, Joachim Pfarr, a mathematical physicist, was the first visiting fellow of our Center for Philosophy of Science whom I brought to Pitt in 1977.

My rapport with Mittelstrass and Wolters became especially warm in 1983, during the three weeks I spent in Konstanz inaugurating a series of six annual lectures by distinguished foreign scholars, the so-called Konstanz Dialogues, instituted by the then Rektor Horst Sund with the support of Mittelstrass. It was Mittelstrass who had established Konstanz's Center for Philosophy and Philosophy of Science (*Zentrum Philosophie und Wissenschaftstheorie*) with which our Pittsburgh Center for Philosophy of Science developed major collaborative programs, especially at the initiative of Jerry Massey when he was director of the Pitt Center.

Moreover, Mittelstrass was the creator and principal editor of the monumental four-volume *Enzyklopädie Philosophie und Wissenschaftstheorie* (*Encyclopedia of Philosophy and Philosophy of Science*), which began appearing in 1980 and is now available in a second edition (volume 4, 2009 or 2010; further volumes thereafter). It contains a very thorough, detailed entry on my career and writings.

My gratifying dealings with German colleagues have broadened. In 2002, I was honored to receive an invitation from Paul Hoyningen-Huene of the University of Hannover to give the prestigious three Leibniz Lectures in its beautifully rebuilt Leibniz-Haus. In late June 2003, I delivered them: the first two were titled "The Poverty of Theistic Cosmology," while the third dealt with an extension of my 2002 article "Critique of Psychoanalysis." An elaborated version of the first two lectures has appeared under their title in the *British Journal for the Philosophy of Science*, December 2004. It is reprinted as chapter 1 of part 1 in the present book.

On its first page, the editor, Dr. Peter J. Clark, graciously appended the following editorial note:

Fifty-one years ago, Professor Grünbaum published his first paper in the *British Journal for the Philosophy of Science* [*not* his first paper *ever*], in the issue for 1953. It was entitled "Whitehead's Method of Extensive Abstraction" (*British Journal for the Philosophy of Science*, vol. 4, pp. 215–26). The editor wishes to acknowledge Grünbaum's extraordinary achievement in philosophy of science and in particular the debt that this journal owes to so distinguished and productive an author.[116]

My 2004 "The Poverty of Theistic Cosmology" is a twin, as it were, of my 1995 essay "The Poverty of Theistic Morality," as well as of a 2004 expanded German version "Das Elend der theistischen Moral," which appeared in *Moral als Gabe: Zur Ambivalenz von Moral und Religion* (*Morality as a Gift: On the Ambivalence of Morality and Religion*), edited by B. Boothe and P. Stoellger.[117] But these writings were preceded by a series of my other publications contra theism, only one of which I shall mention here, because it helped launch the new philosophy of religion journal *Philo*: "Theological Misinterpretations of Current Physical Cosmology," *Philo* 1, no. 1 (Spring/Summer 1998): 15–34.

I have my heart in countering what I see as theological misappropriations of big bang cosmogony. Thus, I was pleased to learn that this topic was avowedly one of the promptings of an invitation I received in October 2005 from the senior editor in chief of the *Harvard Review of Philosophy*, Céline LeBoeuf, to contribute an article to her journal. Her letter of October 14, 2005, read in part:

> It would be hard to overestimate the importance of your contributions to the philosophy of science, for your work has always demonstrated great depth, whether discussing the philosophical foundations of space-time or falsifiability as a criterion for determining scientific theories. Your recent work on the problems of theism and theistic interpretations of current cosmological models is truly exciting, and I believe that any article of yours on this subject would make a valuable contribution to the *Review*. While I have only recently become acquainted with your work on the philosophy of psychiatry, I already appreciate how you have pursued in this domain your earlier reflections on the rationality of scientific theories, and an article on this subject would likewise be welcome. Of course, we would be delighted to receive a submission from you on any topic with which you are currently engaged.

In 1973, I had a visit in Pittsburgh from the originally Hungarian philosopher Imre Lakatos and his then graduate student Spiro Latsis, who was working under him on his doctoral dissertation at the London School of Economics. Sadly, Lakatos died soon thereafter in February 1974. Latsis and Lakatos were planning a colloquium on research programs in physics and economics—in Lakatos's technical sense of "research programs"—to be held in September 1974 in the breathtakingly beautiful setting of Nafplion, Greece, hosted by the Latsis Foundation.

They came to enlist my participation in the Nafplion colloquium in the face of my being loath, as were a number of other colleagues, to visit Greece while the authoritarian, if not fascist, military government was in power in Athens. But, by September 1974, the seven-year-old regime of the colonels had been ousted and the colloquium took place as scheduled.

Spiro Latsis himself has recounted some of this history leading up to my involvement in the colloquium in an essay, "Nafplion Revisited," where he wrote:

> In Pittsburgh we visited Adolf Grünbaum and Herbert Simon. Adolf—though clearly no Popperian rationalist—was an early supporter of the Nafplion venture and besides, he cured a bad cold I had managed to acquire during the trip. For me, it was the start of a long friendship with Adolf and his wife, Thelma.[118]

Surely, I had not quite "cured" his cold, but only taken him to an agreeable dinner.

Soon after Lakatos's very sudden death, Spiro Latsis endowed a memorial prize of ten thousand pounds sterling in the philosophy of science to honor his teacher Lakatos. It was to be awarded annually for an outstanding book, preferably by a younger author. At the outset, I became a member of the steering committee, which uses the recommendations of the designated readers of nominated books (the "selectors") as a basis to award the prize, which has become very prestigious. I have remained a member of the steering committee ever since, having been chosen to be Spiro Latsis's own representative on it. The late John Watkins of the London School of Economics (LSE) was its "convener" for decades, and my longtime, warm friend John Worrall, also of the LSE, has succeeded him in that important role.

Before long, Spiro and his wife, Dorothy, invited Thelma and me to their lovely country home in Gstaad, Switzerland, not very far from their regular apartment in Geneva. Soon I learned that the Latsis Foundation had annually awarded a prize to a young scholar (under forty) at each of the four Swiss universities. But instead of having four sep-arate award ceremonies at these respective institutions, it was decided by 1994 that they would be merged into a single occasion, to be held in Geneva on October 7, 1994.

Spiro and Dorothy asked me to deliver the keynote address at that event, and I was very glad to do so. But since they knew that I suf-fered from jet lag after overnight transatlantic flights, they very thoughtfully arranged for a daytime flight for Thelma and me on the supersonic Concorde plane from New York to Paris in two and a half hours. The skillfully planned and very impressive October 7, 1994, awards ceremony occasioned an article in the *Journal de Genève* of that date.

Soon afterward, in 1995, philosopher of science Elie Zahar of the Department of Philosophy at the LSE decided to retire. Thereupon, John Watkins, the then chair of the department, wrote to me: "The graduate students here want to organize a one-day conference devoted to Elie Zahar, who is taking early retirement; and the person they want most of all to attend it [as a speaker] is you." This time, I was again flown to Europe on the Concorde, presumably at the expense of the Latsis Foundation, but of course to London, and I was very happy to deliver a lecture at the July 6, 1995 Zahar festivity. Since then, the Concorde is no longer in service because it was a commercial failure.

The Latsises own a palatial estate, Casa Estella, at Cap d'Antibes in Juan Les Pins on the French Riviera. They most generously issued a standing invitation to Thelma and me to stay at its guest house, making it—as Dorothy Latsis said—our "home away from home." We have availed ourselves of this munificent expression of their friendship intermittently for many years, enjoying their bountiful hospitality, even when they themselves were not there. Casa Estella is run splen-didly by the amiable Hans Manhard, who has perennially looked after our needs there very ably and cordially.

Indeed, the generosity of Dorothy Latsis's own hospitality can be gauged from the following episode. In corresponding about our next visit to Casa Estella in 2007 with Verena Lenz, a very warm and genial factotum at the Latsis office in Geneva, she informed me that it would not be available to us until September, because the now large Latsis family would occupy it in the latter half of August. I had told Ms. Lenz that after the August 9–15 meeting in Beijing, my next professional engagements were in mid-September in Prague and in Cologne.

But she appreciated that I was loath to go back all the way to Pittsburgh from Beijing with Thelma before traveling to Prague and Cologne. So much so that she generously took the initiative of mentioning my reluctance to Dorothy, who then magnanimously invited Thelma and me to stay by ourselves after we leave Beijing at her spacious luxury apartment in the Chalets du Palace adjoining the famous Palace Hotel in Gstaad, Switzerland, until we proceed to Prague. During August and September 2008, we again stayed in Gstaad, and then at Casa Estella before I presided over the Eighth Ontology Congress in San Sebastian, Spain (September 29–October 4, 2008).

I have largely deferred until now an account of some of the more personal aspects of my academic life at Pitt, starting in 1960 and extending into the present. In 1959, when I was still at Lehigh University, I had read an article in *Time* magazine about the very ambitious plans forged there by the then chancellor Edward Litchfield and about the establishment of the Andrew W. Mellon professorships in the humanities and natural sciences. Thus, when I received an exploratory telephone call from Pitt's Dr. Charles Peake, then the vice chancellor for the academic disciplines, I had an open mind. Yet, let me say, without Peake's formidable talents as an academic visionary, his magisterial know-how as an administrator, his infectious enthusiasm, and his personal magnetism, I would not have explored a Pitt appointment. It happened that I had become aware, during a 1959 lecture I gave at the then Carnegie Institute of Technology (later Carnegie Mellon University) at the instigation of its famed Herbert Simon, that the *then* Pitt philosophy department was, academically, a cemetery.

As long as Peake was in office, until his retirement as provost in 1971, my life at Pitt was really good and indeed creatively exhilarating, as I became the principal architect of a newly constituted Department of Philosophy that was rapidly achieving national and international recognition. Peake and his wife, Margo, became true friends to Thelma and me. After they retired to Gainesville, Florida, our warm friendship blossomed further in lively correspondence until he died in 1993 at age eighty-seven. As requested by the university archives, I deposited our extensive correspondence therein.

Almost from the start of the Center's quest for external funding, Bernard Kobosky, then vice chancellor for public affairs, was a vigorous advocate for the Center for many years in his effective dealings with local foundations. His efforts on our behalf were vital, as were those of Tom Detre later on.

When I first arrived at Pitt, Edison Montgomery ("Monty") was the treasurer. But over the years, he held a series of other important offices, including at least one in the School of Medicine. He played an especially valuable role for me during the crisis involving the departure of Chancellor Litchfield, when it was rumored ominously that, for financial reasons, the trustees would allow the closing of the university. By giving me reliable back-channel information in the face of the alarming rumor, Monty helped sustain my morale and zest for the tasks at hand.

Soon after I had become acquainted with the supporting personnel in Charlie's office, I met Robert Dunkelman there, who eventually became the secretary of the university until his very recent retirement. Over the decades, he was most supportive and helpful to me in more ways than I could enumerate. I knew that I could always count on him when I came to him with a problem. The serene equanimity he displayed added to the pleasure of dealing with him.

When I arrived in 1960, Mrs. Elizabeth McMunn was the only staff member in the philosophy department, and she was there only part-time. But with increased departmental activity, such as the accession of the highly prolific Nicholas Rescher, whom I had just brought from Lehigh in 1961, the appointment of a full-time secretary became imperative.

When I offered Liz the same part-time arrangement she had had, but working with me in the Center, she accepted and then stayed, even when it soon became a full-time job. Since 1961 until her retirement in 1991, she ran my shop in exemplary fashion with outstanding initiative and imagination, as a Jane-of-all-trades: she was the administrative assistant and budget manager in the Center, the secretary for all of my scholarly work and classroom teaching, as well as an exacting editor of my writings.

Not only I, but those of my colleagues with whom she dealt, recognized her superior contribution. Thus, when Wesley Salmon learned that she was retiring, he volunteered the tribute that "Liz is a most difficult, if not impossible act to follow." The level of her involvement in the university community is shown by the active correspondence she conducted with Charlie Peake after his retirement, quite independently of me. That exchange of letters is likewise deposited in the university archives.

When the dynamic Chancellor Litchfield left Pitt in the mid-1960s, because its trustees were reportedly displeased by his financial management and baronial lifestyle, they appointed a committee of six faculty members to advise them in the selection of a successor. I was asked to serve on that committee. The then chairman of the board of trustees, Gwyllem Price, who was also chairman of the board of the Westinghouse Corporation in Pittsburgh, proposed a candidate for the vacancy who was president of a southern university at the time.

When this applicant met with our faculty advisory committee, he blithely boasted that if a professor at his university disagrees with him politically or administratively, he simply fires the professor, whether tenured or not. The trustees had thought of him, I gathered, because they deemed him potentially able to pick the pockets of the Pittsburgh billionaires on behalf of the university.

We realized, of course, that if appointed, this autocrat would be a disaster for the university and would produce an exodus of some of the best members of our faculty. Hence, when we met with the relevant trustees, we minced no words on this score. Fortunately, Henry Hillman, a key trustee and renowned Pittsburgh philanthropist, was of one mind with us and came to the rescue. Said he, "Let him go back to South Carolina and eat corn," which warmed the cockles of my heart! We told Mr. Hillman that, henceforth, we will think of him as one of us and will speak of him as "Professor Hillman."

Soon we endorsed the trustees' selection of Wesley Posvar, a one-time Rhodes Scholar at Oxford, and a graduate of the Military Academy at West Point. At the time, he chaired the political science department at the Air Force Academy. His wife, Mildred Miller, a leading mezzo-soprano at the Metropolitan Opera in New York, who was commuting to there all the way from Colorado, relished the prospect of a shorter commute from Pittsburgh. Posvar was in office for a quarter of a century, during which time his title was changed from "chancellor" to "president," in keeping with the practice at most universities, although it then reverted to "chancellor" again.

During his long tenure, my relations with him were uniformly excellent, and indeed, he and Mildred were patently very cordial and gracious to both Thelma and me.

When I reached age seventy in mid-May of 1993, the mandatory retirement of university faculty at that age was still in force, though it was abolished nationally by law as of January 1, 1994, only seven and a half months later. But, at the initiative of Jerry Massey, who was Center director at the time, yet entirely without any involvement of myself, Wes Posvar gave telling evidence of his valuation of my services by very gladly giving me a lifetime tenured appointment, as of my seventieth birthday in 1993. Thus in September 1990 and several years in advance of my seventieth year, very soon after the then provost Donald Henderson informed me that this appointment was in the works, Chancellor Posvar wrote me:

> I am delighted and pleased to reaffirm certain understandings that have been previously reached with you [by Provost Henderson] and to inform you of our desire to continue your association with the University after reaching age 70 on May 15, 1993.
>
> 1. We are waiving the University mandatory retirement rule in your case and hereby reappoint you indefinitely effective May 16,

1993, as Andrew Mellon Professor of Philosophy with tenure in the Faculty of Arts and Sciences. You will serve as long as you wish and remain physically fit.

2. Your duties as Andrew Mellon Professor may be fulfilled in any department in the Faculty of Arts and Sciences. . . .

I want you to know that the administration of the University is immensely proud to have you on the Pitt faculty. We earnestly hope that you will remain on the faculty for many years to come (September 19, 1990).

Little did I know at the time that Chancellor Posvar's September 1990 reappointment of me with tenure as of May 16, 1993—the day after my seventieth birthday—was to be bitterly resented by one Rudolph Weingartner. He had been provost from the fall of 1987 until his resignation only eighteen months later in the spring in March 1989. Thus, he was clearly no longer in office as provost between the spring of 1989 and late September 1990, when my reappointment was being considered.

Nonetheless, Weingartner firmly believed that Posvar ought to have consulted him about it and became ferociously indignant with wounded pride, claiming in his 2003 autobiography that Posvar had done an "end-run" around him.[119] But in any case, as I learned later on, by the time the chancellor was deciding to reappoint me, he had lost all confidence in Weingartner's academic judgment. Weingartner was implacable and also roundly disparaged me thereafter.

I now need to be candid about the development of my life in the Pitt philosophy department, notably after the "golden era" of the first two decades (1960–1980). First, let me relate some of the happy experiences I had, before I turn to the disappointing ones. It would be very misleading and less than honest to pretend that there were only happy ones.

On the felicitous side of the ledger, I already had occasion to mention the three-day 1990 "Colloquium in Honor of Adolf Grünbaum," which was cosponsored by the Philosophy department along with the Center, the HPS department, the psychiatry department, the Faculty of

138 AN AUTOBIOGRAPHICAL-PHILOSOPHICAL NARRATIVE

Arts and Sciences, the Schools of Health Sciences, and the provost. It bears repetition that it was superbly organized, with much expenditure of effort, by Jerry Massey, who had initiated it, and Mary Connor, who assisted him vigorously, to mark my thirty years of service to Pitt.

Moreover, at the initiative of the Department of Philosophy, notably of its senior members John Haugland and Robert Brandom, and with the active participation of both the Center and the HPS department, Provost Jim Maher approved a major renovation of space on the elegant seventeenth floor of the Cathedral of Learning in room 1701 to create the beautifully furnished "Adolf Grünbaum Philosophical Reading Room," with a door plaque carrying that inscription. Providing a convenient refuge for students and faculty, the reading room was formally dedicated on March 13, 1998. As a surprise to me, my daughter, Barbara, and son-in-law, Ron, had commissioned a painted likeness of me, which was produced from a photograph and is hanging on one of its walls. Very heartwarmingly, their twin sons, Ben and Eli, our two beloved grandsons, now twenty years old, also were there for the dedication.

Once again, the philosophy department made common cause with both the Center and the HPS department, as well as with the provost, in sponsoring an "Adolffest" to mark my eightieth birthday in 2003. It was initiated and organized by my cherished friend Jim Lennox in his capacity as Center director, featuring a day of academic papers, framed by statements from the chancellor and the provost, besides an evening reception. Though my birthday was on May 15, the festivity was held earlier on April 12, because the spring term was then still in session.

Yet clouds had begun to gather and darken twenty years before in the early 1980s after more than two decades of congenial life for me in the philosophy department. One revealing instance is to the point.

The department had several "core courses" in certain of its specialized fields: metaphysics and epistemology, ethics, logic, philosophy of language, and philosophy of science. All incoming graduate students were required to take these core courses during their first two years to put them at ease with their fellow students at the same entry level, thus counteracting their intimidation by more advanced graduate students in other courses.

Out of the blue, early in September 1985, the departmental graduate committee recommended without explanation that the core course in the *philosophy of science* be dropped, while all the other four be retained on the roster. Thus, graduate students who might have their first exposure to philosophy of science in the core course, and who might then want to pursue it further, would instead be channeled willy-nilly into other specialties. Yet no plausible curricular rationale for this dubious recommendation was offered at all. Indeed, one faculty member of the graduate committee, Joseph Camp, who had no competence at all in the philosophy of science, remarked fatuously to me that he would "cover" philosophy of science in his own core course on metaphysics, with the back of his hand, as it were.

It became apparent that the policy of the graduate committee was unmistakably designed to demote the status of the philosophy of science in the department, despite or because of its great renown in the profession. Such was also the judgment of the then Andrew Mellon Professor of Anthropology, Jack Roberts, who alerted me, as a friend, to what was brewing.

Therefore, one of my colleagues in the department, a chaired professor with indubitable credentials in the philosophy of science, reacted to the scheme of the graduate committee by informing them that "it would take only a stroke of the pen" for him to transfer his primary appointment to the HPS department. Other distinguished departmental colleagues of great renown in the philosophy of science also expressed their opposition vigorously. For my part, I wrote the graduate committee (memo of September 10, 1985): "I am dismayed by your proposal, if only because it is clearly a step toward the dismantling of philosophy of science in our department, especially in the wake of Peter [Carl] Hempel's retirement."

On the heels of my memo, David Gauthier, the then chairman of the department, an ethicist, commendably wrote to the graduate committee: ". . . it is not my intention to preside over the dismantling of philosophy of science in our department" (memo of September 10, 1985). In an important follow-up memo, he elaborated sagaciously:

> 1. The core course program seems to me a good one, . . .
>
> 2. I do not believe that there is a canonical set of courses that any department should adopt. I believe that our own set should reflect the interface between our sense of what is philosophically central and our own particular strength as a department.
>
> 3. I therefore am convinced that the Philosophy of Science course should be maintained. *This department built its reputation on the philosophy of science, and it would be both institutionally and philosophically foolish for us to diminish its place in our program.* (memo of September 16, 1985; italics added)

The Philosophy of Science core course remained on the books. And there was nary a word from the graduate committee as to the alleged important curricular problem that the recommended dropping of only that particular course would have solved; nor did they explain whether they would endeavor to solve the phantom problem by means of some alternative to dropping that core course, which was no longer up for grabs. Apparently, they did not mind that their aims were transparent. Naturally, I was deeply disappointed by this episode, the first in a series that permanently left a very sour taste in my mouth.

This ploy was one of several gambits, starting in the early and mid-1980s, which exposed me to—what seemed to me only later— unmistakably envy-driven aggression from several members of the department. Yet, oddly, I was not well equipped myself to discern thinly veiled antagonism or even outright personal hostility until it became patent or was pointed out to me by genuine friends.

Thus, when my 1984 *Foundations* book and I received the aforementioned front-page coverage in the science section of the *New York Times* (January 15, 1985), there was nary a word from any colleague that this unusual publicity was also a feather in the department's cap, not just in my own. Thus, there was *no espirit de corps* (sense of common purpose), which was very alienating

One reason for my myopia, I believe, was that after my traumatic experiences in Nazi Germany, I craved a departmental and university environment in which my colleagues acted like a *benevolent, extended family*. This ardent hope became the source of my belief that they all wished me well, thinking naively that they felt indebted to me, especially because they had repeatedly asked me to run interference for the department with the university administration, which I had done very successfully.

Thus, instead of prudently adopting a wait-and-see attitude toward people I encounter, I generally tend to assume that they are well meaning until and unless they turn out to be patently malevolent. But once I become convinced that someone is an antagonist, I heed Leo Durocher's caveat: "Nice guys finish last," and I defend myself.

But all in all, I have paid a considerable price for having been so naively trusting. In fact, a number of disloyal colleagues were people whom I had benefited by bringing them to Pitt in the first place. Yet, alas, their conduct lent substance to Oscar Wilde's adage that "no good deed goes unpunished."

The alienation I felt from this and several other painful and disappointing experiences I had had in the philosophy department in the early and mid-1980s led me, for the first time, to have the idea of resigning from it, while retaining my supradepartmental Mellon professorship, my then secondary memberships in both the HPS and psychiatry departments, and of course, my chairmanship of the Center.

The centrifugal tendencies in the department persisted, even gaining momentum slowly but surely. Thus, I became increasingly discontented with the operation of the graduate program and with its very ethos. Indeed similarly, in an open, detailed letter (January 22, 2003), written to the department faculty by its graduate students, they complained, in effect, quite specifically that this program was multiply *dysfunctional*; significantly, in a secret ballot, forty-three of the students affirmed their agreement with the letter, five abstained, but none disagreed, although some of them got cold feet thereafter, as they conveyed in a letter to the *University Times*, probably because they feared faculty disapproval.

By 1999, I made no bones about the fact that I was on the verge of resigning. But a beloved colleague in the department, Tamara Horowitz, who expected to become chairperson, asked me to hold off to give her a chance to turn things around when she became chairperson. It was her vigorous agreement with me that motivated me to speak to Weingartner about the declining standards in the graduate program, when he had become chairman, but only to be blithely and dismissively rebuffed by him: as he reported nonchalantly, his own PhD dissertation had not even been read at Columbia University before he received his degree there! I was appalled. Tragically, Tamara died of a malignant brain tumor before she could take the intended action.

In the spring of 2003, Stephen Engstrom was chair of the department. He was highly collegial and very understanding of the various reasons for my disenchantment. As I told him, my alienation was escalating. Yet he made a genuine and strong effort to address my grievances by remedial action.

But, after he had taken soundings individually from the senior faculty, I concluded that there just were not enough people in the department who were motivated to effect the requisite changes, such as substantially reforming the graduate program. Hence, at that stage, I decided to resign from the department, but I deferred my letter of resignation to April 18, 2003, some days after the "Adolffest" of April 12.

In response to my disaffected resignation, Steve Engstrom wrote me very graciously:

On behalf of the Department of Philosophy, I am writing to express the great sadness and regret with which we read your letter of 18 April 2003, in which you informed us that you are resigning from the Department at the end of this term. I want to emphasize that in saying this I am not speaking just for myself or for a few of my colleagues. All of us regard your alienation from the Department in recent years as one of the most unfortunate developments in its history. And speaking for myself, I am disappointed that I was not able to do more to counteract it.

We shall not forget the principal and founding role you played in establishing here one of North America's premier departments of philosophy, both through your own eminent intellectual contributions and through the many distinguished philosophers you attracted to our University.

We will, of course, do our part in taking the practical steps that, as you point out in your letter, your resignation will make necessary. It will be painful for us to remove your name from the Department's roster, but please know that we do wish you well and hope that, despite this separation, the collegial relations you have with members of the Department will survive and flourish for many years to come Signed, "Sincerely, Steve." (April 24, 2003)

Alas, since 2003, developments in the philosophy department have borne out the wisdom of my resignation from it.

Once I had decided to resign, Provost James Maher readily agreed to change the title of my Mellon chair to Andrew Mellon Professor of *Philosophy of Science*. In view of my disaffiliation from the philosophy department, my erstwhile secondary appointment in HPS has since been changed to Primary Research Professor. I am now happy as a lark in *all* of my affiliations at Pitt, and I remain entirely convinced that my resignation from a department that I had nurtured was salubrious for me.

My current contentment at Pitt, and my feeling genuinely at home there, is attributable not only to the congeniality and friendship of my colleagues in the HPS department and the Center but is also a major credit to the personalities and modus operandi of the members of the upper Pitt administration.

Both as Mellon Professor and as chairman of our Center, I report to Provost James Maher, just as I had to then vice chancellor and provost Charles Peake during my first dozen years (1960–1972). I find Jim enormously amiable, competent, and ever constructive, both personally and academically. I am delighted to work with him. Mark Nordenberg, the chancellor, is a very affable and successful helmsman, who radiates good fellowship. It is an indication of both Jim's and Mark's esteem for members of the faculty that they are conscientiously and courteously responsive to letters from us. They make me feel *at home* at Pitt.

I also respect the competence of N. John Cooper, the dean of the School of Arts and Sciences, whose very able championship of time-honored academic values I prize. As a senior member of the HPS department, I too am a beneficiary of his leadership, since the department reports to him.

I greatly look forward to doing more creative work and writing, unencumbered by drains from discontents.

One of my ongoing projects is a two-volume work globally titled *Philosophy of Science in Action*, which—I hope—is an alluring title. It is a collection of my papers, both freshly written and reprinted, which is scheduled to be published, in due course, by Oxford University Press in New York. But the reprinted papers are being equipped with headnotes, either to update their literature or to indicate where subsequent reflection on their content has taken me.

APPENDIX:
MISCELLANEOUS PROFESSIONAL RECOGNITIONS

Let me conclude with various standard sorts of biographical data that I did not have occasion to mention above. They show that I have been quite fortunate in the professional recognitions I have received.

I reproduce here, in the third person, the two-page biographical statement that I supply to publicity departments when they request such information:

BIOGRAPHICAL STATEMENT:
ADOLF GRÜNBAUM

Adolf Grünbaum's writings deal with the philosophy of physics, the theory of scientific rationality, the philosophy of psychiatry, and the critique of theism.

His twelve books include *Philosophical Problems of Space and*

Time (second edition, 1973), *Modern Science and Zeno's Paradoxes* (second edition, 1968), and *The Foundations of Psychoanalysis: A Philosophical Critique* (1984). In 1993, he published *Validation in the Clinical Theory of Psychoanalysis: A Study in the Philosophy of Psychoanalysis*. Oxford University Press in New York City will publish two volumes of his collected papers under the overall title *Philosophy of Science in Action*. The first volume is devoted to his writings in the philosophy of physics, the second volume to his work on other topics. He has contributed over four hundred articles to anthologies and to philosophical and scientific periodicals.

His offices include the presidency of the American Philosophical Association (Eastern Division) and of the Philosophy of Science Association (two terms).

He has been elected to two other, but interconnected presidencies: for 2004–2005, he was the president of the Division of Logic, Methodology and Philosophy of Science of the International Union of History and Philosophy of Science, which is the worldwide umbrella organization of the various national associations or societies in the philosophy of science, on the one hand, and of the history of science, on the other. Upon completing this presidency, Grünbaum automatically became the president for 2006–2007 of the International Union for the History and Philosophy of Science.

He is a Fellow of the American Academy of Arts & Sciences, a member of the Académie Internationale de Philosophie des Sciences, a Fellow of the American Association for the Advancement of Science, and a Laureate of the International Academy of Humanism. In 1985, he delivered the Gifford Lectures in Scotland as well as the Werner Heisenberg Lecture to the Bavarian Academy of Sciences in Munich. In mid-2003, he delivered the three Leibniz Lectures at the University of Hannover, Germany.

He is the recipient of a 1985 "Senior U.S. Scientist" Humboldt Prize, and of Italy's 1989 Fregene Prize for science (Rome, Italy). All four prior recipients of this prize, which is awarded by the Italian Parliament, were Nobel laureates in one of the natural sciences. In May

1990, Yale University awarded him the Wilbur Lucius Cross Medal for "outstanding achievement." In May 1995, he received an Honorary Doctorate from the University of Konstanz in Germany. In May 1998, the venerable University of Parma in Italy awarded him its Silver Medal in recognition of his "prestigious career." In 1989, he received the first-ever Master Scholar and Professor Award from the president of the University of Pittsburgh. Furthermore, in March 1998, that university dedicated the Adolf Grünbaum Philosophical Reading Room.

Currently, he is the Andrew Mellon Professor of Philosophy of Science, Primary Research Professor in the Department of History and Philosophy of Science, Research Professor of Psychiatry, and chairman of the Center for Philosophy of Science, all at the University of Pittsburgh. In April 2003, he resigned from the Department of Philosophy there, while retaining his lifetime Mellon Chair and all his other appointments.

In 1983, a *Festschrift* for him, edited by R. S. Cohen and L. Laudan, appeared under the title *Physics, Philosophy and Psychoanalysis: Essays in Honor of Adolf Grünbaum*. It was reprinted in 1992. Another *Festschrift*, *Philosophical Problems of the Internal and External Worlds: Essays on the Philosophy of Adolf Grünbaum*, edited by J. Earman, A. Janis, G. Massey, and N. Rescher, and containing the proceedings of the October 1990 International Colloquium in Honor of Adolf Grünbaum, which was held at the University of Pittsburgh, was published in 1993.

In 1992, Mr. Harvey Wagner, chief executive officer of the Teknekron Corporation, and his wife, Leslie Wagner, honored Professor Grünbaum for his role as an undergraduate teacher of Mr. Wagner's by a gift of one million dollars to the Center for Philosophy of Science at the University of Pittsburgh.

In mid-October 2002, the Center for Philosophical Education in Santa Barbara, California, held a two-day conference titled "The Adolf Grünbaum Symposium in Honor of the Works of Professor Adolf Grünbaum." Its proceedings are published in this volume, *Phi-*

losophy of Religion, Physics, and Psychology: Essays in Honor of Adolf Grünbaum, edited by A. Jokić. And this *Festschrift* is very affectionately dedicated to Harvey and Leslie Wagner.

A more complete biographical account can be found under the "Grünbaum" entry in the twenty-first edition of *Who's Who in the World*, the fifty-fourth edition of *Who's Who in America*, the eighth edition of *Who's Who in American Education*, and the 2006 *Encyclopedia Judaica*.

ACKNOWLEDGMENTS

This text has had the benefit of very helpful comments from Thomas Detre, Robert S. Cohen, and my wife, Thelma.

Adolf Grünbaum
University of Pittsburgh
2510 Cathedral of Learning
Pittsburgh, PA 15260-6125
TEL: (412)624-5738
FAX: (412)648-1068
E-mail: grunbaum@pitt.edu

GLOSSARY OF ACRONYMS

APA   American Philosophical Association
BBS   *Behavioral and Brain Sciences*
CPS   Center for Philosophy of Science
HPS   Department of History and Philosophy of Science
PPST   *Philosophical Problems of Space and Time*
STR   Special Theory of Relativity
VF   *Voprosy Filosofii*

REFERENCES

1. G. Wills, *Papal Sin, Structures of Deceit* (New York: Doubleday, 2000), p. 21.

2. *World Book Encyclopedia* (1966), 14: 668.

3. I. Fisher, "Pope Described as Recovering after Surgery" *New York Times*, February 26, 2005, pp. A1–A6.

4. A. Grünbaum, (1993). *Validation in the Clinical Theory of Psychoanalysis: A Study in the Philosophy of Psychoanalysis* (Madison, CT: International Universities Press), chap. 4.

5. I. Jakobovitz, *London Times*, May 9, 1987.

6. A. Grünbaum, "The Poverty of Theistic Morality," in *Science, Mind and Art: Essays in Honor of Robert S. Cohen*, ed. K. Gavroglu et al., Boston Studies in the Philosophy of Science, vol. 165 (Dordrecht, Netherlands: Kluwer Academic Publishers, 1995), pp. 203–42.

7. R. J. Neuhaus, "Can Atheists be Good Citizens?" *First Things* (August/September 1991): 17–21.

8. Ibid., p. 17.

9. A. Grünbaum, *The Foundations of Psychoanalysis: A Philosophical Critique* (Berkeley: University of California Press, 1984), introduction and sections 1, 2, and 3. See also A. Grünbaum, "Reply to Symposium on the Grünbaum Debate," *Psychoanalytic Dialogues* 10, no. 2 (2000): 338–41.

10. B. Russell, *Our Knowledge of the External World* (London: George Allen & Unwin Ltd., 1926), p. 183.

11. A. Grünbaum, *Modern Science and Zeno's Paradoxes* (Middletown, CT: Wesleyan University Press, 1967).

12. A. Grünbaum, *Modern Science and Zeno's Paradoxes* (London: George Allen & Unwin, 1968).

13. A. Grünbaum, "Are Infinity Machines Paradoxical?" *Science* 159, no. 3810 (1968): 396–406; A. Grünbaum, "Can an Infinitude of Operations Be Performed in a Finite Time?" *British Journal for the Philosophy of Science* 20, no. 3 (1969): pp. 203–18.

14. A. Grünbaum, "The Poverty of Theistic Cosmology," *British Journal for the Philosophy of Science* 55, no. 4 (2004): 561–614; reprinted in this volume.

15. A. J. Ayer, *Language, Truth and Logic* (New York: Dover Publications, 1952).

16. A. Grünbaum, *Philosophical Problems of Space and Time* (New York: A. A. Knopf, 1963).

17. A. Grünbaum, *Philosophical Problems of Space and Time*, 2nd ed. (Dordrecht, Netherlands: D. Reidel Publishing, 1973).

18. L. Sklar, review of *Philosophical Problems of Space and Time*, 2nd ed., *Journal of Philosophy* 74, no. 8 (1977): 494–500.

19. A. Grünbaum, "Operationism and Relativity," *Scientific Monthly* 79, no. 4 (1954): 228–31.

20. A. Grünbaum, "Operationism and Relativity," in *The Validation of Scientific Theories*, ed. P. Frank (New York: Collier Books, 1961), pp. 83–92. This article had also appeared in a 1957 edition of this book on pp. 84–94.

21. A. Grünbaum, "Logical and Philosophical Foundations of the Special Theory of Relativity," *American Journal of Physics* 23, no. 7 (1955): 450–64.

22. H. Putnam, "An Examination of Grünbaum's Philosophy of Geometry," in *Philosophy of Science: The Delaware Seminar*, vol. 2, ed. B. Baumrin (New York: Interscience Publishers, 1963), pp. 205–55.

23. A. Grünbaum, *Geometry and Chronometry in Philosophical Perspective* (Minneapolis: University of Minnesota Press, 1968), chap. 3, pp. 195–371; A. Grünbaum, "Reply to Putnam," in Boston Studies in the Philosophy of Science, vol. 5, ed. R. S. Cohen and M. Wartofsky (Dordrecht, Netherlands: D. Reidel Publishing, 1968), pp. 1–150.

24. Grünbaum, "Reply to Putnam."

25. For the convenience of readers who may be interested, I list them here: "Free Will and the Laws of Human Behavior," *VF* no. 6 (1970): 62–74; "Freud's Theory: The Perspective of a Philosopher of Science," *VF* no. 4 (1991): 90–106; "Origin versus Creation in Physical Cosmology," *VF* no. 2 (1995): 48–60; "A Century of Psychoanalysis," *VF* no. 7 (1997): 85–98; "A New Critique of Theological Interpretations of Physical Cosmology," *VF* no. 4 (2002): 67–88; "The Poverty of Theistic Cosmology," an essay that, because of its length was spread over three successive issues, *VF* no. 8 (2004): 99–114; *VF* no. 9 (2004): 149–62; *VF* no. 10 (2004): 114–24; a Russian translation of my multiply reprinted twenty-page article "Critique of Psychoanalysis," first published in the 2002 *Freud Encyclopedia*, was published in *VF* no. 3 (2007): 105–29.

26. A. Grünbaum, "The Poverty of Theistic Cosmology," p. 561.

27. A. Grünbaum, "Soviet Atheism and Psychoanalysis Under Perestroika," *Free Inquiry* 9, no. 2 (1989): 52–53.

28. A. Grünbaum, "Science and Ideology," *The Scientific Monthly* 79, no. 1 (1954): 13–19.

29. K. Kovalchick, *Celebrating 40 Years: A History* (Pittsburgh: University of Pittsburgh Press, 2002), p. 2.

30. A. Grünbaum, "The Nature of Time," in *Frontiers of Science and Philosophy*, ed. R. G. Colodny, University of Pittsburgh Series in the Philosophy of Science, vol. 1 (Pittsburgh: University of Pittsburgh Press, 1962), pp. 147–88.

31. R. C. Alberts, *Pitt: The Story of the University of Pittsburgh 1787–1987* (Pittsburgh: University of Pittsburgh Press, 1986), p. 283.

32. W. Salmon, *Space, Time and Motion*, 2nd rev. ed. (Minneapolis: University of Minnesota Press, 1980).

33. W. Salmon, ed., *Zeno's Paradoxes* (Indianapolis: Bobbs-Merrill, 1970), p. viii.

34. W. Salmon, "Comments," in *Experience, Reality and Scientific Explanation: Essays in Honor of Merrilee and Wesley Salmon*, ed. M. C. Galavotti and A. Pagnini (Boston: Kluwer Academic Publishers, 1999), p. 211.

35. Salmon, *Space, Time and Motion*, p. 237, n. 6.

36. W. Salmon, "Psychoanalytic Theory and Evidence," in *Psychoanalysis, Scientific Method, and Philosophy*, ed. S. Hook (New York: New York University Press, 1959).

37. A. Grünbaum, *Validation in the Clinical Theory of Psychoanalysis*, p. 361.

38. R. S. Cohen, and L. Laudan, eds., *Physics, Philosophy and Psychoanalysis: Essays in Honor of Adolf Grünbaum* (Dordrecht, Netherlands: D. Reidel Publishing, 1983; reprint, 1992).

39. A. Grünbaum, "Free Will and Laws of Human Behavior," in *New Readings in Philosophical Analysis*, ed. H. Feigl, W. Sellars, and K. Lehrer (New York: Appleton-Century-Crofts, 1972), pp. 605–27.

40. A. Grünbaum, and W. C. Salmon, eds., *The Limitations of Deductivism* (Berkeley: University of California Press, 1988).

41. A. Grünbaum, "Freud's Theory: The Perspective of a Philosopher of Science," in a Centenary Celebration Volume of the *Presidential Addresses of the American Philosophical Association, 1900–2000*, ed. R. T. Hull (forthcoming). This is a revised version of my original presidential address in *Proceedings and Addresses of the American Philosophical Association* 57, no. 1 (1983): 5–31.

42. R. Bernstein, "Philosophical Rift; A Tale of Two Approaches," *New York Times*, December 29, 1987, p. A1.

43. J. Earman et al., eds., *Philosophical Problems of the Internal and External Worlds: Essays on the Philosophy of Adolf Grünbaum* (Pittsburgh, PA/Konstanz, Germany: University of Pittsburgh Press/University of Konstanz Press, 1993).

44. C. Daly, review of *Philosophical Problems of the Internal and External Worlds: Essays on the Philosophy of Adolf Grünbaum*, ed. J. Earman et al., *Canadian Philosophical Review* 15, no. 3 (1995): 167.

45. A. Hanly, review of *Foundations of Psychoanalysis*, by A. Grünbaum, *Journal of the American Psychoanalytic Association* 36 (1988): 521–28.

46. P. A. Schilpp, ed., *The Philosophy of Karl Popper*, vol. 2 (Chicago: Open Court Publishers, 1974), p. 1004.

47. K. Popper, *Realism and the Aim of Science* (Totowa, NJ: Rowman & Littlefield, 1983), chap. 2, p. 164, n. 1.

48. A. Grünbaum, *Validation in the Clinical Theory of Psychoanalysis*, chap. 2.

49. A. Grünbaum, *The Foundations of Psychoanalysis*, chap. 1, section B.

50. A. Grünbaum, *Validation in the Clinical Theory of Psychoanalysis*, chap. 2., pp. 49–68. See also A. Grünbaum, "Popper's Fundamental Misdiagnosis of the Scientific Defects of Freudian Psychoanalysis and of Their Bearing on the Theory of Demarcation," a paper delivered at the "Rethinking Popper" conference held at the Czech Academy of Sciences, Prague, September, 10–14, 2007, which appeared in *Psychoanalytic Psychology* 25, no. 4 (October 2008) and in the proceedings volume of the Prague conference, published by Springer in 2009.

51. M. Scriven, "The Experimental Investigation of Psychoanalysis," in *Psychoanalysis, Scientific Method and Philosophy*, ed. Sidney Hook (New York: New York University Press, 1959), pp. 226–51.

52. P. S. Holzman, "Psychoanalysis: Is the Therapy Destroying the Science?" *Journal of the American Psychoanalytic Association* 33, no. 4 (1985): 733.

53. R. R. Holt, *Freud Reappraised: A Fresh Look at Psychoanalytic Theory* (New York & London: Guilford Press, 1989), p. 327.

54. Ibid., p. 328.

55. R. R. Holt, "Some Reflections on Testing Psychoanalytic

Hypotheses," open peer commentary on Adolf Grünbaum, *The Foundations of Psychoanalysis: A Philosophical Critique*, in *Behavioral and Brain Sciences* 9, no. 2 (1986): 242–44.

56. P. A. Schilpp, *The Philosophy of Karl Popper*, p. 984.

57. A. Grünbaum and W. Salmon, eds., *The Limitations of Deductivism*, p. 17, n. 20.

58. A. Grünbaum, *The Foundations of Psychoanalysis*, p. xii.

59. R. Wollheim, *Sigmund Freud* (New York: Viking Press, 1971), pp. 12–13.

60. S. Freud, "A Short Account of Psychoanalysis," in *The Standard Edition of the Complete Psychological Works of Sigmund Freud*, vol. 19, ed. and trans. J. Strachey (London: Hogarth Press, 1955; original work published in 1924), p. 194.

61. Ibid.

62. J. Breuer and S. Freud, "On the Psychical Mechanism of Hysterical Phenomena: Preliminary Communication," in *The Standard Edition of the Complete Psychological Works of Sigmund Freud*, vol 2, edited by J. Strachey pp. 3–17 (London: Hogarth Press, 1955; original work published in 1893).

63. C. Hanly, review of *The Foundations of Psychoanalysis*, by A. Grünbaum, *Journal of the American Psychoanalytic Association* 36 (1988): 524.

64. R. Holt, "Some Reflections on Testing Psychoanalytic Hypotheses," open peer commentary in *Behavioral and Brain Sciences* 9, no. 2 (1986): 243.

65. B. Rubinstein, "Psychoanalysis and the Philosophy of Science: Collected Papers of Benjamin Rubinstein, M.D.," in *Psychological Issues*, Monograph 62–63, ed. R. Holt (Madison, CT: International Universities Press, 1997).

66. A. Grünbaum, "The Reception of My Freud-Critique in the Psychoanalytic Literature," *Psychoanalytic Psychology* 24, no. 3 (2007): 545–76.

67. F. Crews, *Skeptical Engagements* (New York: Oxford University Press, 1986).

68. F. Crews, "The Future of an Illusion," *New Republic*, January 21, 1985, p. 28.

69. A. Grünbaum, "Précis: *The Foundations of Psychoanalysis: A Philosophical Critique*," *Behavioral and Brain Sciences* 9, no. 2 (1986): 217–28; A. Grünbaum, "Author's Response: Is Freud's Theory Well-Founded?" *Behavioral and Brain Sciences* 9, no. 2 (1986): 266–84.

70. A. Grünbaum, *The Foundations of Psychoanalysis*, p. 161.

71. B. Von Eckhardt, "Adolf Grünbaum: Psychoanalytic Epistemology," in *Beyond Freud: A Study of Modern Psychoanalytic Theorists*, ed. J. Reppen (Hillsdale, NJ: Analytic Press, 1985), pp. 353–54.

72. A. Grünbaum, *The Foundations of Psychoanalysis*, p. 1.

73. A. Grünbaum, "The Hermeneutic Versus the Scientific Conception of Psychoanalysis: An Unsuccessful Effort to Chart a *Via Media* for the Human Sciences," in *Einstein Meets Magritte, An Interdisciplinary Reflection: The White Book of Einstein Meets Magritte*, ed. D. Aerts (Dordrecht, Netherlands: Kluwer Academic, 1999), pp. 219–39.

74. A. Grünbaum, "The Poverty of the Semiotic Turn in Psychoanalytic Theory and Therapy," in *Between Suspicion and Sympathy, Paul Ricoeur's Unstable Equilibrium, a Festschrift in Honor of Paul Ricoeur's 90th Birthday*, ed. A. Wiercinski (Toronto, ON: The Hermeneutic Press, 2003), pp. 602–19.

75. M. Eagle, review of *The Foundations of Psychoanalysis*, by A. Grünbaum, in *Philosophy of Science* 53 (1986): 65–88.

76. H. Eysenck, review of *The Foundations of Psychoanalysis*, by A. Grünbaum, *Behaviour Research and Therapy* 23, no. 1 (1985): 89–90.

77. P. Robinson, "Adolf Grünbaum: The Philosophical Critique of Freud," in *Freud and His Critics* (Berkeley: University of California Press, 1993) chap. 3, pp. 180–81.

78. Ibid.

79. P. Holzman, introduction in A. Grünbaum, *Validation in the Clinical Theory of Psychoanalysis*, pp. 117–18.

80. A. Grünbaum, "Remarks on Miller's review of *Philosophical Problems of Space and Time*," *Isis* 68, no. 243 (1977): 447–48.

81. Grünbaum, *Validation in the Clinical Theory of Psychoanalysis*, pp. xvii–xxii.

82. A. Grünbaum, "Critique of Psychoanalysis," in *The Freud Encyclopedia*, ed. E. Erwin (New York/London: Routledge, 2002), pp. 117–36.

83. Breuer and Freud, "On the Psychical Mechanism of Hysterical Phenomena: Preliminary Communication," 1893, pp. 6–7.

84. S. Freud, "An Autobiographical Study," *The Standard Edition of the Complete Psychological Works of Sigmund Freud,* vol. 20, edited by J. Strachey (London: Hogarth Press, 1959; original work published in 1925), p. 45.

85. Breuer and Freud, "On the Psychical Mechanism of Hysterical Phenomena: Preliminary Communication," pp. 6–7.

86. A. Grünbaum, "Critique of Psychoanalysis," pp. 125–26.

87. A. Grünbaum, "The Placebo Concept in Medicine and Psychiatry," *Psychological Medicine* 16 (1986): 19–38.

88. A. Grünbaum, "The Placebo Concept in Medicine and Psychiatry," in *Non-Specific Aspects of Treatment*, ed. N. Shepherd and N. Sartorius, World Health Organization (Toronto, ON: Hans Huber Publishers, 1989), pp. 7–38.

89. P. S. Sullivan, review of "The Placebo Concept in Medicine and Psychiatry," by A. Grünbaum, in *Philosophical Psychopathology*, edited by G. Graham and G. L. Stephens (Cambridge, MA: MIT Press, 1994), pp. 285–324; *Behavior and Philosophy* 24, no. 2 (1996): 177.

90. R. Wollheim, "Desire, Belief, and Professor Grünbaum's Freud," chap. 6 in *The Mind and Its Depths* (Cambridge, MA: Harvard University Press, 1993).

91. A. Grünbaum, "Freud's Permanent Revolution: An Exchange, a Reply to Thomas Nagel," *New York Review of Books* 41, no. 14 (August 11, 1994): 54–55.

92. E. Erwin, *A Final Accounting* (Cambridge, MA: MIT Press, 1996) pp. 95, 104–106, 124–25, and 130–36.

93. J. Lear, "The Shrink Is In," *New Republic*, December 25, 1995, p. 22.

94. P. Caws, "Psychoanalysis as the Idiosyncratic Science of the Individual Subject," *Psychoanalytic Psychology* 20, no. 4 (2003): 624.

95. A. Grünbaum, "Is Sigmund Freud's Psychoanalytic Edifice Relevant to the 21st Century?" *Psychoanalytic Psychology* 23, no. 2 (2006): 258.

96. A. Grünbaum, *The Foundations of Psychoanalysis*, chap. 4.

97. Lear, "The Shrink Is In," p. 22.

98. *World Book Encyclopedia* (1966), 14: 103.

99. A. Grünbaum, "Is Sigmund Freud's Psychoanalytic Edifice Relevant to the 21st Century?" pp. 99–109. See esp. pp. 107–108.

100. E. Erwin, "Philosophers on Freudianism: An Examination of Replies to Grünbaum's *Foundations*," in J. Earman et al., *Philosophical Problems of the Internal and External Worlds*, pp. 409–58.

101. E. Erwin, "Psychoanalysis: Past, Present, and Future," *Philosophy and Phenomenological Research* 57, no. 3 (1997): 671–96.

102. A. Grünbaum, "Author's Response: Is Freud's Theory Well Founded?" *Behavioral and Brain Sciences* 9 (1986): 266–80.

103. A. Grünbaum, letter to the editor, "A Reply to Jonathan Lear's 'The Shrink is In,'" *New Republic*, January 29, 1996, p. 5.

104. A. Grünbaum, *Validation in the Clinical Theory of Psychoanalysis*, chap. 4; Grünbaum, *The Foundations of Psychoanalysis*, chap. 8.

105. P. E. Meehl, "Commentary: Psychoanalysis as Science," *Journal of the American Psychoanalytic Association* 43, no. 4 (1995): 1021.

106. R. Roudinesco, *Pourquoi la Psychanalyse?* (Paris: Fayard, 1999).

107. R. Roudinesco, *Why Psychoanalysis?* (An English translation of her 1999 French original) (New York: Columbia University Press, 2001), pp. 71–74.

108. A. Green, "Against Lacanism," *Journal of European Psychoanalysis* 2 (1995–1996): 169–85.

109. N. Weill, review of *The Foundations of Psychoanalysis*, by A. Grünbaum, *LeMonde*, December 27, 1996, p. 12.

110. A. Barberousse, review of *The Foundations of Psychoanalysis*, by A. Grünbaum, *Revue de L'Association Henri Poincaré* 7, no. 6 (June 1997): 27–32.

111. O. Flanagan, review of *Validation in the Clinical Theory of Psychoanalysis*, by A. Grünbaum, *London Times Literary Supplement*, October 29, 1993, p. 3.

112. A. H. Esman, review of *Validation in the Clinical Theory of Psychoanalysis*, by A. Grünbaum, *American Journal of Psychiatry* 152, no. 2 (1995): 283.

113. S. Mitchell, "The Analyst's Knowledge and Authority," *Psychoanalytic Quarterly* 62, no. 1 (1998): 4–5.

114. A. Grünbaum, "Is Simplicity Evidence of Truth?" in *Philosophy of Science*, ed. A. O'Hear (London: Cambridge University Press, 2007) and in Supplement No. 61 of *Philosophy*, pp. 261–73. A revised, enlarged version appeared in *American Philosophical Quarterly* 45, no. 2 (2008): 179–89.

115. R. Farr, editorial review of *Freud, Conflict and Culture: Essays on His Life, Work and Legacy*, retrieved September 28, 1998, from www .Amazon.com.

116. A. Grünbaum, "The Poverty of Theistic Cosmology," this volume.

117. A. Grünbaum, "Das Elend der theistischen Moral," in *Moral als Gabe: Zur Ambivalenz von Moral und Religion*, ed. B. Boothe and P. Stoellger (Würzburg: Königshausen and Neumann, 2004), pp. 143–75.

118. S. Latsis, "Nafplion Revisited," in *Inflation, Institutions and Information: Essays in Honour of Axel Leijonhufvud*, ed. D. Vas and K. Velupillai (Philadelphia, PA: Trans-Atlantic Publications, 1996), p. 28.

119. R. Weingartner, *Mostly About Me* (Bloomington, IN: Author House Publishing, 2003), p. 351.

Part One

THEISTIC COSMOLOGY

1.
THE POVERTY OF THEISTIC COSMOLOGY*

Adolf Grünbaum**

ABSTRACT

Philosophers have postulated the existence of God to explain (I) why any contingent objects exist at all rather than nothing contingent, and (II) why the fundamental laws of nature and basic facts of the world are exactly what they are. Therefore, we ask: (a) Does (I) pose a well-conceived question that calls for an answer? and (b) Can God's presumed will (or intention) provide a cogent explanation of the basic laws and facts of the world, as claimed by (II)? We shall address both (a) and (b). To the extent that they yield an unfavorable verdict, the afore-stated reasons for postulating the existence of God are undermined.

As for question (I), in 1714, G. W. Leibniz posed the primordial existential question (hereafter "PEQ"): "Why is there something contingent at all, rather than just nothing contingent?" This question has

---

*Editorial note: Fifty-one years ago, Professor Grünbaum published his first paper in the *British Journal for the Philosophy of Science* [*not* his first paper *ever*], in the issue for 1953. It was entitled "Whitehead's Method of Extensive Abstraction" (*British Journal for the Philosophy of Science* 4: 215–26). The editor wishes to acknowledge Grünbaum's extraordinary achievement in philosophy of science and in particular the debt that this journal owes to so distinguished and productive an author.

**This essay originated in the first two of my three Leibniz Lectures, delivered at the University of Hannover, Germany, June 25–27, 2003. © British Society for the Philosophy of Science, 2004.

two major presuppositions: (1) A state of affairs in which nothing contingent exists is indeed genuinely possible ("the null possibility"), the notion of nothingness being both intelligible and free from contradiction; and (2) *De jure*, there should be nothing contingent at all, and indeed there would be nothing contingent in the absence of an overriding external cause (or reason), because that state of affairs is "most natural" or "normal." The putative world containing nothing contingent is the so-called null world.

As for (1), the logical robustness of the null possibility of there being nothing contingent needs to be demonstrated. But even if the null possibility is demonstrably genuine, there is an issue: Does that possibility require us to explain why it is not actualized by the null world, which contains nothing contingent? And, as for (2), it originated as a corollary of the distinctly Christian precept (going back to the second century) that the very existence of any and every contingent entity is utterly dependent on God at any and all times. Like (1), (2) calls for scrutiny. Clearly, if either of these presuppositions of Leibniz's PEQ is ill founded or demonstrably false, then PEQ is aborted as a nonstarter, because in that case, it is posing an ill-conceived question.

In earlier writings (Grünbaum 2000, 5), I have introduced the designation "SoN" for the ontological "spontaneity of nothingness" asserted in presupposition (2) of PEQ.

Clearly, in response to PEQ, (2) can be challenged by asking the counterquestion, "But why should there be nothing contingent, rather than something contingent?" Leibniz offered an *a priori* argument for SoN. Yet it will emerge that *a priori* defenses of it fail, and that it has no empirical legitimacy either. Indeed physical cosmology spells an important relevant moral: As against any *a priori* dictum on what is the "natural" status of the universe, the verdict on that status depends crucially on empirical evidence. Thus PEQ turns out to be a nonstarter, because its presupposed SoN is ill founded! Hence PEQ cannot serve as a springboard for creationist theism.

Yet Leibniz and the English theist Richard Swinburne offered divine creation ex nihilo as their answer to the ill-conceived PEQ. But, being predicated on SoN, their cosmological arguments for the existence of God are fundamentally unsuccessful.

The axiomatically topmost laws of nature (the "nomology") in a scientific theory are themselves unexplained explainers, and are thus thought to be true as a matter of brute fact. But theists have offered a theological explanation of the specifics of these laws as having been willed or intended by God in the mode of agent causation to be exactly what they are.

A whole array of considerations are offered in section 2 to show that the proposed theistic explanation of the nomology fails multiply to transform scientific brute facts into specifically explained regularities.

Thus, I argue for the poverty of theistic cosmology in two major respects.

1. WHY IS THERE SOMETHING RATHER THAN NOTHING?

1.1 Refined statement of Leibniz's primordial existential question (PEQ)

Leibniz's 1714 essay "Principles of Nature and of Grace Founded on Reason" (1714; 1973b, sec. 7, 199) is the locus classicus of the question *"Why is there something rather than nothing?"* In the German translation of his French original, this question is *"warum es eher Etwas als Nichts gibt"* (ibid., 13).

We shall speak of this query as "the primordial existential question" and will use the acronym "PEQ" to denote it for brevity. But we must refine the statement of PEQ to preclude a *trivialization*, which Leibniz certainly did not intend when he asked this question. As we shall see, he believed that God is "a necessary being, bearing the reason of its existence within itself" in order to provide a *non-circular* "sufficient reason" for "the existence of the [*contingent*] universe" (ibid., sec. 8, 199). But if there is a *necessary* being, there can be no question why it exists, rather than not, because such a being could not possibly fail to exist. Therefore, it would clearly *trivialize* Leibniz's cardinal PEQ, if it were asked concerning a "something" comprising one or more entities whose existence is logically or metaphysically *necessary*.

Hence, the scope of the term "something" in his PEQ must obviously be restricted to entities whose existence is logically *contingent*; entities whose *non*existence is *logically possible*. And similarly for the scope of the term "nothing." Accordingly, we can formulate Leibniz's nontrivial construal of PEQ as follows: "Why is there something contingent at all, rather than just nothing contingent?" Philip Quinn (2003) has usefully characterized that articulation of PEQ as an "explanation-seeking contrastive why-question." He calls it "contrastive" because it features the contrastive locution "rather than."

William Craig (2001, sec. 2, 375–78) is oblivious to the *non*trivial construal of PEQ above. Thus, in the paper in question, published in this journal and titled "Professor Grünbaum on the 'Normalcy of Nothingness' in the Leibnizian and Kalam Cosmological Arguments," which is directed against my earlier essay in this journal (Grünbaum 2000), Craig obfuscates and eviscerates Leibniz's primordial question, which drives Craig to an exegetical falsehood as follows: "It must be kept in mind that for Leibniz (in contrast to Swinburne) [. . .] a state of nothingness is logically impossible" (ibid., 377). But Craig's assertion here is a red herring precisely because, for *both* Leibniz *and* Swinburne, a state of affairs in which there is nothing *contingent* is indeed logically possible. If Craig is to be believed and Leibniz had regarded a state of nothingness to be logically impossible, then his PEQ would have been tantamount to asking fatuously: Why is there something rather than a specified logically *impossible* state of affairs? This alone, it appears, is a reductio ad absurdum of Craig's exegesis of Leibniz.

As we shall see in some detail in Section 1.9, Swinburne (1991, 128) deems the existence of God to be logically *contingent*, and therefore he *excludes* God from a state of affairs in which there is nothing

164 THE POVERTY OF THEISTIC COSMOLOGY

contingent. But Swinburne, as well as Leibniz, was all too aware that, if there are entities that exist *necessarily*, then even a state in which there is nothing contingent cannot exclude such entities. Moreover, Leibniz (1714; 1973b, sec. 8, 199) had inferred that God exists necessarily qua sufficient reason for the "existence of the [contingent] universe." Hence Leibniz deemed the existence of God to be compossible with a state featuring nothing contingent, whereas Swinburne denied that compossibility, having concluded that God exists only contingently.

It is crucial to note at the outset that PEQ rests on important presuppositions. If one or more of these presuppositions is either ill founded or demonstrably false, then PEQ is aborted as a *non*starter, because it would be posing a *non*issue (pseudoproblem). And, in that case, the very existence of something contingent, instead of nothing contingent, does *not* require explanation. In earlier writings (Grünbaum 1998, 16; 2000, 5, 19), I have used the rather pejorative term "pseudoproblem"—"Scheinproblem" in German—to reject "a question that rests on an ill-founded or demonstrably false presupposition" (2000, 19). But, since the term "pseudoproblem" was given currency by the Vienna Circle, I immediately issued the caveat that, in my own use of it, "I definitely do not intend to hark back to early positivist indictments of 'meaninglessness'" (ibid.). Terminology aside, PEQ will indeed turn out to be a nonstarter because one of its crucial presuppositions is demonstrably ill founded. As we shall see, that presupposition is a corollary of a distinctly Christian doctrine, which originated in the second century CE.

What are the most important presuppositions of PEQ? Clearly, one of them is that the notion of a state of affairs in which absolutely nothing *contingent* exists is both *intelligible* (meaningful) and *free from contradiction*. Let us call such a putative state of affairs "the null possibility," as the English philosopher Derek Parfit does (1998a, 420). And let us speak of a supposed world in which there is nothing contingent as the "null world."

Yet it is vital to recognize that the null possibility is *not* shown to be logically genuine by the premise that each contingent entity, taken individually, might possibly not exist. After all, this premise is entirely compatible with the *denial* of the null possibility. Indeed, the familiar fallacy of composition is being committed if one infers that all entities, taken *collectively*, might possibly fail to exist merely because each contingent entity, taken *individually*, might possibly fail to exist.

In just this way, both Derek Parfit (1998b, 24) and the English theist Richard Swinburne (1996, 48) seem to have fallaciously inferred the logical robustness of the null possibility after enumerating a finite number of actual entities, *each* of which individually may possibly fail to exist. And their commission of the fallacy of composition then blinds them to their obligation to justify the null possibility as logically sound before posing PEQ. Alternatively, they may just have taken the null possibility for granted peremptorily. Thus, Parfit (1998b, 24) gave the following version of PEQ:

> [W]hy is there a Universe at all? It might have been true that nothing [contingent] ever existed: no living beings, no stars, no atoms, not even space or time. When we think about this ["null"] possibility [Parfit 1998a, 420], it can seem astonishing that anything [contingent] exists.

In this statement, Parfit presumably construed the term "nothing" to mean "nothing *contingent*," as Leibniz did. Evidently, Parfit inferred the null possibility without ado, declaring: "It might have been true that nothing ever existed." But he gave no cogent justification for avowing this logical possibility to be genuine: he just assumed peremptorily that the *nihilistic proposition* "There is nothing," or "The null world obtains," is both intelligible and free from contradiction. Instead of providing a conceptual *explication* of the null possibility, Parfit has evidently offered a mere *open-ended enumeration* of the absence of familiar ontological furniture from the null world: "no living beings, no stars, no atoms, not even space or time." Thereupon, he enthrones PEQ on a pedestal (ibid., column 1): "No question is more sublime than why there is a Universe [i.e., some world or other]: why there is anything rather than nothing." Besides presupposing that the null possibility is logically robust, Parfit's motivation for PEQ tacitly *pivots* on the supposition that, *de jure*, there should be nothing contingent.

1.2 Is it imperative to explain why there isn't just nothing contingent?

Parfit told us "When we think about this ['null'] possibility, it can seem astonishing that anything exists." And assuming such an astonished response, he feels entitled to ask why the null possibility does *not* obtain, i.e., why there is something after all, rather than just nothing. But I must ask: Why should the *mere contemplation* of the null possibility reasonably make it "seem astonishing that anything exists"?

If some of us were to consider the logical possibility that a person might conceivably metamorphose spontaneously into an elephant, for example, I doubt strongly that we would feel even the slightest temptation to ask why that *mere logical possibility is not realized*. But what if someone were to reply that, in such a case, we are not puzzled because, as we know empirically with near certainty, people just don't ever turn into elephants? Then I would retort: Indeed, and what could possibly be more commonplace empirically than that something or other does exist? On the other hand, consider, as just a thought experiment, that per impossibile, a person actually metamorphoses into an elephant. If we were suddenly to witness such a spontaneous transformation, we would all be aghast, and we would ask urgently: Why, oh why, did this monstrous transformation occur?

Why then, I put it to Parfit, should anyone reasonably feel astonished at all that the null possibility, if genuine, has remained a mere logical possibility and that something does exist *instead*? In short, why *should* there be just nothing, merely because it is logically possible? This *mere* logical possibility, I claim, does *not suffice* to legitimate Parfit's demand for an explanation of why the null world does *not* obtain, an explanation he seeks as a philosophical anodyne for his misguided ontological astonishment.

1.3 Must we explain why any and every de facto unrealized logical possibility is not actualized?

To justify a negative answer to this question, let us inquire quite generally: For *any* and *every de facto* unrealized logical possibility, is it well conceived to demand an explanation of the fact that it is *not* actualized? As we know, Leibniz's principle of sufficient reason (PSR) has been used to answer affirmatively that every fact has an explanation. Yet, as we shall see in section 1.71, Leibniz himself did not regard that principle as itself an adequate justification for his PEQ, because he also relied on its presupposition SoN to convey that the existence of something contingent is not to be expected at all, and therefore calls for explanation. But even his PSR is demonstrably unsound.

To appraise his principle of sufficient reason, consider within our universe the grounds for the demise of Laplacean determinism in quantum theory. This *empirically* well-founded theory features irreducibly stochastic probability distributions governing such phenomena as the spontaneous radioactive disintegration of atomic nuclei, yielding emissions of alpha or beta particles and/or gamma rays. In this domain of phenomena, there are not only logically but also nomologically (i.e., law-based) possible particular events that *could* but do *not* actually occur under specified initial conditions. Yet it is impermissibly legislative ontologically to insist that merely because these events are thus possible, there *must* be an explanation entailing their specific nonoccurrence, and similarly, of course, for stochastically governed, actually occurring events. This lesson was not heeded by Swinburne (1991, 287), who avowed entitlement to *pan*explainability, declaring: "We expect all things to have explanations." In our exegesis of Leibniz in section 1.71 below, we shall deal further with his PSR.

The case of quantum theory shows that an empirically well-grounded theory can warrantedly discredit the tenacious demand for the satisfaction of a previously held ideal of explanation, such as Leibniz's principle of sufficient reason. To discover that the universe does not accommodate rigid prescriptions for explanatory understanding is not tantamount to scientific failure; instead, it is to discover positive reasons for identifying certain coveted explanations as phantom. And to reject the demand for them is legitimate in the face of Charles Saunders Peirce's heuristic injunction not to block the road to inquiry, for this rejection does not abjure the search for a new, better theory in which the original explanatory quest may appear in a new light.

The demise of the PSR at the hands of microphysics spells a moral for Parfit's question why the null possibility does not obtain: the *mere* logical possibility of the null world—assuming it to be genuine—does *not suffice* to legitimate Parfit's demand for an explanation of why the null possibility does not obtain, rather than something contingent.

Nonetheless, Richard Swinburne declared ([1991], 283): "It remains to me, as to so many who have thought about the matter, a source of extreme puzzlement that there should exist anything at all." And, more recently, he opined (Swinburne 1996, 48): "It is extraordinary that there should exist anything at all. Surely the most natural state of affairs is simply nothing: no universe, no God, nothing." It is here, incidentally, that Swinburne apparently commits the fallacy of composition, as Parfit did, in trying to vouchsafe the null possibility by an enumeration of contingent entities, each of which, taken individually, may possibly fail to exist.

1.4 Is a world not containing anything contingent logically possible?

We need to be mindful of a further imperative to demonstrate that the null possibility hypothesized by PEQ is logically authentic, if indeed it is: some philosophers have explicitly denied the intelligibility of a kindred possibility. Thus, Henri Bergson has argued relatedly against nothingness: "The idea of absolute nothingness has not one jot more meaning," he tells us (1974, 240; originally published in 1935), "than a square circle." True enough, Richard Gale (1976, 106–13) has given a number of detailed reasons for rejecting Bergson's claim of unintelligibility. Yet Gale's own proposed explication of the hypothetical claim that "Nothing exists" is itself so qualified as to drive him to the following unfavorable conclusion: "it is not [logically] possible for there to be [absolutely] Nothing" (ibid., 116).

To state the nub of his reasoning, let me again use the locution "null world" to speak of a putative world in which the null possibility in fact obtains. Then we can say that the null world is devoid of space-time, no less than of all other contingent objects. But according to Gale's account (1976, 115–16), the receptacle of space and time (extension and duration), along with the "positive" properties or "forms," "are the ontological grounds for the possibility of there being Nothing." Hence Gale contends that "there is no possibility of *their* not existing. Put differently, it is not possible for there to be [absolutely] NOTHING, for there must at least be the [spatio-temporal] Receptacle and the forms" (ibid., 116). Thus, Gale diverges from Parfit's view that space and time exist only contingently: for Gale, they exist necessarily and hence exist even in the null world; but for Parfit, they are excluded from the null world, qua existing only contingently. Therefore, it is puzzling that, in the face of this exclusion, Parfit used the seemingly *temporal* term "ever" when he told us that "it might have been true that nothing ever existed."

But, as Edward Zalta has pointed out (private communication), it is unclear how Gale's avowal of space and time as existing *necessarily*, and Bergson's indictment of meaninglessness, are relevant to the issue of the intelligibility of the null possibility. That possibility pertains to contingent existents, not to necessary ones. After all, as we saw, Leibniz's null world contains necessarily existing entities like his God, while being devoid of all contingent ones. Hence Gale's argument has not gainsaid the pertinent sort of null possibility. And, as for Bergson, he is addressing the hypothesis that "absolutely nothing exists," rather than the hypothesis that nothing *contingent* exists. And the latter may be meaningful, even if the former is not.

But are there positive arguments that establish the meaningfulness of the null possibility? The reader is referred to philosophical misgivings or challenges issued by Edward Zalta of Stanford University, which place the burden of proof on those who deem PEQ to be well conceived (private communication).

In any case, it should be borne well in mind that the provision of a viable explication of the null possibility is surely not *my* philosophical responsibility, but rather belongs to the protagonists of PEQ, who bear the onus of legitimating their question. In the absence of assurance that the null possibility is logically authentic, PEQ might well be aborted as a nonstarter for that reason alone.

How, then, are we to understand more deeply the *tenacity* with which PEQ has been asked not only by some philosophers but even in our culture at large? An illuminating set of answers is afforded, it seems, by delving critically into three kinds of impetus for this ontological question, as follows: (1) historically based assumptions going back to the second century of the Christian era, which served to inspire PEQ; (2) explicitly *a priori* logical justifications of PEQ put forward by Leibniz, Parfit, Swinburne, and Robert Nozick, and (3) hypothesized emotional sources articulated by Arthur Schopenhauer in his magnum opus *The World as Will and Representation* (*Die Welt als Wille und Vorstellung*).

Let us consider these three sorts of impetus for PEQ seriatim:

1.5 Christian doctrine as an inspiration of PEQ

On Maimonides' reading of the opening passage of the Book of Genesis, the Mosaic God created the world *out of nothing*. Yet there is recent biblical exegesis contending that this doctrine of creation ex nihilo was not avowed in the Book of Genesis. Though the doctrine may have had a prehistory, it was first widely held by Christian theologians, beginning in the second century CE, as a *distinctly Christian* precept (May 1978). Thus, in an exegetical essay on "Genesis's account of creation" in the Old Testament, the Jewish scholar Norbert Samuelson wrote a few years ago (Samuelson 2000, 128): "'this [Hebraic] cosmology presupposes that initially God is not alone. Prior to God's act of creation [. . .] the earth, [and] water are the stuff from which God creates." But Christian writers regard their specific conception of divine creation ex nihilo as a philosophical advance over the account in the Book of Genesis, if only because they held that an omnipotent God had no need for preexisting materials to create the universe. Thus, as one such writer noted rather patronizingly, "The abstract notion of nothing does not seem to have been reached by the Israelite mind at that time" (Loveley 1967, 419). And, evidently, the notion of nothingness was essential to generate PEQ.

According to traditional Christian ontological doctrine, the *very existence* of any and every contingent entity other than God himself is utterly dependent on God at any and all times. Let us denote this fundamental Christian axiom of total ontological dependence on God by "DA," for "Dependency Axiom." Clearly, DA entails the following cardinal maxim: "without God's [constant creative] support [or perpetual creation]," the world "would instantly collapse into nothingness" (Hasker 1998, 695; cf. also Edwards 1967, 176). This assumption played a crucial role in subsequent philosophical history. Thus, in later centuries, precisely this hypothesis DA was avowed, as we shall see, by such philosophers as Thomas Aquinas and Descartes, among a host of others.

Evidently, DA in turn entails that, in the absence of an external cause, the spontaneous, natural, or normal state of affairs is one in which nothing contingent exists at all. As will be recalled, in earlier writings (Grünbaum 2000, 5), I have denoted the assertion of this ontological spontaneity of nothingness by "SoN." As before, we shall usually speak of the putative state of affairs in which no contingent objects exist at all as "the null world," a locution that is preferable to the term "nothingness." In that parlance, SoN asserts the ontological spontaneity of the null world.

As we see, the fundamental Christian ontological axiom DA of total existential dependence on God *entails* SoN. In other words, logically the truth of SoN is a *necessary condition* for the truth of the fundamental ontological tenet of Christian theism. In this clear sense, SoN is a *presupposition* of DA, which will turn out to be a heavy doctrinal burden indeed. SoN is "a heavy doctrinal burden," because, as we shall see, it is *completely baseless*.

According to SoN, the actual existence of something contingent or other—qua deviation from the supposedly spontaneous and natural state of nothingness—automatically requires a *creative external cause ex nihilo*, a so-called ratio essendi. And such a supposed creative cause must be distinguished, as Aquinas emphasized, from a merely transformative cause: transformative causes produce changes of state in contingent things that *already exist in some form*, or the transformative causes generate new entities from previously existing objects, such as in the building of a house from raw materials.

Furthermore, in accord with the traditional Christian commitment to SoN, creation ex nihilo is required at every instant at which the world exists in some state or other, whether it began to exist at some moment having no temporal predecessor in the finite past, or has existed forever. More precisely, having presupposed SoN, traditional Christian theism makes the following major claim: in the case of any contingent entity E other than God himself, if E exists, or begins to exist *without* having a *transformative* cause, then its existence must have a creative cause *ex nihilo*, **rather than being externally UNCAUSED.**

Yet, as some scholars have pointed out, "To the ancient Indian and Greek thinker the notion of *creation* [*ex nihilo*] is unthinkable" (Bertocci 1968; 1973, 571). Thus, in Plato's *Timaeus*, there is no *creation ex nihilo* by the Demiurge, who is held to transform chaos into cosmos, although that notion is very vague. Indeed, as John Leslie (1978, 185) has pointed out informatively: "To the general run of Greek thinkers *the mere existence of things* [or of the world] *was nothing remarkable*. Only their *changing* patterns provoked [causal] inquisitiveness" (italics added). And he mentions Aristotle's views as countenancing the acceptance of "reasonless existence."

It is a sobering fact that, before Christianity molded the philosophical intuitions of our culture, those of the Greeks and of many other world cultures (Eliade 1992) were basically different ontologically. No wonder that Aristotle regarded the material universe as uncreated and eternal. In striking contrast, *SoN is deeply ingrained in the traditional Christian heritage*, even among a good many of those who reject Christianity in other respects. And the Christian climate lends poignancy to Leslie's conjecture that "When modern Westerners have a tendency to ask why there is anything at all, rather than nothing, possibly this is *only* because they are heirs to centuries of Judaeo-Christian thought" (ibid.; italics added). So much for the Christian historical contribution to PEQ via its SoN doctrine.

1.6 Henri Bergson

Early in the twentieth century, Henri Bergson was alert to the often-beguiling, if not insidious, role of SoN in metaphysics, and he aptly articulated that assumption as inherent in PEQ. In 1935, speaking of occidental philosophy, Bergson (1974, 239–40) lucidly wrote disapprovingly concerning PEQ as follows:

> [P]art of metaphysics moves, consciously or not, around the question of knowing why anything exists—why matter, or spirit, or God, rather than nothing at all? But the question presupposes that reality fills a void, that underneath Being lies nothingness, that *de jure* there should be nothing, that we must therefore explain why there is *de facto* something.

Bergson's concise formulation of SoN as a presupposition of the Primordial Existential Question is that "*de jure* there should be nothing." But as a rendition of this cardinal presupposition of PEQ, his formulation that *de jure* there *should* be nothing is significantly incomplete: it needs to be amplified by the further claim that there indeed *would* be nothing in the absence of an overriding external cause or reason! Thus, let us bear in mind hereafter that *SoN makes the following very strong ontological assertion*: **De jure, there *should* be nothing contingent at all rather than something contingent, and indeed, there *would* be just nothing contingent in the absence of an overriding external cause (reason).**

In a chapter devoted to PEQ, Robert Nozick (1981, 122) notes, as Bergson had, that this inveterate question is predicated on SoN, a presupposition avowing, in his words, "a presumption in favour of nothingness." As he puts his view there: "To ask 'why is there something rather than nothing?' assumes that nothing(ness) is the natural state that does not need to be explained [causally], while deviations or divergences from nothingness have to be explained by the introduction of special causal factors."

Importantly, SoN can be challenged by the *counterquestion*: "But why *should* there be nothing contingent, rather than something contingent?" And, indeed, why *would* there be nothing contingent in the absence of an overriding external cause (reason)? In effect, Leibniz (1714; 1973b, sec. 7, 199) endeavored to disarm this challenge, as we are about to see, when he tried to legitimate SoN—albeit unsuccessfully—as part of a *twofold a priori* justification of PEQ.

Since PEQ is predicated on SoN, PEQ will be undermined in due course by the failure of *a priori* defenses of SoN, *and* by the unavailability of any empirical support for it either. Hence I am not persuaded by Nicholas Rescher's (1984, 19) claim that PEQ—which he calls "the riddle of existence"—"does not seem to rest in any obvious way on any particularly problematic presupposition." Although the defects of SoN are indeed not obvious, SoN will, in fact, turn out to be a "particularly problematic presupposition" of PEQ.

1.7 A priori justifications of PEQ by Leibniz, Parfit, Swinburne, and Nozick

A number of writers have used ideas such as simplicity, nonarbitrariness, naturalness, and probability in an attempt *to justify PEQ a priori*. In each case, the argument seems to be that a state of affairs in which nothing contingent exists has a crucial property (simplicity, nonarbitrariness, naturalness, high probability, and so forth) that we would *a priori* expect the world to have. This *a priori* expectation is then presumed to validate the second presupposition of PEQ, which entails that the null world would be actual in the absence of an external cause. Typically, the property in question (e.g., simplicity) has not even been sufficiently articulated; but even if its character is taken for granted, the governing principle (e.g., the world should be simple) has not been justified, as we can see now and elsewhere (Grünbaum 2008).

1.7.1 Leibniz

In stark contrast to Bergson, both Leibniz (1714; 1973b, sec. 7, 199) and Swinburne (1991, 283–84) maintained that SoN is *a priori true.* And their reason was that the null world is simpler, both ontologically and conceptually, than a world containing something contingent or other. This very ambitious assertion poses two immediate questions: (a) Is the null world really *a priori* simpler, and indeed the *simplest* world ontologically as well as conceptually? And (b) even assuming that the null world *is* thus simpler, does its supposed maximum dual simplicity *mandate ontologically* that there *should* be just nothing *de jure*, and that, furthermore, there *would* be just nothing in the absence of an overriding cause (reason), as claimed by SoN?

As for question (a), re the maximum twofold simplicity, the Swedish philosophers Carlson and Olsson (2001, 205) speak of the null world as "the intrinsically simplest of all possibilities," and they add that they "have not seen it questioned." But as for the question of *conceptual* simplicity, there is one *caveat*, which needs to be heeded.

To see why, note first that Leibniz couched his original 1697 statement of PEQ in terms of "worlds" when he demanded "a full reason why there should be any world rather than none" (1697; 1973a, 136). This formulation suggests that, conceptually, the very notion of the null world may well range—by *negation* or *exclusion*—over all of the possible contingent worlds or objects other than itself that are *not* being actualized in it. But this collection of unrealized, noninstantiated contingent worlds is *super-denumerably infinite* and is of such staggering complexity that it boggles the mind!

As we have remarked, the champions of the maximum simplicity of the null world have not given us a demonstrably viable explication of the notion of the null world as being logically authentic or robust. Therefore, they cannot claim to have *ruled out* that, conceptually, this notion is highly complex, instead of being the simplest. So much for the caveat pertaining to the purported maximum *conceptual* simplicity of the null world.

Beyond this caveat, we do not need to address the ramified issues raised by (a) in dealing with Leibniz's defense of SoN except to say that, to my knowledge, the purported conceptual and *ontological* maximum simplicity of the null world has not been demonstrated by its proponents.

But let us assume, just for the sake of argument, that Leibniz and Swinburne could warrant *a priori* the maximum conceptual and ontological simplicity of the null world, as avowed by Leibniz, when he declared (1714; 1973b, sec. 7, 199): "'nothingness' is simpler and easier than 'something.'" It is of *decisive importance*, I contend, that *even if the supposed maximum ontological simplicity of the null world were warranted a priori, that presumed simplicity would **not** mandate the claim of SoN that de jure **the thus simplest world must be spontaneously realized ontologically** in the absence of an overriding cause.* Yet, to my knowledge, neither Leibniz nor Swinburne nor any other author has offered any cogent reason at all to posit such an ontological imperative.

Let us quote and then comment on the context in which Leibniz (1714; 1973b, secs. 7 and 8, 199), formulates his PEQ and then seeks to justify it at once by relying carefully on both of the following two premises: (1) his principle of sufficient reason (PSR), and (2) an *a priori* argument from simplicity for the presupposition SoN inherent in PEQ:

7. Up till now we have spoken as *physicists* merely; now we must rise to *metaphysics*, making use of the *great principle*, commonly but little employed, which holds that *nothing takes place without sufficient reason*, that is to say that nothing happens without its

being possible for one who has enough knowledge of things to give a reason sufficient to determine why it is thus and not otherwise. This principle having been laid down, the first question we are entitled to ask will be: *Why is there something rather than nothing?* For "nothing" [the null world] is simpler and easier than "something"[1]. Further supposing that things must exist, it must be possible to give a reason *why they must exist just as they do and not otherwise.*

8. Now this sufficient reason of the existence of the universe cannot be found in the series of contingent things, that is to say, of bodies and of their representations in souls. [. . .] Thus the sufficient reason, which needs no further reason, must be outside this series of contingent things, and must lie in a substance which is the cause of this series, or which is a necessary being, bearing the reason of its existence within itself; otherwise we should still not have a sufficient reason, with which we could stop. And this final reason of things is called *God*.

(Incidentally, note that although the English translation by Parkinson and Morris of the first sentence of Leibniz's section 8 speaks of the sufficient reason "of" the existence of the universe, I shall hereafter replace their "of" by "for," since the German translation of Leibniz's [1956] text uses the term "für.") These two major passages in Leibniz's sections 7 and 8 invite an array of comments:

(1) Right after enunciating his principle of sufficient reason (PSR), Leibniz poses PEQ "*Why is there something rather than nothing?*" as "the first question we are entitled to ask." And immediately after raising this question, he relies on simplicity *to justify* its presupposition SoN that, *de jure*, there should be nothing contingent at all, rather than something contingent: "For 'nothing' [the null world] is simpler and easier than 'something'." But, in

---

1. It merits mention that, in the German translation and French original version of Leibniz's 1714 essay (1956, 13 and 12, respectively), the word "nothing" in the sentence "For 'nothing' is simpler and easier than 'something'" is rendered by the respective nouns "das Nichts" and "le rien."

the class of logically contingent entities to which this claim of greater simplicity pertains, "nothing" (the null world) and "something" (a world featuring something) are mutually exclusive and jointly exhaustive. Thus Leibniz is telling us here, in effect, that the null world is the *a priori simplest* of all, besides being "the easiest." But, alas, he does not tell us here in just what sense the null world is "the easiest."

This thesis of the intrinsically greatest *a priori* ontological and conceptual simplicity of the null world has been a veritable mantra in the literature, which is why Carlson and Olsson (2001, 205) wrote that they "have not seen it questioned."

(2) It is vital to appreciate that Leibniz explicitly went beyond his PSR to justify his PEQ on the heels of enunciating PSR and posing PEQ: fully aware that PEQ presupposes SoN, he clearly did *not* regard PEQ to be justified by PSR alone, since he explicitly offered a *simplicity argument* to justify the presupposed SoN immediately after posing PEQ. Most significantly, he is *not* content to rely on PSR to ask *just* the truncated question: "Why is there something contingent?" Instead he uses SoN in his PEQ to convey his dual thesis that (i) the existence of something contingent **is not to be expected at all**, and (ii) *its actual existence therefore cries out for explanation in terms of the special sort of **noncontingent** causal sufficient reason he then promptly articulated in his section 8.*

Thus, the soundness of Leibniz's justification of his PEQ evidently turns on the cogency of his PSR as well as of his *a priori* argument for SoN. As for the correctness of his PSR, recall our preliminary objections to it in section **1.3**, which were prompted there by Parfit's erotetic musings. The modern history of physics teaches that PSR, which Leibniz (1714; 1973b, sec. 7) avowedly saw as metaphysical, cannot be warranted *a priori* and indeed is untenable on empirical grounds. The principle asserts that every event—in Leibniz's parlance, anything that "takes place" or "happens" ["*geschieht*," i.e., "*sich ereignet*" in the German translation of his original French text]—has an *explanatory* "reason [cause] sufficient to determine *why it is thus and not otherwise*" (italics added). Leibniz's inclusion of the locution "and not otherwise" is presumably intended to emphasize an important point: his PSR guarantees the existence of a sufficient reason not only for the actual occurrence of a given specific event E, but also for the actual *non*occurrence of any and every *specific* event that is *different* from **E** in some respect or other. Quite reasonably, therefore, PSR has been taken to avow the existence of a reason sufficient to explain any and every *fact* pertaining to an individual event. For example, one such fact might be the occurrence of the complex event of Ronald Reagan *not* having been killed when Hinckley shot at him.

In sum, PSR is untenable: irreducibly stochastic laws in quantum physics tell us that some events have no individual explanations but occur as a matter of brute fact. And, assuming with Leibniz that there is no infinite regress of explanations, the history of science strongly supports the view that, no matter how we axiomatize our body of knowledge, every such axiomatization will feature some contingent fundamental laws or other that are unexplained explainers in that axiomatization and codify brute facts. And, as for Leibniz's *a priori* argument from simplicity for SoN, we saw earlier in this section **1.7.1** that it does not pass muster.

(3) To set the stage for a further instructive commentary on the subtle deficits of PSR, recall from Section **1.3** Swinburne's own formulation of SoN, which reads in part (1996, 48; italics added): "surely *the most natural* state of affairs is simply nothing." As will be shown in section **1.7.3**, this formulation of SoN entails the following consequence: it would be natural—though *not* "most natural"—for *our* world or universe U_0 *not* to exist rather than to exist. Let us denote that corollary of SoN by "SoN(U_0)." Now, though Leibniz's PSR turned out to be untenable, suppose just for

argument's sake that we were to grant him his PSR for now. Then someone may be tempted to believe that its explanatory demand could suffice after all, *without* a separate additional argument for SoN's corollary SoN(U_0), to legitimate the question "Why does our universe U_0 exist, rather than not?" a question which presupposes that corollary. In short, the issue is whether PSR can *single-handedly* license the injunction to explain the existence of our universe in particular.

To evaluate the claim that it can, note that Leibniz (ibid., sec. 8) called for an explanatory "sufficient condition for the existence of the universe [U_0]." But, importantly, this explanatory demand generates at least three *distinct* questions, which differ both from each other and from PEQ:

(Q_1): "Why does U_0 exist, rather than not?"

(Q_2): "Why does U_0 exist, rather than just nothing contingent?"

(Q_3): "Why does U_0 exist, featuring certain laws L_0, rather than some different sort of universe U_n, featuring logically possible different laws of nature L_1?"

Moreover, note that each of these questions, no less than PEQ, is predicated on a presupposition of its own, which is asserted by the *alternative* stated in its "rather than" clause. The relevant presupposition is that, *de jure*, the corresponding alternative should obtain and indeed might well or would obtain in the absence of an overriding cause (reason). And the question then calls for an explanation of the *deviation* from the supposedly *de jure* alternative.

In question Q_1, the alternative A_1 is that U_0 should not exist. Hence Q_1 is predicated on SoN(U_0). But its alternative, A_1, is noncommittal as to whether *other* universes likewise should *not* exist. Thus, A_1 differs from the alternative A_2 in Q_2 that nothing contingent at all should exist, which asserts SoN. A_2, in turn, differs from the alternative A_3 in Q_3 that something else exists in lieu of U_o. Thus Q_1, Q_2, and Q_3 are different questions whose answers therefore may well be different. Furthermore, observe that Q_2, besides differing from Q_1 and Q_3, also differs from PEQ: whereas Q_2, asks specifically why our U_o exists, rather than nothing contingent at all, PEQ asks why *something or other* exists, as against nothing at all.

In effect, Q_3 demands an explanation of why *our* world U_o exists, as contrasted with a logically possible different sort of universe featuring different laws of nature. Interestingly, Leibniz (1714; 1973b, sec. 7, 199) called for an answer to *that* question only after having assumed that: "Further, supposing that things must exist, it must be possible to give a reason *why they must exist just as they do* and not otherwise" (italics in original). By way of anticipation, note that, as section 2 will show, the volitional theological answer to question Q_3 by such theists as Leibniz, Swinburne, and Quinn completely fails to deliver on their explanatory promises.

We can now deal specifically with the afore-stated issue: if Leibniz's PSR were granted, could it *single-handedly* legitimate question Q_1 ("Why does U_o exist, rather than not?") *without* having to justify Q_1's presupposition SoN(U_o) by an additional argument? This question will now enable us to demonstrate the inability of PSR to serve solo as a warrant for Q_1, just as it turned out above to be incapable of licensing PEQ without a further argument justifying the latter's presupposition SoN. As will be recalled, Leibniz himself recognized that limitation of PSR by his recourse to the supposed greater simplicity of the null world as the ontological underwriter of the presupposition SoN of his PEQ.

In what sense, if any, might his PSR underwrite his demand for a 'sufficient reason for the existence of the universe [U_o]'? This injunction is clearly more specific than the imperative to supply a sufficient reason for the existence of something contingent or other. In his statement of PSR, Leibniz asserts the existence of a sufficient reason for what "*takes place*" or "happens." And, very importantly, this reason for the occurrence of events, he tells us, is "sufficient to determine *why it is thus and not otherwise*" (1714; 1973b, 199; italics added). Evidently, his contrasting alternative to "*what takes place*" is "*otherwise.*" But, crucially, that alternative demonstrably fails, however, to be *univocal!*

An example that, alas, has become quotidian since the two attacks on the World Trade Center in New York will illustrate the considerable ambiguity of Leibniz's notion of "otherwise." When we explain ordinary sorts of events, we typically know instances of their occurrence as well as instances of their nonoccurrence. Furthermore, we have evidence concerning the conditions relevant to both kinds of instances, and often have information as to their relative frequency. That information often tells us which of these sorts of events, if either, is to be expected routinely.

Indeed, when we explain the occurrence of a given event, the *contrasting non*occurrence of that event can *take different forms*: we may wish to explain why the event occurred rather than some *specified* other sort of event that might have occurred *instead*; or we may just ask why the given event occurred rather than not.

Thus, when a presumably well-built skyscraper collapsed, we may ask why it did, rather than withstand an assault on it, though not without considerable damage. Or we may ask why a presumably well-built skyscraper collapsed, rather than just staying intact. The latter contrasting alternative of staying intact constitutes the "*natural*" career of well-built skyscrapers in the absence of an overriding external cause.

In the context of Leibniz's inquiry into "the sufficient reason for the existence of the universe [U_0]," the relevant event or happening is the existence of the universe through time. Hence we must ask in that context: What becomes of his call for a sufficient reason that determines "why it [an event] is thus and not otherwise"?

The ambiguity of this request for a sufficient reason is shown by the different contrasting alternatives in questions Q_1, Q_2, and Q_3, questions that pertain to the existence of U_0. One such alternative to the existence of U_0 is simply that it does not exist, another that nothing contingent exists, but yet another alternative is that some other sort of universe U_n exists *instead* of U_o. Yet the demand to explain the existence of U_o, as issued by PSR, does *not* dictate *which* one of these alternatives is presupposed by it as "otherwise"! Thus, PSR fails to license the question Q_1 single-handedly by failing to single out its presupposed $SoN(U_o)$ as against the very different presuppositions of questions Q_2 and Q_3.

For precisely this portentous reason, it turns out that to ask "Why does U_o exist?" is *to ask an incomplete question*! Hence Leibniz's PSR is incapable of showing that if the existence of U_o is to be explained, it must be explained qua deviation from its nonexistence, as against qua deviation from some other alternative. In short, PSR itself does not license Q_1 as against Q_2 or Q_3. The tempting belief that it does so single-handedly is a will-o'-the-wisp.

The failure of his PSR to underwrite the particular question Q_1 "Why does U_o exist, rather than not?" also emerges from one of the reasons for rejecting $SoN(U_o)$ as ill founded. As we saw, when we call for an explanation of such events as the collapse of a skyscraper, we do so against the background of having observed specified instances of their *non*occurrence. Indeed, these nonoccurrences may well be so very frequent that we are warranted in taking them to be "natural." But obviously, yet very importantly, we have never ever observed an event constituted by the *non*existence of our U_o, let alone found evidence that its nonexistence would be "natural." Nor does the empirically supported big bang cosmology feature such a temporal event (Grünbaum 1998, sec. 5, 25–26). Accordingly, it is ill founded to regard the *non*existence of our U_o to be "natural," such that PSR could then warrant the question why U_o exists, rather than not.

(4) As we saw in section **1.1** on refining Leibniz's PEQ to preclude its unintended trivialization, its articulated version states: "Why is there something *contingent* at all, rather than just nothing *contingent*?" Alas, this rather straightforward construal of PEQ was lost on Craig. As we saw in section **1.1**, his bizarre, misguided reading of Leibniz turns PEQ into the ludicrously fatuous question: Why is

there something rather than a specific logically impossible state of affairs? And it will be recalled that, contrary to Craig, for Leibniz no less than for Swinburne, the null world—being devoid of all contingent existents, though clearly not of any necessary existents—is indeed logically possible. Yet Craig (2001, 377) denied this exegetical fact and offers a non-sequitur:

> It must be kept in mind that for Leibniz (in contrast to Swinburne) God's existence is logically necessary, so that [*sic!*] a state of nothingness is logically impossible. Hence, Leibniz *cannot* be assuming that a state of absolute nothingness is the natural or normal state of affairs, as Swinburne does.

But here Craig has irrelevantly fabricated for himself a notion of "a state of (absolute) nothingness" that is indeed logically impossible, because he himself incoherently banished all *necessary* existents from it, rather than only all contingent existents, as in Leibniz's null world. Thus, Craig clearly offers a non sequitur in claiming that "for Leibniz (in contrast to Swinburne) God's existence is logically necessary, so that a state of nothingness is logically impossible," a baseless conclusion. Moreover, once Craig had concocted his own incoherent state of "absolute nothingness," which is logically impossible by excluding necessary existents, he manufactured an incoherent *pseudo* version of SoN that features such phantom nothingness. Thereupon, he informs us irrelevantly and misleadingly that Leibniz "cannot" be claiming such absolute nothingness to be "the natural or normal state of affairs, as Swinburne [allegedly] does."

But surely, nobody of just ordinary intelligence, let alone Swinburne, would avow Craig's incoherently devised pseudo version of SoN! Swinburne's null world excludes only all *contingent* existents, while including, of course, any necessary existents. Thus, his conception of SoN plainly does not pertain to Craig's inane artifact of "absolute nothingness," with which Craig saddled him.

Unfortunately, Craig used an *ignoratio elenchi* and an exegetical jumble to reject my attribution of SoN to Leibniz, although SoN is a patent presupposition of Leibniz's PEQ, as Bergson and Nozick explicitly recognized.

(5) In section **1.3** above, and under item (3) of the current section **1.7.1**, there is a preliminary mention of Swinburne's alternative formulation of SoN, which reads in part (1996, 48; italics added): "surely *the most natural* state of affairs is simply nothing," a state devoid of all and only *contingent* existents. And as we shall see in Section **1.7.3**, this formulation of SoN entails the following consequence: it would be natural—though not "most natural"—for our world or universe U_0 of inanimate matter, biological organisms, and Homo sapiens *not* to exist, rather than to exist, a corollary of SoN that we have denoted by "SoN(U_0)." Since Leibniz evidently presupposed SoN in his PEQ, he is likewise committed to its corollary SoN(U_0).

But the existence of our U_0 is a *deviation* from its purportedly "natural" *non*existence, as avowed by SoN(U_0). And any deviation from naturalness calls for *explanation* in terms of a suitable cause or reason. No wonder that, in the opening sentence of Leibniz's afore-cited section 8, he demanded that the answer to his PEQ provide a "sufficient reason" for "the existence of the universe [U_0]."

Yet, true to form, Craig (2001, 378) tells us blithely:

> Even with respect to the physical universe, moreover, Leibniz did not hold that the natural or normal state of affairs is the non-existence of the physical universe, for he held (notoriously) that God's creation of the world is, like God Himself, necessary.

But, as Nicholas Rescher has argued in a thorough chapter significantly titled "Contingentia Mundi: Leibniz on the World's Contingency"(2003, 45):

> From the earliest days of his philosophizing Leibniz insisted upon the contingency of the world. It was always one of his paramount aims to avert a Spinozistic necessitarianism, and he regarded the

contingency of the world's constituents and processes as an indispensable requisite towards this end, one in whose absence the idea of divine benevolence would be inapplicable.

To the further detriment of Craig's exegesis, Rescher explains (ibid., 46):

Leibniz distinguishes between two different modes of necessity. The one is the *metaphysical* necessity of that whose opposite is logico-conceptually impossible. And the other is the *moral* necessity of that whose opposite is ethically unacceptable. Only the former is absolute and categorical, the latter is standard-relative and dependent upon ultimately evaluative ethical considerations.

Rescher elaborates crucially (ibid., emphasis added):

God's choice of the best available alternative for actualization [i.e., "God's creation of the world," in Craig's words], while indeed a certain fact regarding God, is only *morally* and not metaphysically necessary—*and thereby contingent* [footnote omitted].

And while God's moral perfection as creator of this best of worlds is itself a morally necessary truth, it is emphatically not metaphysically necessary. (ibid., 49)

As shown by Rescher's account of Leibniz's views, Craig used an exegetically false premise to deny incorrectly that Leibniz was committed to SoN(U$_o$). It would seem that the remainder of Craig's misguided gloss on Leibniz does not merit discussion.

1.7.2 Derek Parfit

Parfit has gone beyond Leibniz and Swinburne in laying down alleged *a priori* ontological imperatives. Claiming that the null world is not only the *a priori* simplest but also the "least arbitrary," Parfit believes that SoN is warranted all the more qua presupposition of PEQ. And hence he thinks that the supposed minimum arbitrariness of the null world confers even greater urgency on PEQ (1998b, 25). Thus, he asks, "Why is there a Universe at all? Why doesn't reality take its simplest and least arbitrary form: that in which nothing ever exists?" But why, one must ask, should ontological reality spontaneously be "least arbitrary," even assuming that the null world is *a priori* "least arbitrary" in Parfit's intended sense? He develops his reasoning (1998a) as follows. First he declares quite generally (p. 420): "If some possibility would be less puzzling, or easier to explain, we have more reason to think that it obtains." And then, echoing Leibniz's undemonstrated belief that the null world is the *a priori* simplest both conceptually and ontologically, Parfit offers that world as his paradigm example (p. 421): "Is there some global possibility whose obtaining would be in no way puzzling? That might be claimed of the Null Possibility [. . .]. Perhaps, of all the global possibilities, this would have needed the least explanation. It is much the simplest and it seems the easiest to understand."

Parfit's view that the null possibility "seems easiest to understand" may well have been suggested by Leibniz's dictum (1714; 1973b, sec. 7, 199) that "'nothing' is simpler and easier than 'something'," a claim that Leibniz offered unsuccessfully to justify SoN qua presupposition of his PEQ. But the null possibility may well not be "easiest to understand": as we saw in section **1.7.1**, conceptually the very notion of the null world may well range—by *negation* and *exclusion*—over a mind-boggling, nondenumerable infinitude of possible contingent worlds. This complexity seems to have been tacitly discerned by Parfit (1998a, 420) in his implicitly *open-ended* enumeration of objects that are *excluded* from the null world.

In any case, Parfit has to retract his view that the null possibility is the least problematic (ibid.): "Even if this possibility would have been the easiest to explain, it does not obtain. Reality does not take its simplest and least puzzling form."

Why then did Parfit claim, in the first place, that we have more reason to think that the allegedly simplest and "least arbitrary" null world would obtain than that some other unremarkable possible world would be actualized? Without ado, he appeals to some concept of "coincidence," saying (Parfit 1998a, 424):

> Coincidences can occur. But it seems hard to believe. We can reasonably assume that, if all possible worlds exist, that is *because* that makes reality as full as it could be.

> Similar remarks apply to the Null Possibility. If there had never been anything, would that have been a coincidence? Would it have merely happened that, of all the possibilities, what obtained was the *only* possibility [i.e., the *one* possibility] in which nothing exists? That is also hard to believe. Rather, if this possibility had obtained, that would have been because it had that feature.

Yet now Parfit no longer claims that some remarkable global feature F of some possible world W (e.g., maximum simplicity) warrants *the presumption that W would obtain*, rather than some unremarkable world. Instead, now we learn that *if* W does obtain and features some remarkable global property F, then we can infer explanatorily that W was ontologically mandated by the fact that F was ontologically *self-realizing*! Relying on such purported mandatoriness, Parfit says (ibid., 426–27): "Our world may seem to have some feature that would be unlikely to be a coincidence. We might reasonably suspect that our world exists, not as a brute fact but because it has this feature."

But how can Parfit tell whether a given global feature F of our world or of some other possible world in fact is "unlikely to be a coincidence"? He told us that if every conceivable world were actualized in a so-called "plenary universe," it would be very unlikely to be a coincidence. And if the null possibility were to obtain, that too would be very unlikely to be coincidental. On the other hand, we learn (Parfit 1998a, 424), if a universe of 57 worlds were to exist, its cardinality of 57 "could hardly" be self-actualizing and hence *would be coincidental*. In the case of the plenary universe and of the null world, Parfit (ibid.) is struck alike by the fact that "of all the global possibilities, the one that obtains" is just that particular one. But, clearly, in the 57-worlds universe as well, only one of the "countless global possibilities" is instantiated.

Why then, one must ask, would the obtaining of the *extremities* of the full range of possibilities *not* be coincidental and therefore cry out for being explained, whereas the actualization of one of the possibilities *in between* would be just coincidental? Why, indeed, isn't the 57-member world-ensemble just as *a priori improbable* as either the plenary universe or the null world?

Apparently, Parfit determines which instantiations are not coincidental, and which ones are, by *tacitly* appealing to some metric of *a priori* probabilities for the actualization of possible worlds that he has not even begun to articulate, let alone to justify. *A fortiori*, he has not shown that the then *a priori* improbable actualizations call for *ontological* explanation just because they are *a priori* improbable. For these reasons alone, it seems, his bizarre ontology of potentially self-actualizing explanatory global properties does not enlarge our philosophical horizons.

Furthermore, Swinburne (1998, 428) objects cogently to Parfit's self-actualizing scenario for our world:

> Parfit's suggestion that there might be some non-causal explanation of the existence of the Universe involves his claiming that there is some kind of principle at work in producing the Universe, which is never operative in producing more limited effects within the Universe. But then we have absolutely no reason for supposing that that kind of principle is ever at work, or that such a principle explains anything at all. [footnote omitted]

Note, incidentally, that when Parfit purports to explain that our universe U_o exists by recourse to its possession of some remarkable self-actualizing property P, what he is, in effect, claiming to explain is why U_o is actualized, *rather than* some other universe U_n that *lacks* P.

Parfit even generalizes his notion of self-actualizing remarkable global features into a comprehensive doctrine of cosmological explanation (Parfit 1998a, 424):

If some possibility obtains because it has some feature, that feature selects what reality is like. Let us call it the *Selector*. A feature is a *plausible* Selector if we can reasonably believe that, were reality to have that feature, that would not merely happen to be true.

There are countless features which are not plausible Selectors. Suppose that fifty-seven worlds exist. Like all numbers, 57 has some special features. For example, it is the smallest number that is the sum of seven primes. But that could hardly be *why* that number of worlds exist.

I have mentioned certain plausible Selectors. A possibility might obtain because it is the best, or the simplest, or the least arbitrary, or because it makes reality as full as it could be, or because its fundamental laws are as elegant as they could be. There are, I assume, other such features, some of which we have yet to discover.

For each of these features, there is the *explanatory* possibility that this feature is the Selector. That feature then explains why reality is as it is.

Among the *a priori* selectors, which Parfit countenanced as "plausible," he included the minimization of *a priori* arbitrariness: "A possibility might obtain because it is [. . .] the simplest, or the least arbitrary" (1998a, 424). Thus when he inquired (1998b, 25) why the null possibility does *not* obtain, he asked why reality does *not* "take its simplest and least arbitrary form."

Parfit's *a priori* view that the minimization of arbitrariness is *ontologically legislative* had been championed years earlier by Peter Unger (1984), who took a leaf out of Robert Nozick's (1981, 128) fanciful "fecundity principle," which richly populated the universe with an infinite "plenitude" (Unger 1984, 45) of isolated worlds on avowedly altogether *a priori* grounds (Unger 1984, sec. 10, 45–49). Though Unger admits to being "uncertain" as to what he means by the conception of a "highly unarbitrary" world (p. 47), he reaches a very gloomy *empirical* verdict concerning the realization of *non*arbitrary features in our world (Unger 1984, 48–49):

> Well, then, what *is* the available empirical evidence, and what does it indicate about the actual world? As empirical science presents it to us, is the world we live in, the world of which we are a part, a world notable for its lack of natural arbitrariness? Far from it, the actual world, our evidence seems to indicate, is full of all sorts of fundamental arbitrary features, quirks that seem both universal for the world and absolutely brute.

> [. . .]

> According to available evidence, and to such a theory of our actual world as the evidence encourages, the actual world has nowhere near the lack of arbitrariness that rationalist intuitions find most tolerable. To satisfy the rationalist approach, our evidence tells us, we must look beyond the reaches of our actual space and time, beyond our actual causal network. For there to be a minimum of arbitrariness in the universe entire, indeed anything anywhere near a minimum, we might best understand the universe as including, not only the actual world, but infinitely many other concrete worlds as well.

Thus, neither Parfit nor Unger have supplied any cogent reason for believing that *a priori* nonarbitrariness—insofar as that notion is clear at all—is *ontologically legislative* in the sense of mandating what sort of world is actualized.

As an ontological injunction, the minimization of arbitrariness seems to be just a case of apriorism run amok.

1.7.3 Richard Swinburne and Thomas Aquinas vis-à-vis SoN

Interestingly, Swinburne's alternative formulation of SoN, already mentioned in section **1.7.1** above, can serve as a point of departure for arguing, as a lesson from the history of science, that SoN stands or falls *on empirical but not on a priori* grounds. Swinburne's alternative formulation of SoN reads in part (1996, 48): "Surely the most natural state of affairs is simply nothing." Let us develop the implications to which Swinburne commits himself by his thesis that the null world is "Surely the most natural state of affairs." The articulation of the corollaries of his version of SoN will then enable us to make an *empirical* judgment of its validity.

Assume, for argument's sake, that the null world were actualized. And consider the set W of logically possible contingent worlds (or objects) that *fail to be instantiated* or realized in the null world. Then Swinburne's version of SoN makes the following claim: it is *most natural* that *none* of the possible member-worlds or objects in W are actualized. But if this collective failure of actualization is "most natural," what is the ontological status, in regard to "naturalness," of the individual member-worlds of W, taken singly or distributively? As we are about to see, Swinburne's SoN entails that, for each individual member of W, taken singly, it is *natural*—though *not* "most natural"— that it *not* be actualized rather than that it exist de facto.

To see that this conclusion follows from his version SoN, note that if it *were* natural for *some one* member world of W to *exist actually rather than not*, then the state of affairs in which *no* member is actualized—i.e., the null world—could *not* be the *most* natural of all. Notice, incidentally, that the stated inference from SoN does *not* commit the fallacy of division, since it reasons from the superlative attribute "most natural" of the collection W to the merely positive attribute "natural" of its individual members.

Thus, it is a corollary of Swinburne's version of SoN that it is natural for our world or universe U_o of inanimate matter, biological organisms, and Homo sapiens not to exist, rather than to exist. As before, let us denote this particular corollary by "SoN(U_o)." A like corollary pertains to each individual member-world of a putative mega-universe that features universes in addition to ours in a world ensemble. But the actual existence of our U_o is a deviation from its allegedly *de jure* "natural" nonexistence, as avowed by SoN (U_o).

Hence its existence calls for explanation in terms of a suitable external cause or reason. Indeed, precisely because Swinburne claims that there *should* be nothing instead of our universe U_0, he is driven to ask *why* our cosmic abode U_0 does exist, even though, *naturalistically*, it supposedly should not.

By the same token, he, like other theists before him, appeals to SoN(U_0) to demand an explanation of how our world is *kept in existence*, instead of reverting to its supposedly natural state of not existing (1996, 49). And, again in concert with other theists, he therefore holds the view that "God keeps the universe in being, whether he has been doing so forever or only for a finite time" (quoted in Quinn 1993, 593). This is the Christian doctrine of perpetual divine creation, which is labelled *creatio continuans*.

Thomas Aquinas is one of the early major theists who simply assumes SoN tout court in his metaphysics of essence and existence (1948; *Summa Theologiae* Ia 50, art. 2, ad 3). Logically contingent existing entities, he holds, are "composed" of both essence and existence. The ontologically spontaneous state of affairs, in his view, is à la SoN that *no* logically contingent objects exist, but that their essences constitute *potentialities* for actualization. Yet since logically contingent objects do exist, Aquinas's assumption of SoN prompts him to conclude that there must be a *ratio essendi*, a creative "act of being" which actualizes their essences. And that act, we have learned, is performed by God. Moreover, the divine bestowal of being has to take place at every instant of their existence. Thus, he concluded: "the being of every creature [i.e., every logically contingent thing] depends on God, so that not for a moment could it subsist, but would fall into nothingness were it not kept in being by the operation of Divine power" (quoted in Quinn 1993, 593).

Aquinas's peremptory assumption of SoN as a presupposition of his argument above for perpetual divine creation is further evident from his view that God could *annihilate* objects by *merely ceasing to conserve them*, without any destructive act, since they would then *spontaneously* lapse into nonexistence (see Quinn 1993, 593–94).

But, now suppose, for the sake of argument, that one conceptualizes any logically contingent object with Aquinas as being "composed" of its essence and existence, which is quite dubious anyway. Even then, it is baseless to assume that the essence of the object must be *ontologically prior*, as it were, to its existence *such that SoN is true*. Evidently, Aquinas *does not justify SoN* but begs the question by taking it for granted in the service of his creationist theological agenda.

William Craig (2001, 383) concedes that "Grünbaum is correct in seeing the spontaneity of nothingness [. . .] to lie at the heart of the Thomist cosmological argument [for the existence of God]." But then Craig objects (ibid., n. 6): "it is up to Grünbaum to explain why he rejects the Thomistic metaphysics that underlies the argument, which he has not even begun to do." Although Craig himself recognized that Aquinas's cosmological argument is predicated on SoN, he did not see that this argument is a petitio principii, since Aquinas begs the question with respect to SoN. Thus, contrary to Craig, I am entitled to reject Aquinas's cosmological argument as ill founded after having impugned his baseless contention that essence is ontologically prior to existence *such that SoN is true*.

Alas, the sixteenth-century Jewish Kabbalist Meir ben Gabbai encumbers his affirmations of SoN with primitive word magic. Yet without even a hint of intellectual disapproval, Nozick (1981, 122, note*, second paragraph) reports that according to ben Gabbai, "only God's continuing production of the written and oral Torah maintains things in existence." Speaking of the alleged dire ontological consequences of any interruption at all of God's continuing production of the written and oral Torah, that Kabbalist wrote: "were it [i.e., this divine production of Hebrew words] to be interrupted for even a moment, all creatures would sink back into their non-being." Thus, ben Gabbai relies on divine word magic to replace the ontological role that Aquinas assigned to divine creation ex nihilo. But plainly, the Torah scribes are the ones who keep producing the written Torah. And it is ongoing human verbal communication among the faithful that preserves the Torah orally. Hence, if ben Gabbai is to be believed, human language users are ontologically necessary at all times to prevent the mighty cosmos from lapsing into nothingness in accord with SoN, a patent absurdity that no one should ever have taken seriously.

But, as we are about to appreciate, SoN is altogether ill founded *empirically*, so that *any* cosmological argument or doctrine that is predicated on it is likewise empirically unwarranted. And since *a priori* defenses of SoN are seen to have failed, it will then emerge as an induction from various episodes in the history of science that SoN stands or falls on *empirical grounds*.

1.7.4 The "natural" status of the world as an empirical question

Consider the corollary of SoN pertaining to our own world U_0, i.e., $SoN(U_0)$, in its own right. As we recall, the latter corollary asserts that it is *natural* for our world U_0 *not* to exist, rather than to exist. As against any *a priori* dictum on what is the "natural" status of our world, the verdict on that status will now be turning out to depend crucially on empirical evidence. Two cosmological examples will spell this empirical moral:

(a) The *natural evolution* of one of the big bang models of the universe countenanced by general relativistic cosmology is a clear cosmological case in point. This model, the so-called Friedmann universe, is named after the Russian mathematician Alexander Friedmann. In the context of the general theory of relativity (GTR), the big bang, dust-filled "Friedmann universe" has the following features (Wald 1984, 100–101):

(i) It is a spatially closed spherical universe (a "3-sphere"), which expands from a big bang to a maximum finite size, and then contracts into a crunch

(ii) It exists altogether for only a finite span of time, such that no instants of time existed prior to its finite duration or exist afterward (Grünbaum 1998, 25–26)

(iii) As a matter of natural law, its total rest mass is conserved for the entire time period of its existence, so that, during that time, *there is no need for a supernatural agency to prevent it from lapsing into nothingness* à la Aquinas, or as in René Descartes's scenario in his Third Meditation (1967, 168).

Evidently, the *"natural"* dynamical evolution of the Friedmann big bang universe *as a whole* is specified by the *empirically supported* general relativistic theory of cosmology. And if there is a world ensemble of big bang worlds, the "natural" evolution of the members of this mega-universe would likewise be based on such physical laws as are hypothesized to prevail in them. But the epistemic warrant for these presumed laws likewise cannot dispense with empirical evidence, if they are to become warrantedly known to us. Thus, the "natural" behavior of big bang worlds is not vouchsafed *a priori*.

(b) The same epistemic moral concerning the empirical status of cosmological naturalness is spelled by the illuminating case of the now largely defunct Bondi and Gold steady-state cosmology of 1948, if only because its account of the hypothesized steady state of the expanding universe as being natural owes its demise to contrary empirical evidence.

Their 1948 steady-state theory (Bondi 1960) features a spatially and temporally infinite universe in which the following steady-state cosmological principle holds: as a matter of natural law, there is large-scale conservation of matter *density*. Note that this conservation is *not of matter*, but of the *density* of matter over time. The conjunction of this constancy of the density with the Hubble expansion of the universe then entails a rather shocking consequence: throughout space-time, and without any matter-generating agency, matter (in the form of hydrogen) pops into existence completely naturally in violation of matter-energy conservation (Bondi 1960, 73–74, 140, 152). Hence the steady-state world features the accretion or formation of *new* matter as its *natural*, normal, spontaneous behavior. And although this accretive formation is ex nihilo, it is clearly *not* "creation" by an external agency. Apparently, if the steady-state world were actual, it would impugn the ontology of the medieval Latin epigram "Ex nihilo, nihil fit."

Its spontaneous matter accretion occurs at the rate required by the constancy of the matter density amid the mutual galactic recession, and it populates the spaces vacated by the mutual galactic recession. Thus, in the hypothesized Bondi and Gold world, the spontaneous accretion of matter would be explained deductively as *entirely natural* by the conjunction of two of its fundamental physical postulates. But the rate of this spontaneous cosmic debut of new matter is small enough to leave the received matter-energy conservation law essentially intact locally (terrestrially).

Precisely because the new matter is held to originate *spontaneously* in the steady-state world, it is salutary to use the *agency-free* term "matter *accretion*" to describe this hypothesized process. And we must shun the use of the misleading *agency-loaded* term "matter *creation*," because the noun "creation" denotes an act of causing something to exist by an agency *external* to its object. Thus the notion of creation calls for a creator.

It was therefore quite misleading that the cosmologist Hermann Bondi, who was a dedicated secular humanist, equated the problem of the origin of the universe with the alleged "problem of creation" and declared that the steady-state theory solves "the problem of creation" *scientifically*, whereas "other theories," such as the big bang theory, do not, but hand it over to metaphysics (Bondi 1960, 140). Yet since SoN is turning out to be groundless, the purported problem of creation emerges to be a nonissue.

The steady-state theory owes its demise to the failure of its predictions and retrodictions to pass observational muster in its competition with the big bang cosmology. This episode again teaches us that *empirically-based scientific theories are our sole epistemic avenue to the "natural" behavior of the universe at large*, though only fallibly so, of course, since such theories are liable to be replaced by others in the light of further empirical findings.

In earlier writings (Grünbaum 1998, sec. 3, 22–23; 2000, 5–7), I have given other examples from the history of physics (Aristotelian and Newtonian mechanics) and from the history of biology (spontaneous generation of life from nonliving substances), showing how evolving empirically-based theories in these domains provided *changing* conceptions of the "natural," spontaneous behavior of *subsystems* of the universe. By the same token, these episodes illustrate *how misguided it is to persist in asking for an external cause of the deviations of such systems from the pattern that an empirically discredited theory erroneously affirms to be "natural."*

Thus, as I emphasized furthermore in previous writings (Grünbaum 1973, 406–407; 1998, secs. 2, 3, 4), the history of empirical science has legitimated the *theory-relative rejection* of certain why-questions. Bas van Fraassen (1980, 111–12) referred to my legitimation of such rejections and has aptly encapsulated their upshot: "the important fact for the theory of explanation is that not everything in a theory's domain is a legitimate topic for why-questions; and that *what is* [legitimate], *is not determinable a priori*" (italics added).

Thus it is fitting that we should ask: What is the empirical verdict on $SoN(U_0)$, a corollary of Swinburne's SoN, which asserts that "It is natural for our universe not to exist, rather than to exist"? Its proponents have offered no empirical evidence for it from cosmology, let alone for SoN itself, believing mistakenly, as we saw, that it can be vouchsafed *a priori* à la Leibniz.

But what if they were to counter the injunction to supply empirical evidence, objecting that it demands the impossible. Impossible, it might be said, because, in principle, there just can be no evidence from within our actual world that might show that it is the "natural" state of our cosmos *not* to exist in the first place. To this, my retort is twofold: it is not obvious that this epistemic predicament is genuine; but even if it were, it would only redound to the baselessness of $SoN(U_0)$, rather than tell against the legitimacy of demanding evidence for any and every averral of cosmic naturalness, including $SoN(U_0)$.

Philip Quinn (2003) agrees that, as he put it, "current scientific theories and the empirical evidence on which they rest provide little or no support for SoN." But he then chides me, declaring that, "the *de facto* history of science falls far short of establishing the strong modal conclusion that this issue [of the credentials of SoN] *cannot* be settled *a priori* because only empirical evidence could have a bearing on it." And he opines that it is "*scientistic*"—in von Hayek's (1952) pejorative sense of being philosophically imperialistic—to maintain, as I do, that "only empirical evidence of the sort that supports scientific theories could have a bearing on the acceptability of SoN."

Quinn's complaint of scientism prompts several responses from me:

(a) There is an important *asymmetry*, I submit, between an empirical and an aprioristic adjudication of the truth of SoN, because of *their very different track records* in making determinations of naturalness. *A priori* defences of SoN seem to have failed, as we have already seen. But the *scientific* record of determining *what transpires naturally* is *brilliant*, not only in physics but also in biology. Witness the history of the theories as to the *spontaneous generation of life* from inorganic materials to which we alluded before, starting with Pasteur but including Oparin and Urey in the twentieth century. Furthermore, the rich history of the *disintegration of Kantian apriorism* in regard to the external world spells a strong caveat against the expectation of an *a priori vindication* of SoN.

(b) As Quinn noted, I appeal inductively to the substantial evidence from the history of science to infer that settling the merits of SoN *a priori* is pie in the sky. But I allow, of course, that this induction is *fallible*. Hence I would surely be open to correction, if someone were unexpectedly to come up with a cogent *a priori* argument for SoN. But I trust I can be forgiven if I do not expect that to happen at all. And I do not see that I am being rash in my epistemological attitude.

Indeed, I deem it very important that Quinn makes both of the following concessions, the first of which bears repetition from before:

(i) "It seems to me that current scientific theories and the empirical evidence on which they rest provide little or no support for SoN" (2003).

(ii) "If only empirical evidence can settle the issue of whether SoN is true [as claimed by Grünbaum] and it [SoN] is not well supported by empirical evidence, then SoN—[qua presupposition of PEQ]—is, indeed, ill founded and the *pseudoproblem charge* [against PEQ] *has been established* [as claimed by Grünbaum]." [ibid., italics added]

(c) I attach great importance to pointing out why one should generally be left rather unmoved by the charge of scientism in metaphysics and epistemology, not only in this context, but also in others. It is easy enough to raise the red flag of scientism, whenever scientific reasoning impugns hypotheses or presumed ways of knowing that someone wishes to immunize against scientific doubts. But, as I see it, he who would level the charge of scientism *responsibly* incurs a major obligation: to come up with a *positive vindication* of the principles or methods that this critic wishes to *contrapose* to the scientific ones. In the present case, I would ask for *positive* reasons to expect that *a priori* methods can settle the merits of SoN. But neither Quinn nor, to my knowledge, anyone else has given any such reason.

As an example of an irresponsible use of the charge of scientism, I mention the hypothetical case in which scientific findings are adduced against the demonic possession theory of insanity, only to be rejected as "scientistic." The charge would be that the scientific ontology is too impoverished to have room for Satan or other demons, who spring the confines of its allegedly narrow horizons.

1.7.5 Robert Nozick

Robert Nozick has offered some distinctive views on SoN and PEQ. He (1981, 126) notes correctly that SoN is presupposed by the question "Why is there anything at all, rather than nothing?" a presupposition he deems to be "a very strong assumption" in the following sense: "to ask this question [i.e., PEQ] is to presume a great deal, namely, that nothingness is a natural state requiring no explanation, while all deviations from nothingness are in need of explanation." But the explanation of *deviations from the* natural state does not preclude that the natural state *also* may have an explanation of its own. Indeed, as shown by our scientific cosmological illustrations, Nozick was not entitled to suppose that the hypothesized "natural" state would *itself* simply require "no explanation," if it were to obtain.

Strangely, Nozick goes on to shroud the natural state of the world in blanket agnosticism, declaring tout court (ibid., 126): "The first thing to admit is that we do not know what the natural state is." But surely the *fallibility* of the evolving verdicts of our empirically supported scientific theories as to the natural behavior of the universe is not tantamount to our wholesale ignorance of the natural state of affairs.

Indeed, if Nozick were right that the natural behavior of the world is unknown to us tout court, then none of those who endorse PEQ as an authentic question, himself included, could even entertain the claim of its presupposition SoN that the *null world* is the most natural cosmic state. And that loss would then abort PEQ even before it is posed.

Yet despite his agnostic disclaimer concerning SoN, Nozick is undaunted in tackling PEQ *a priori* after first noting that this question is deeply, if not uniquely, problematic for the following reason (ibid., 115):

The question [i.e., PEQ] appears impossible to answer [footnote omitted]. Any factor introduced to explain why there is something will itself be part of the something to be explained, so it (or anything utilizing it) could not explain all of the something—it could not explain why there is *anything* at all. Explanation proceeds by explaining some things in terms of others, but this question seems to preclude introducing anything else, any explanatory factors.

In effect, Leibniz (1697; 1973a, 136–37; 1714; 1973b, sec. 8, 199) had anticipated Nozick's objection here by arguing that if PEQ is to have an answer, the sufficient reason for the existence of something could not be provided by a series of *contingent* somethings, because they would form an infinite explanatory regress; instead, he contended, the required sufficient reason (cause) terminates the regress by existing *necessarily*.

Though Nozick has posed a forbidding difficulty for PEQ, he insists that PEQ is "not to be rejected" (1981, 116) and writes:

> This chapter [i.e., his chap. 2, on PEQ] considers several possible answers to the question [PEQ]. My aim is not to assert one of these answers as correct (if I had great confidence in any one, I wouldn't feel the special need to devise and present several); the aim, rather, is to loosen our feeling of being trapped by a question with no possible answer—one impossible to answer yet inescapable. [. . .] The question cuts so deep, however, that any approach that stands a chance of yielding an answer will look extremely weird. Someone who proposes a non-strange answer shows he didn't understand this question. Since the question is not to be rejected, though, we must be prepared to accept strangeness or apparent craziness in a theory that answers it.
>
> Still, I do not endorse here any one of the discussed possible answers as correct. It is too early for that. Yet it is late enough in the question's history to stop merely asking it insistently, and to begin proposing possible answers. Thereby, we at least show how it is possible to explain why there is something rather than nothing, how it is possible for the question to have an answer.

Alas, the hospitality then displayed by Nozick to avowed "extreme weirdness" and "apparent craziness" does not stop short of countenancing explanations vitiated by gross logical improprieties or crude abuses of language. And he is plainly not offering them tongue-in-cheek. Let us cite one of his proposed answers to PEQ, although it will turn out to be a mere farce.

Recall that the null world, which is assumed to obtain *de jure* by PEQ, also excludes the existence of time. Yet Nozick (ibid., 123) will now offer us a *temporal* scenario *a priori* from which, he claims, one could conclude that "there is something rather than nothing because the nothingness there once was nothinged itself, thereby producing something [thereafter]." Nozick depicts the grotesque scenario as follows:

> Is it possible to imagine nothingness being a natural state which itself contains the force whereby something is produced? One might hold that nothingness as a natural state is derivative from a very powerful force toward nothingness, one any other forces have to overcome. Imagine this force as a vacuum force, sucking things into nonexistence or keeping them there. If this force acts upon itself, it sucks nothingness into nothingness, producing something or, perhaps, everything, every possibility. If we introduced the verb "to nothing" to denote what this nothingness force does to things as it makes or keeps them nonexistent, then (we would say) the nothingness nothings itself. (See how Heideggerian the seas of language run here!) Nothingness, hoisted by its own powerful petard, produces something.

When Nozick speaks of "the nothingness there once was," he means, I take it, that at one time, the null world obtained. He envisions further that the null world itself contains "a very powerful force toward nothingness." Even this much already seems incoherent, since the null world is presumably *devoid* of all forces, physical fields, or forms of energy. If there were such a force, we learn, it would annihilate (destroy) any preexisting things permanently, without residue.

Nozick describes this putative action of the force metaphorically and misleadingly by speaking of the force "sucking things into nonexistence and keeping them there." But there are no things to be destroyed in the null world. Hence, if *per impossibile*, a thing-consuming "vacuum force" were operative in the null world after all, on what does Nozick think it can act? Its function, he tells us would be to act "upon itself," presumably to suspend or annihilate itself as an agency of potential destruction. But clearly, his putative "force toward nothingness" is not *itself* identical with nothing or "nothingness."

Hence even if the annihilating force can intelligibly act on itself, it would *not* be annihilating "nothing" or the null world as the object of its self-destruction. Besides, "nothing" (the null world) is not a thing-like substance such as a fluid or a gas that could be "sucked out" (evacuated), let alone *itself* be annihilated, leaving "something" in its wake!

Thus, Nozick is misformulating his own scenario by beguilingly saying that, when acting on itself, the force "sucks nothingness [*sic!*] into nothingness, producing something or, perhaps, everything, every possibility." Needless to say, Nozick's metaphysical Potemkin village is impervious to appraisal by empirical evidence. Alas, in his attempt to propose this answer to PEQ, Nozick's imagination seems to have gone berserk, leaving only bewildered indignation in its wake.

Nozick also takes logical liberties in another tack he explores toward dealing with PEQ. But now he is considering *undermining* the question, instead of proposing an answer to it (ibid., 130): "Why is there something rather than nothing? There isn't. There's both." Here he invokes a so-called fecundity assumption, which asserts (p. 128) that "all possibilities are realized" in the following sense: "All the possibilities exist [are realized] in independent noninteracting realms, in 'parallel universes'" (p. 129). But, as shown by our discussion of the null possibility, the obtaining of the null world, which Nozick declares to be *compossible* with the existence of a superabundance of different actualized universes, does logically *exclude* the realization (actualization) of any and all logically possible contingent worlds *other than itself*. Thus, the null world cannot be one of the "noninteracting realms" alongside parallel universes that constitute other actualized possibilities. Therefore, Nozick's fecundity principle cannot serve to undermine PEQ, although that question has turned out to be a nonstarter, because it presupposes the truth of the baseless SoN.

Yet, the Nobel laureate physicist Steven Weinberg (1993, 238) entertains Nozick's (1981, 128) "fecundity principle" even to the extent of declaring: "If this principle is true, then our own quantum-mechanical world exists but so does the Newtonian world of particles orbiting endlessly and so do worlds that contain nothing at all." So,

Weinberg countenances a *plurality* of "noninteracting" *null* worlds! That seems unintelligible.

So much for Nozick's *a priori* treatment of PEQ.

1.8 Hypothesized psychological sources of PEQ

It would be appropriate to consider possible emotional inspirations of PEQ, if we are to understand the tenacity with which it has been asked.

As Charles Larmore has emphasized, the unflinchingly pessimistic Arthur Schopenhauer held, contrary to Kant, that it is not reason as such that drives us to pose questions such as PEQ (Schopenhauer 1966, vol. 1, appendix). In a chapter on "Man's Need for Metaphysics," Schopenhauer wrote (1958, vol. 2, ch. 17, p. 161; italics added):

> undoubtedly it is the knowledge of death, and therewith the consideration of the suffering and misery of life, that give the strongest impulse to philosophical reflection and metaphysical explanations of the world. *If our life were without end and free from pain, it would possibly not occur to anyone to ask why the world exists, and why it does so in precisely this way, but everything would be taken purely as a matter of course.*

Elaborating further on "Man's Need for Metaphysics," Schopenhauer declared (ibid., 171):

> In fact, the balance wheel, which maintains in motion the watch of metaphysics that never runs down, is the clear knowledge that this world's nonexistence is just as possible as its existence [. . .]. What is more, in fact, we very soon look upon the world as something whose non-existence is not only conceivable, but even preferable to its existence [. . .]. Accordingly, philosophical astonishment is at bottom one that is dismayed and distressed.

But to the detriment of Schopenhauer's diagnosis of the emotional inspiration of PEQ, he leaves fundamentally *unexplained* why that

question has apparently been posed *only*—or at least principally—by
the heirs of the distinctly Christian doctrine SoN. After all, the thinkers
in other cultures who did *not* raise it were just as conscious of death
and the miseries of life as the legatees of traditional Christian doctrine.

Yet it would be interesting to investigate empirically the motives
of philosophers who embrace PEQ as an *authentic* question, so as to
learn to what extent, if any, such philosophers are driven by emotions
of the sort conjectured by Schopenhauer. It is perhaps not implausible
that our deeply instilled fear of death has prompted some of us to
wonder why we exist so precariously. And some of us may then have
extrapolated this precariousness, more or less unconsciously, to the
existence of the universe as a whole. But whatever the emotional inspiration of PEQ, no such motivation can vindicate it as a *philosophically*
viable question, since its presupposed SoN is baseless.

Disappointingly, after declaring PEQ to be the most fundamental
question of metaphysics, Heidegger (1953, 1) *psychologized* it away
as inspired by existential anxiety, thereby essentially echoing
Schopenhauer's ideas on the psychology of PEQ.

But PEQ dies hard. In 1999, it was the focus of a massive book of
over 750 pages by the Swiss philosopher Ludger Lütkehaus. Published
in German, its title in English becomes *Nothing: Farewell to Being,
End of Anxiety*. Let Lütkehaus speak for himself in stating the aim of
his opus (1999, 29) in his German original.

> *Es [dieses Buch] versucht, die Präokkupationen eines Denkens zu
> revidieren, das seinsfixiert, "ontozentrisch" in seinen Werthierar-
> chien, "ontomorph" in seinen Begriffen und Vorstellungen und
> bedingungslos "ontophil" in seinen Antrieben ist.*

> *Den symptomatischsten Ausdruck hat dieses Denken in seinem para-
> noiden, 'nihilophoben' Verhältnis zum Nichtsein, zum "Nichts"
> gefunden. Nicht die "Seinsvergessenheit," wie es die Todtnauberger
> Schule beklagt—die Nichtsvergessenheit bezeichnet das wahre
> "Schwarze Loch" seiner ontologischen Amnesie. Nichtsvergessen-
> heit, Nichtsangst und Seinsgier bilden sein "ontopsychologisches,"*

"ontopathologisches" Syndrom. Und gerade damit arbeitet dieses Denken der Vernichtung und Selbstvernichtung zu. Das ist die— vielleicht tragische—Ironie der so seinsfixierten westlichen Seins- geschichte.[2]

After this avalanche of words, I have no idea just what Lütkehaus would have each of us do to overcome our alleged "ontopathological syndrome." Should we forsake all *joie de vivre*?

1.9 PEQ as a failed springboard for creationist theism: the collapse of Leibniz's and Swinburne's theistic cosmolog- ical arguments

We are now ready to appraise the theistic creationist answers given to PEQ by Leibniz and Swinburne respectively as part of their *cosmolog- ical argument* for the existence of God.

Swinburne has argued cogently against Leibniz that, if there is a God, his existence is logically contingent no less than that of the uni- verse. As he reasoned carefully (1991, chap. 7, p. 128):

> it seems coherent to suppose that there exist a complex physical uni-
> verse but no God, from which it follows that it is coherent to suppose
> that there exist no God, from which in turn it follows that God is not
> a logically necessary being. If there is a logically necessary being, it
> is not God [footnote omitted].

And having deemed the existence of God to be logically contingent, the sweep of Swinburne's version of SoN *excludes* God along with our con-

---

2. In my English translation, it reads: "This book is an attempt to revise the preoccupation of a mode of thinking that is fixated on being, 'ontocentric' in its hierarchy of values, 'ontomorphic' in its concepts and ideas, and unconditionally 'ontophilic' in its motivations. This mode of thought has found its most symptomatic expression in its paranoid, 'nihilophobic' relation to non-being, to noth- ingness. Not the forgetting of being, as deplored by the Todtnauberg School, but rather the forgetting of nothingness is the genuine 'black hole' of its ontological amnesia. The forgetting of nothingness, fear of nothingness, and ontological greed constitute its 'ontopsychological,' 'ontopathological' syn- drome. And precisely thereby this way of thinking conduces to annihilation and self-annihilation. Thus fixated on being, this is the perhaps tragic irony of the Western history of being."

tingent universe from the null world. Thus, as we recall, Swinburne formulated SoN as follows: "Surely the most natural state of affairs is simply nothing: no universe, no God, nothing" (1996, 48). But, on the basis of his SoN, he had demanded, in response to PEQ, a suitably potent external divine cause to explain the existence of the universe *qua deviation from nothingness*. And he issued this explanatory demand for a creator ex nihilo as a challenge to atheists (ibid., 48–49; 2).

But clearly, what is sauce for the goose is sauce for the gander: if God does exist contingently, as Swinburne claims, then the contingent existence of the deity also constitutes a *deviation* from the allegedly most natural state of nothingness! Thus, on Swinburne's version of SoN, the existence of God requires causal explanation *in answer to PEQ* no less than the existence of the universe does. Hence Swinburne is not entitled to take the existence of God for granted, as he does, to explain the existence of the universe *in answer to PEQ*. To point out against Swinburne that, on his version of SoN, God and the universe *alike* require causal explanation is *not* to saddle him with an infinite regress of explanations.

How, then, does he deal with the following inescapable challenge from his version of SoN? If he is going to give an answer to PEQ, as he does, he needs to explain why God exists, *rather than just nothing contingent*, fully as much as he needs to explain why our universe exists, rather than just nothing contingent. Yet Swinburne is *oblivious* to this major challenge as emanating from his SoN!

Thus, unencumbered by this explanatory debacle, Swinburne opines onesidedly (1996, 2): "the view that there is a God [. . .] explains the fact that there is a universe at all." And, again in accord with his SoN, he claims furthermore (ibid., 49) that God also keeps "the many bits of the universe" in existence.

But in regard to the imperative to explain why God exists rather than just nothing, Swinburne (ibid.) is driven to concede that "inevitably we cannot explain" it. Thus he claims (1991, 127) that "The choice is between the universe as [explanatory] stopping point and God as stopping point." Yet, Swinburne defaulted on his explanatory debt when he conceded that the existence of God "inevitably" defies explanation. He had assumed that debt by embracing his version of SoN, which excludes God from the null world and turns the existence of the deity into a deviation from the "most natural" state of nothingness. Hence, contrary to Swinburne, on his own premises, God does not qualify as an explanatory "stopping point" after all.

Thus Swinburne has indeed deservedly incurred a jibe akin to the one Schopenhauer famously issued against those who demand a creative cause of the existence of the universe, but then suspend a like demand to explain the existence of God: Swinburne has treated SoN like a hired cab that he dismissed, just when it reached his intended theological destination.

But let there be no misunderstanding of my use of Schopenhauer's simile of the hired cab against Swinburne. I am emphatically *not* maintaining generally that a theological hypothesis T can be explanatory only if T itself can, in turn, be explained; instead, I am contending that Swinburne hoists himself with his own petard: in answer to PEQ, his recourse to SoN to call for a theistic explanation of the very existence of the universe *boomerangs*, because his SoN likewise requires a causal explanation of the existence of God, which, he tells us explicitly, is not to be had.

Leibniz (1714; 1973b, secs. 7 and 8, 199) and Swinburne (1991, 121– 30) have offered the prima facie most persuasive of the traditional "first cause" cosmological arguments for the existence of God as creator of the universe ex nihilo. Though they differ as to whether God exists necessarily (Leibniz) or contingently (Swinburne), the common core of their cosmological arguments can be encapsulated as follows: (i) The null world, which is devoid of all contingent existents, is the simplest (ontologically), (ii) SoN is true: *De Jure*, the null world should obtain qua being the simplest, and indeed it would obtain as the most "natural" or normal state of affairs in the absence of an external cause (or "reason), (iii) But the de facto existence of our universe of contingent objects is a massive *deviation* from the null world mandated by SoN, (iv) This colossal existential deviation from ontological "normalcy" cries out for explanation by a suitably potent cosmic cause, making an answer to PEQ imperative. The required cause is a creator ex nihilo. Hence the God of theism exists.

Thus it is clear that the theistic creationist answers given to PEQ by both Leibniz and Swinburne are each predicated on a version of SoN in the face of contingently existing things. Their versions differ somewhat: Leibniz tells us that, in the absence of an overriding reason (cause), the nihilistic state of affairs is ontologically imperative, because it is "simpler and easier" than the state of something contingent, whereas Swinburne claims that the null world is "the most natural state of affairs." But I have been at pains to argue that both of these versions of SoN are baseless, each for reasons of its own. Yet they both avow the central claim of SoN that *de jure* there *should* be nothing contingent and that there would indeed be nothing contingent in the absence of an overriding external divine creative cause.

But the ill-founded SoN is clearly a presupposition of PEQ. Therefore, the cosmic ontological question PEQ is a nonstarter by posing a pseudo-issue. Yet the purported imperative to answer precisely this global question is the basis for Leibniz's (secs. 7 and 8) and Swinburne's (1991, chap. 7) cosmological arguments for the existence of God. Thus, **PEQ cannot serve as a springboard for creationist theism.** *Hence Leibniz's and Swinburne's cosmological arguments are fundamentally unsuccessful.*

2. DO THE MOST FUNDAMENTAL LAWS OF NATURE REQUIRE A THEISTIC EXPLANATION?

We now consider theistic arguments that have been offered, not about the existence of contingent objects, but about the explanation of the natural laws that are exhibited by their behavior, laws which are sometimes called the "nomological structure" of the world, the "nomic structure" or, briefly, its "nomology." Theists have claimed to explain the nomology as having been willed or intended by God in the mode of agent causation to be exactly what it is. We shall speak of this supposed theistic explanation as "the theological volitional explanation of the nomology." And the principal issue in this section **2** will be whether creationist theism succeeds in explaining the specific content of the nomological structure, as claimed by its advocates.

2.1 The ontological inseparability of the laws of nature from the furniture of the universe

Unless there is an infinite regress of explanations, every explanatory theory will feature some set of *unexplained explainers*. Yet any theory can be axiomatized in *alternative* ways. For example, in Euclid's synthetic plane geometry, the famous parallel postulate (number 5) can be interchanged with the theorem that the sum of the interior angles of a rectilinear triangle is 180 degrees as follows: this angle sum theorem now becomes the 5th postulate, while the previous parallel postulate now becomes a theorem. Similarly, Newtonian dynamics has been alternatively axiomatized by means of the so-called calculus of variations. Thus, the fundamentality of a postulate is *not* absolute, but depends on the axiomatization. Hereafter, when we speak of "the most fundamental laws" of the nomology, this characterization is to be understood to within just such axiomatic relativity.

In a scientific theory pertaining to the laws of nature and featuring unexplained explainers, the most fundamental of these laws (in the given axiomatization) will hold as a matter of brute fact. But, as we just saw in section **1.9**, in a theistic system, the existence of God is its avowed unexplained explainer and thus is *its* brute fact. Hence, as we noted, Swinburne declared (1991, 127): "The choice is between the universe as [explanatory] stopping point and God as stopping point."

The nomology consists of the lawlike regularities exhibited by the physical, biological, and biopsychological constituents of the universe. It is of cardinal importance here that these nomic patterns *inhere* in the behavior of the world's furniture and do not exist independently alongside it. Theists and atheists can agree that the laws do not hover over the universe, as it were, in some separate realm. Swinburne (1991, 43) rightly rejected this hypostatization: "Talk about laws of nature is really only talk about the power and liabilities of bodies."

In short, the laws are inextricably intertwined with the material *content* of the universe. Hence we can speak of their intrinsic entanglement as "the *ontological inseparability*" of the nomology from the world's furniture.

But just this inextricability has a very important corollary pertaining to a posteriori teleological arguments for the existence of a designer God, a corollary that seems to have been overlooked heretofore. If the theistic God is to endow the laws of nature with teleological features—such as permitting the formation of intelligent life—then he must do so precisely BY MEANS OF creating the material content of the world ex nihilo as his handle on the laws. Thus, the designer role which the theist attributes to the deity cannot be fulfilled by God without his being the creator ex nihilo. *Yet the a posteriori argument for a designer God cannot **also** shoulder the probative burden of warranting divine creation ex nihilo, an onus that was borne by the received cosmological argument.* Indeed, if the teleological argument had the probative resources to argue not only for a *designer God* but also for a *creator God* ex nihilo—the latter being the conclusion of the cosmological argument—then the teleological argument would make the cosmological one SUPERFLUOUS! But it does not. Instead, the two arguments complement one another as follows: The proponent of the design argument for the existence of God is engaged in *adding* to the cosmological conclusion, which is that God the creator exists, the *further* a posteriori conclusion that a *designer* God exists, who is also a creator God. Hence, an argument for God *the* **designer** that uses as a premise the existence of God **the creator** does not "beg the question" as to the existence of a designer God.

Accordingly, the assertion of divine creation ex nihilo, which is the *conclusion* of the failed cosmological argument for the existence of God, is now seen to be a tacit *premise* of the traditional teleological argument that seemingly *goal-directed* features of the world call *a posteriori* for a cosmic *designer*. In other words, if God is to implement his inferred role as cosmic designer of the nomology, he must be assumed to be the creator ex nihilo of the substantive fabric and texture of the universe.

But absent a successful cosmological argument for the occurrence of divine creation ex nihilo, the teleologist must bear the enormous additional probative burden of somehow warranting the very framework of creation ex nihilo in which the teleological arguments are inevitably anchored. In short, the teleologist is in dire need of some kind or other viable, cogent *substitute* for the received cosmological argument, an argument whose most persuasive versions (by Leibniz and Swinburne) turned out to be genuinely flawed in section **1.9**. Yet, no such substitute is extant.

2.2 The probative burden of the theological explanation of the world's nomology

The theological volitional explanation of the nomology, which we shall develop in the next section, has to shoulder a multiple heavy probative burden as follows:

(a) Since the theist purports to explain the laws of nature as the product of divine intention, the ontological inherence of the laws of nature in the cosmic furniture commits him/her to the claim that God brought the nomic structure into existence *by means of creating ex nihilo the world's furniture* from which that structure is inseparable; thus just like the theistic argument for a *designer* God, the theological volitional explanation of the nomic structure is in dire need of a successful substitute for the failed cosmological argument for divine creation ex nihilo.

(b) According to theism, God is the creator ex nihilo of all logically contingent existing entities, whenever they exist, though of course he does not create himself. If that claim were true, then God would *automatically also* be the simultaneous creator ex nihilo of such laws of nature L as govern the content of the universe, precisely because the nomic structure L is intrinsic to the furniture of the universe.

214 THE POVERTY OF THEISTIC COSMOLOGY

(c) Yet we must heed a caveat: it would *not* follow from the reliance of the theistic volitional explanation of the nomology on creation ex nihilo that if the universe is *not* the product of divine creation ex nihilo, then no sort of supernatural agency—such as the phantasmic demiurge in Plato's *Timaeus*—might have been the craftsman of the laws L by *transforming prior chaos* into a cosmos. But, the traditional theist is unwilling to countenance a divine cosmic craftsman who merely transforms a preexisting chaotic world into a nomic universe, holding that an omnipotent God had no need for preexisting substances to create a universe. Hence the notion of a mere cosmic transformative craftsman is unavailable to the theist, and hence would not enable the theistic explanation of the nomology to dispense with the *equivalent* of a cosmological argument for creation ex nihilo.

But, as we saw in section **1.9**, the received cosmological argument for divine creation ex nihilo is erected in response to PEQ on the quicksand of SoN. And since this cosmological argument is thus *ill founded*, neither the divine volitional explanation of the nomic structure, nor the aforementioned teleological argument for the existence of God can build on it *cumulatively*; instead, they must then bear the enormous *additional* probative burden of somehow warranting the very *framework* of creation ex nihilo from which the theistic volitional explanation of the world's nomology and boundary conditions is inseparable. But no such warrant is in sight.

Philip Quinn (2003, 592) has tried to parry my claim that the commitment to a divine volitional explanation of the nomic structure confronts the theist with the probative burden of providing the equivalent of a successful cosmological argument for creation ex nihilo. Quinn denies that burden and chides me for the "error" of "underestimating the number of sources from which justification for the existence of the God of theism can be derived." And he explains that the contributions from these various sources "can combine to yield a cumulative case argument."

Yet the issue before us is specifically the warrant, if any, for the theist's purported explanation of the world's nomology as having been intended by God in the mode of agent causation to be exactly what it is. And how does Quinn think his envisioned cumulative argument can *dispense* with the *specific* demonstration that the theistic volitional explanation of the nomology must show the creation of the nomic structure to be part and parcel of the divine creation ex nihilo of the material content of the world? Any such specific demonstration, it seems, does bear the same heavy probative burden as the received cosmological argument, which failed.

But, serious though it is, the need for a successful substitute for the failed cosmological argument is *merely one of a whole array of defects of the theistic explanation of the nomology.* We shall turn to these other failings after first articulating the proposed volitional theistic explanation of the nomology as developed by Swinburne and Quinn.

2.3 The theistic explanation of the cosmic nomology

In a 1993 *Festschrift* for me, Philip Quinn set the stage for advocating a theological explanation of the nomology, which purportedly transforms scientific *brute* facts into specifically explained regularities. Quinn says (1993, 607):

> The conservation law for matter-energy is logically contingent. So if it is true, the question of why it holds rather than not doing so arises. If it is a fundamental law and only scientific explanation is allowed, the fact that matter-energy is conserved is an inexplicable brute fact. [. . .] If there is a[ny] deepest law, it will be logically contingent, and so the fact that it holds rather than not doing so will be a brute fact.

Quinn now proceeds (ibid.) to draw two inferences from the *scientific* brute fact status of the most fundamental laws of nature, assuming that there are such "ultimate" laws. He writes:

> There are, then, genuine explanatory problems too big, so to speak, for science to solve. If the theistic doctrine of creation and conservation is true, these problems have solutions in terms of *agent-causation. The reason why there is a certain amount of matter-energy and not some*

other amount or none at all is that God so wills it, and the explana-
tion of why matter-energy is conserved is that God [creatively] *con-*
serves it [as required by SoN]. (italics added)

In the same vein as Quinn, Swinburne (1996, 21–22) characterizes
explanation in terms of agent-causation as "intentional" or "personal."
And speaking of the laws of nature L, Swinburne endeavors to prepare
the ground for that sort of theistic explanation of facts of nature (1991,
125):

Why does the world contain just that amount of energy, no more, no
less? [The laws] L would explain why whatever energy there is
remains the same; but what L does not explain is why there is just
this amount of energy.

Evidently, Quinn and Swinburne presume to *quantify* the "amount of
matter-energy" univocally. But even in elementary Newtonian
mechanics, after integration of the equation of motion to derive its law
of conservation of dynamical energy, the numerical value of the total
energy is dependent on the arbitrarily (i.e., humanly) chosen *zeros* of
the component potential and kinetic energies. How then is divine volition to explain that "there is just this [numerical] amount of energy"?
Does God create to within the zeros?

But let us consider more generally the context of the volitional theological explanations offered by Swinburne and Quinn in answer to
their question of why the actual world's nomology is what it is. This
question singles out the presumed ultimate laws and facts of nature for
explanation. And our two theists demote science for its inability to
answer their question. As they make clear (Swinburne 1991, 125;
Quinn 1993, 607), they consider their theistic volitional explanation of
the ultimate nomology to be a major explanatory advance over scientific brute fact. Yet neither Swinburne nor Quinn *spelled out* their very
ambitious deductive theistic explanation of the nomic structure.

Therefore, I now offer a reconstruction of essentially the deductive
explanatory reasoning that, I have good reason to believe, they origi-

nally had in mind. Quinn (private communication) authenticated my reconstruction in regard to the view he held before 2003. After codifying Swinburne's and Quinn's original versions of the purported theistic explanation of the basic laws of nature, we shall address their more recent accounts.

As for the earlier versions of Swinburne's and Quinn's explanation of the nomology qua product of divine agency, let me significantly refine my earlier formulation of it (Grünbaum 2000, 20). To determine whether their explanation redeems the very ambitious claims they made for it, let us have in mind, for the sake of concreteness, Swinburne's own example, originally endorsed by Quinn, of explaining theologically the supposed specific amount of total energy in the universe, which they depict as a *scientific* brute fact. Or, just for argument's sake, suppose that the nonlinear partial differential equations codifying Einstein's theory of the gravitational-cum-metric field *were* ultimate laws of nature. How, then, did Swinburne and Quinn envision that explanatory recourse to divine agency would *transform* these and all other specific putative scientific *brute* facts into volitionally *explained* facts?

I have schematized their original presumed theistic explanations in a deductive argument, using the familiar term "explanandum" to denote *what is to be explained*, which is asserted in the conclusion of the deductive argument below. But in this schematic reconstruction, the purportedly explained actual specifics of the most basic laws and facts are patently *not* stated in either the conclusion or in the premises, if only because they are not known; instead each of the premises and the conclusion speak of the unspecified laws of nature by means of *place holders*. Yet, whatever these specifics actually are, this explanatory schema is presumably the theistic solution—in Quinn's words— to "genuine explanatory problems too big, so to speak, for science to solve" (Quinn 1993, 607). With these understandings, the supposed volitional explanation becomes schematically:

218 THE POVERTY OF THEISTIC COSMOLOGY

Deductive Theistic Volitional Explanation of the Presumed Ultimate Laws and Facts of Nature

Premise 1. God freely chose (intended or *willed*) that the contents of our world exist and that they exhibit the laws that inhere in them.

Premise 2. Being omnipotent, God was, and is, perpetually able to cause directly (i.e., creatively bring about ex nihilo) the existence of the world's contents , so that they exhibit the laws which inhere in them.

Premise 3. If God chooses that *p*, and is able to cause it to be the case that *p*, then *p*.

Conclusion/Explanandum: The contents of our world exist and exhibit the laws that inhere in them.

This deductive argument invites some elaboration:

Premise 1 is to be understood more explicitly as entailing that God chose or intended or willed the realization of the possible world which is in fact actual so as to be nomologically precisely what it is, rather than the actualization of another possible world featuring alternative fundamental laws or facts, such as a different value of the presumed numerical total energy (Swinburne 1991, 125; Quinn 1993, 607).

Swinburne uses the lowercase letter *e* to denote the *explanandum*, which states the facts to be explained by the explanatory argument. And he articulated the substance of premises 2 and 3 above in the following two statements:

(a) "clearly whatever [the explanandum] *e* is, God, being omnipotent, has the power to bring about *e*. He will do so, **if he chooses to do so**" (1991, 109; bolding added). And the *e* that God chooses to bring about will be *compatible* with the assumed omnibenevolence of his aims (ibid.).

(b) "God, being omnipotent, cannot rely on [mediating] causal processes *outside his control* to bring about effects, so his range of easy control must *coincide* with his range of *direct control* and *include all states of affairs which it is **logically possible** for him to bring about*" (ibid., 295; italics and bolding added).

Claiming that the nomology inherent in the world's content is explained as the product of divine intention, the theist's explanation requires that God brought the nomology into being *by means of creating ex nihilo the cosmic furniture* in which it inheres: Evidently, in the absence of a *cogent*, viable *substitute* for the failed received cosmological argument for divine creation ex nihilo, the theistic volitional explanation of the nomology relies crucially on the *conclusion* of the cosmological argument as its underwriter. Thus, the theistic volitional scenario inherits the epistemic liabilities of that argument, set forth in Section **1**, much as does the a posteriori argument for a *designer* God, as we saw in section **2.1**.

Yet Swinburne claims to offer *a scientized* epistemology for creation ex nihilo (1996, 2):

> *The very same criteria which scientists use to reach their own theories* leads us to move beyond these theories to a creator God who sustains everything in existence. (ibid.; italics added)

Moreover, Swinburne asserts theistic *pan*explainability, declaring (ibid.):

> using those same [scientific] criteria, we find that the view that there is a God explains *everything* we observe, not just some narrow range of data [italics in original]. It explains the fact that there is a universe at all [*via* SoN], [and] that scientific laws operate within it. (cf. also his 1991, chap. 4 on "Complete Explanation")

But, as we shall see further on, in a paper of early 2003, which was cited in section **1.7.4** a propos of the epistemic status of SoN, Quinn parted company with Swinburne and modified his earlier version of the theistic explanation of the nomology. It will turn out that, in this latest version, Quinn distanced himself, in effect, from Swinburne's afore-cited 1996, purportedly *scientized* epistemology of theistic panexplainability.

Yet in Swinburne's reply (2000, 481–85) to my lengthy essay of the same year (Grünbaum 2000), he claimed to offer a clarification (p. 482) of his account of explaining the nomic structure theologically. Alas, on the contrary, this supposed clarification will be seen to muddy the waters. As will emerge under "Objection 4" below, it features a *conflation* of the Bayesian *confirmation* of the hypothesis that God exists, on the one hand, with the volitional theistic *explanation* of the specific content of the basic nomic structure, on the other. Hence, as in the case of Quinn, we shall discuss Swinburne's views in *two* stages, deferring scrutiny of his supposed clarification.

As against the thesis that theism solves "genuine explanatory problems too big [. . .] for science to solve," I now offer a series of further cardinal objections to the purported divine volitional explanation of the nomology. Even if that theistic explanation did not depend on demonstrated creation ex nihilo, these impending additional major objections thoroughly undermine it. In developing these animadversions, let us be mindful of Swinburne's afore-cited claims that theism is of a piece, epistemologically, with scientific theorizing, while transcending science by offering panexplainability and transforming scientific brute facts into specifically explained states of affairs.

2.4 Further major defects of the theological explanation of the fundamental laws of nature

OBJECTION 1. How does Swinburne reason *epistemically* that God *actually chose* to bring about the specific de facto *e* of the explanandum? He surely needs to validate this premise in order to attribute the prevailing *e* causally to divine creative volition. Obviously, that premise is not vouchsafed at all by Swinburne's *conditional* assurance that God will bring about *e*, "*if he chooses to do so*" (italics added). Equally patently, it would beg the question, if one were to answer our question here by claiming that God must have chosen to produce *e*, since *e* is actually the case! In sum, although the premise that God actually chose to produce *e* is explanatorily crucial for Swinburne and Quinn, no independent evidential support for it is in sight.

Indeed, just this epistemic gaping hole *alone fundamentally undermines* Swinburne's purported theistic volitional explanations of the *specific* content of the world's *ultimate* nomic structure, and of such presumed basic facts as the envisioned specific amount of total cosmic energy.

OBJECTION 2. To the detriment of Quinn and Swinburne, the volitional theological explanation of the nomology features *a built-in sort of ex post facto defect*, which prevents the evidence in the explanandum *e* from providing a check on the validity of the explanatory theistic premises! And this liability is not only anathema in the epistemology of scientific theories, *but is unacceptable in any sort of explanation based on evidence.*

Thus, let us consider a hypothetical situation in which the steady-state world of the 1948 Bondi and Gold theory were actual, a world which we had occasion to discuss in section **1.7.4**. If that world were actual, the theistic explanatory premises would be that omnipotent God willed the law of the constancy of mass *density* as well as the Hubble expansion of the galaxies. Alternatively, suppose that the actual world were one exhibiting mass-energy conservation as well as Swinburne's and Quinn's envisioned specific amount of total energy. In the latter event, our theists would explain this *different* state of affairs equally confidently by telling us that the Deity intended and chose to implement mass-energy conservation, rather than the *density* conservation of the steady-state world.

Thus, whichever of the two cosmologies actually materializes, *the evidence in the explanandum e provides no check on the validity of the explanatory premises!* And the crux of this *immunity* from evidential check is achieved formally by the following bizarre device: *Whatever the content of the explanandum in the conclusion, that same explanandum is **identically built** into the premises!*

In this way, the theistic explanation of the nomology is purchased effortlessly in advance on the cheap. But no building of the explanandum identically into the premises is found in the respected sciences, as illustrated by the following explanations in physics and biology:

(1) the Newtonian gravitational explanation of the orbit of the moon;

(2) the deductive-nomological explanations of optical phenomena furnished by Maxwell's equations, which govern the electromagnetic field, or, in a statistical context,

(3) the genetic explanations of hereditary phenotypic human family resemblances.

In short, the range of the explanatory latitude of the theistic volitional explanation is prohibitively permissive, in clear contravention of Swinburne's afore-cited declaration (1996, 2): *"The very same criteria which scientists use to reach their own theories* lead us to move beyond these theories to a creator God who sustains everything in existence" (ibid.; italics added). Moreover, the building of the explanandum identically into the premises is unacceptable *in any sort of explanation based on evidence.* Nor can it be made acceptable by abjuring "scientism"!

OBJECTION 3. As we recall, Swinburne (1991, 125) wrote:

> Why does the world contain just that amount of energy, no more, no less? L [the basic laws of nature] would explain why whatever energy there is remains the same; but what L does not explain is why there is just this amount of energy.

Let us denote by E_0 the putative specific amount of total energy in the universe, which Swinburne and Quinn (1993, 607) each consider well defined, and which they characterized as a *scientific brute fact.* Their point in so doing is to claim that explanatory recourse to divine agency would *transform* this specific unexplained fact, as well as the specific content of scientifically ultimate laws of nature, into volitionally *explained* items. Thus, we recall, Swinburne (1996, 2) wrote grandiosely: "the view that there is a God explains *everything* we observe, not just some narrow range of data."

To make good on his thesis of such explanatory *specificity*, Swinburne would need to be able to justify the following contention: given the hypothesis *h* that God exists in conjunction with assumed relevant background knowledge *k*, the specific pertinent *e* is a *deductive consequence* of the conjunction of *h* and *k*; i.e., the probability of the *explanandum e* on this conjunction is 1. Presumably, Swinburne's example of explaining the specific total amount of energy E_0 theistically is intended to make the general point of such deductive explainability far beyond E_0: it is to illustrate his global contention that the theistic hypothesis *h* "explains *everything* we observe" (italics in original).

But how does Swinburne see himself as vindicating this mind-bogglingly ambitious claim? Surely one is entitled to have expected him *to spell out the details* of the explanatory argument for at least one major case. Very disappointingly, that very reasonable expectation is dashed. Instead, immediately after having avowed theistic volitional *pan*explainability, he greatly weakens his explanatory thesis. Speaking of the universe, he now maintains just that the existence of God explains *generically* that *there are* laws of nature, and we learn that *h* explains much more modestly "that there is a universe at all [via SoN], [and] that scientific laws operate within it." And in his earlier opus, *The Existence of God*, in a summary of several of its chapters, he wrote, again generically (1991, 287): "What science cannot explain [but theism can] is why the laws of nature are of the character they are."

Yet it simply won't do to offer a theistic argument for the likelihood of a generic nomic structure, even if successful, as a substitute for redeeming the vaunted theistic panexplainability of the *specific content* of the fundamental scientific laws and facts, items which science avowedly leaves unexplained as brute facts. *Indeed, it is regrettably misleading philosophically to offer demonstrably hollow panexplainability as an improvement upon the scientific explanatory enterprise!*

OBJECTION 4. For the sake of the discussion, suppose that Swinburne had articulated a formally valid deductive theistic volitional argument for the basic laws and facts *e* of the universe. Importantly, even the provision of such an argument would not suffice to qualify the deduction as *explanatory*.

A deliciously hilarious example cited by Wesley Salmon over thirty years ago (1971, 34) makes this point tellingly by featuring a *pseudoexplanation* of why John Jones did not become pregnant during the past year. The purported cause is that he took birth control pills all year, and the causal hypothesis is that no man who takes such pills ever becomes pregnant. It then follows impeccably that Jones did not become pregnant last year. But plainly, the birth control pills are causally irrelevant here.

An elementary classroom example of a causal pseudoexplanation is that, other things being equal, victims of the common cold who are coffee drinkers recover from it within one month, because drinking coffee is *therapeutic* for the common cold. Note that although this pseudoexplanation is stupendously predictive, it is nonetheless unacceptable causally: as we know, the afflicted cold sufferers recover equally well if they do *not* ingest any coffee at all.

Quite generally, ever since Francis Bacon taught, it has been known that, at least in the case of *causal* hypotheses, the *mere* deducibility of some data from some such hypotheses (together with known initial conditions) does *not* suffice to qualify the hypotheses as explanatory; nor does it qualify the data as supporting evidence for the hypotheses. To believe that it does is to indulge in dubious *hypothetico-deductive pseudo confirmation*. What is being overlooked by such a belief is that, although the causal hypotheses (in conjunction with the known initial conditions) entail the particular data, the hypothesized causal factors are often actually *causally irrelevant* to the data that are to be explained.

If the causal hypotheses are to be explanatory, they need to meet further well-known epistemic requirements, such as furnishing suitable "controls" instantiating actual causal relevance. Thus, in the present case, the theist's claim that God is the creative cause of the existence of the world and thereby the architect of its laws of nature should offer evidence *against* the rival *null* hypothesis that *no* external creative cause ex nihilo is required. If SoN *were* at all evidentially warranted, it could serve to *rule out* that rival null hypothesis. But, as shown in section **1.7**, SoN is baseless and hence unavailable to rule out the rival hypothesis that no creative cause ex nihilo is needed at all.

OBJECTION 5. The premises in the theistic volitional explanation yield that a divine volitional state, though itself uncaused, issued in God's creatively causing ex nihilo the existence of our nomological world. Yet, again, given the demise of SoN, *transformative* causation is the only kind of causation for which we have evidence—be it agent causation or event causation—rather than creative causation ex nihilo.

Thus, as emerges from the preceding considerations in objections 1 to 5 inclusive, Swinburne did not score a point against atheism when he wrote (1991, 287):

> The only plausible alternative to theism is the supposition that the world with all the characteristics I have described just is, has no explanation. That however is not a very probable alternative. We expect all things to have explanations.

But this assertion does not even cohere with Swinburne's claim (1996, 49) that the existence of God has no explanation, as we saw in section **1.9**.

OBJECTION 6. As mentioned in section **2.3**, more recently Swinburne (2000, 481–85) offered a reply to my prior objections of the same year (Grünbaum 2000, 17–29) to his account of theistically explaining the world's ultimate nomic structure and other scientific brute facts. Recall his showcase paradigm example of the putative total cosmic energy E_0 (1991, 125) and, more generally, his claim of theistic volitional pan-explainability (1996, 2) of the *specifics* of "*everything* we observe." Astonishingly, in his reply to me, he *sabotages* his erstwhile grandiose vision of a theistic explanatory edifice as follows (2000, 482):

The hypothesis *h* which I consider to explain the data *e* is not, "there is a God and he causes *e*" (which is what Grünbaum may be supposing on his p. 20), but "there is a God" (as he explicitly recognizes on p. 36). Given *h*, it follows that God can bring about *e*, but how probable it is that he will, depends on whether (in virtue of his perfect goodness) he has good reason to do so. (God's perfect goodness, I claim, follows from his omniscience and his perfect freedom, that is his freedom from influences other than rational considerations.) Quite a bit of my writing is devoted to showing that he does have such good reason—e.g. that simple regularities in nature give to finite beings the power to grow in power and knowledge, etc., and that that is a good thing.

As we know, Bayes's theorem in the calculus of probability, if used to probabilify *hypotheses*, is a device for updating the *evidential* appraisal of a hypothesis on the basis of new, or previously unavailable, or unconsidered evidence. Thus, as Wesley Salmon (2001, 79) has emphasized, "Bayes' theorem belongs to the context of confirmation, not to the context of explanation." And this important distinction is, of course, *not* lessened at all by the fact that, once a hypothesis is sufficiently confirmed, it can qualify *epistemically* to serve as a premise in an explanation.

In my earlier critique of Swinburne (Grünbaum 2000, 35), I cited Salmon's reiteration (2000, 79) of Hempel's caveat that "Explanation-seeking why-questions solicit answers to questions about why something occurred, or why something is the case. Confirmation-seeking why-questions solicit answers to questions about why *we believe* that something occurred or something is the case." And, being mindful that Bayes's theorem belongs to the context of *confirmation*, I wrote (ibid.):

Swinburne [. . .] muddies the waters. He tries to use Bayes' theorem both to probabilify (i.e., to increase the confirmation of) the [hypothesis of the] existence of God, on the one hand, and, on the other, to show that theism offers the best [simplest] explanation of the known facts, assuming that God exists. And his [Swinburne's] account of the notation he uses in his statement of the theorem reveals his failure to heed the Hempel-Salmon distinction.

In his reply to me (Swinburne 2000, 482), Swinburne turned a deaf ear to the relevance of the Hempel-Salmon distinctions. And it was thus lost on him that *I* was explicitly speaking of the theistic hypothesis, which he was trying to *confirm* (incrementally) à la Bayes, when I went on to say (Grünbaum 2000, 36):

> It is vital to be clear on what Swinburne takes to be the hypothesis h in his Bayesian plaidoyer for the existence of the God of theism. He tells us explicitly: "Now let h be our hypothesis—'God exists' (1991, 16)."

But, contrary to Swinburne's reply to me (2000, 482), I absolutely *never, ever* "explicitly recognized" that the hypothesis, which he took to be **sufficient** to **explain** the data e was just the *parsimonious* one h "there is a God" or "God exists." This *confinement* of the explanatory premises to h never even occurred to me, because such a parsimonious hypothesis obviously could not possibly redeem Swinburne's mantra that theism explains the *specifics* of "*everything* we observe" (1996, 2).

After all, as I pointed out emphatically under my objection 1 above, to make good on that omnivorous explainability, it is hopelessly insufficient to declare with Swinburne (2000, 482) that "Given h, it follows that God *can* bring about e, but how probable it is that he will, depends on whether (in virtue of his perfect goodness) he has good reason to do so" (italics added). Nor does it help rescue Swinburne's forlorn all-encompassing explanatory pretensions to point out, as he does, that "he [God] does have such good reason" as, for example, "that simple regularities in nature give to finite beings the power to grow in power and knowledge."

To have even a hope of redeeming his explanatory mantra, Swinburne does indeed require at least the following *conjunctive* theistic hypothesis, which he mentions but rejects (2000, 482): "God exists *and* he chose to cause *e ex nihilo*"—a *stronger* hypothesis which I articulated in the deductive volitional explanation I have set forth above. In short, in effect Swinburne has now *repudiated* his erstwhile signature doctrine of all-encompassing theistic explainability, rather than having offered a relevant cogent rebuttal to me.

For his part, Quinn (2003, 593) has come to appreciate these serious defects in Swinburne's views, so that, by 2003, he developed a quite different conception of the theistic explanation of the ultimate nomic structure and basic facts of the world. Now Quinn mentions three positive answers to the question "Why does the possible world that is in fact actual obtain, rather than another?" and he suggests ([ibid.]) that, presented with three answers to it, the majority of contemporary theists would prefer the explanation that "God had a sufficient reason to actualize it [i.e., the de facto existing world], *but this reason is utterly beyond our ken*" (italics added). Yet this sort of surrogate explanation belongs to fideist rather than natural theology! Therefore, I cannot see why a theist would expect anyone who does not *antecedently* believe in God to embrace theism as *explanatory*, if it features, as this forlorn surrogate explanation does, resort to the old chestnut that God's sufficient reason passes all human understanding.

To be sure, the intellectual humility expressed by it is ingratiating. But that explanation forsakes any conjecture as to God's specific reason for choosing the actual nomic structure, *as against an alternative one*. And yet Quinn's erstwhile *plaidoyer* for a theistic explanation was precisely, like Swinburne's, that it transforms scientific brute facts into specifically *explained* states of affairs. As I have argued, it does nothing of the kind: neither Swinburne nor Quinn have redeemed at all their vaunted promise to explain theologically what science leaves unexplained.

3. CONCLUSION

In parts 1 and 2 of this essay, I have argued for "The poverty of theistic cosmology" in the following *two* respects: neither the theistic answer to the question "Why is there something contingent rather than nothing contingent?" nor the theological explanation of the ultimate nomological architecture of the world withstands evidential scrutiny.

ADDENDUM

This chapter is reprinted, with a few minor improvements, from my article "The Poverty of Theistic Cosmology" in the *British Journal for the Philosophy of Science* 55, no. 4, pp. 561–614.

I have published a much shorter sequel to it under the title "Why Is There a Universe AT ALL, Rather Than Just Nothing?" That sequel is the text of my presidential address of August 9, 2007, at the thirteenth quadrennial World Congress of the Division of Logic, Methodology, and Philosophy of Science, which is one of the two divisions of the International Union of History and Philosophy of Science. My presidential address appears in the volume of proceedings of the Thirteenth Congress, which is titled *Logic, Methodology and Philosophy of Science*, and is edited by Clark Glymour of the United States, Wei Wang of China, and Dag Westerstahl of Sweden. These proceedings are being published by King's College Publications, London, UK, in 2009.

In advance of that publication, the journal *Free Inquiry* has published a variant of its text in two installments: part 1 appeared in the June/July 2008 issue (vol. 28, no. 4, pp. 32–35); part 2 in the August/September 2008 issue, vol. 28, no. 5, pp. 37–41.

As will be recalled from section **1.7.1** on Leibniz, I contended that the hypothesized *a priori* maximum ontological simplicity of the null world does *not* mandate the claim of SoN that, *de jure*, the thus simplest world must be spontaneously realized ontologically in the absence of an overriding cause, because *a priori* simplicity is *not* ontologically legislative.

But Swinburne has given a special theological twist to simplicity in the 2004, second edition of his book *The Existence of God* (Oxford

University Press). There, he tells us (p. 336): ". . . if there is to exist something, it seems impossible to conceive of anything simpler (and therefore *a priori* more probable) than the existence of God." And in his 1997 monograph *Simplicity as Evidence of Truth*, as extended in his 2001 Oxford book *Epistemic Justification*, he has stated empirical conditions under which, he claims, the simpler of two rival hypotheses is most probably true. Might Swinburne then be able to claim that the theistic existential hypothesis is inductively more likely to be true than its atheistic competitor? Elsewhere, I have thoroughly justified a decidedly *negative* answer to this question: See my "Is Simplicity Evidence of Truth?" (Grünbaum, 2008), especially its "A Coda on Atheism Versus Theism" (p. 188).

ACKNOWLEDGMENTS

In thinking about the issues I discussed in this chapter, I greatly benefited from the scholarship, advice, and comments of several valued colleagues as well as of three graduate research assistants. I wish to express my warm gratitude to them.

Philip Quinn, who was my doctoral dissertation student at the University of Pittsburgh nearly four decades ago, was a steadfast and generous interlocutor on the entire spectrum of the issues I treat, as can also be gleaned from within my text.

Teddy Seidenfeld was a great and indefatigable resource in specifically appraising Swinburne's Bayesian argument for the existence of God, an appraisal which I may offer, in due course, in a sequel to the present essay. For now, I can refer the reader to my paper "Is Simplicity Evidence of Truth?" (Grünbaum 2008) for my detailed critique of Swinburne's 1997 Aquinas Lecture "Simplicity as Evidence of Truth," as extended in his 2001 Oxford book *Epistemic Justification*.

My Pittsburgh colleagues Richard Gale, Gerald Massey, Nicholas Rescher, and the late Wesley Salmon served helpfully as sounding boards on one or another subtopic.

My graduate assistants Emily Aiken, Alan Love, and James Tabery ably processed and/or summarized some relevant literature for me. Jim Tabery also went over the drafts of this essay with a fine-toothed comb, making good suggestions to improve my exposition.

REFERENCES

Aquinas, T. 1948. *Summa Theologia, cura et studio Sac. Petri Caramello, cum textu ex recensione Leonina.* Turin: Marietti.

Bergson, H. 1974. *The Two Sources of Morality and Religion*, translated by R. A. Audra and C. Brereton, with the assistance of W. Carter. 1935; Westport, CT: Greenwood Press.

Bertocci, P. 1968, 1973. "Creation in Religion." In *Dictionary of the History of Ideas: Studies of Selected Pivotal Ideas*, vol.1: *Abstraction in the Formation of Concepts to Design Argument*, edited by P. Wiener. New York: Scribner, p. 571.

Bondi, H. 1960. *Cosmology*, 2nd ed., Cambridge Monographs on Physics Series, Cambridge: Cambridge University Press.

Carlson, E., and E. J. Olsson. 2001. "The Presumption of Nothingness." *Ratio* 14, no. 3: 203–21.

Craig, W. 2001. "Professor Grünbaum on the 'Normalcy of Nothingness' in the Leibnizian and Kalam Cosmological Arguments." *British Journal for the Philosophy of Science* 52: 371–86.

Descartes, R. 1967. "Meditation III. Of God: That He Exists." In *The Philosophical Works of Descartes*, vol. 1, translated by E. S. Haldane and G. R. T. Ross. Cambridge: Cambridge University Press, pp. 157–71.

Edwards, P. 1967. "Atheism." In *The Encyclopedia of Philosophy*, vol. 1, edited by P. Edwards. New York: Macmillan and Free Press; London: Collier Macmillan Ltd., pp. 174–89.

Eliade, M. 1992. *Essential Sacred Writings from Around the World*. San Francisco: HarperCollins.

Gale, R. M. 1976. *Negation and Non-Being*. American Philosophical Quarterly Monograph Series, no. 10–0084–6422. Oxford: Blackwell.

Grünbaum, A. 1973. *Philosophical Problems of Space and Time*. 2nd ed. Boston Studies in the Philosophy of Science, vol. 12. Dordrecht: Reidel.

————. 1998. "Theological Misinterpretations of Current Physical Cosmology." *Philo*, 1 no. 1: 15–34.

————. 2000. "A New Critique of Theological Interpretations of Physical Cosmology." *British Journal for the Philosophy of Science* 51: 1–43.

————. 2008. "Is Simplicity Evidence of Truth?" *American Philosophical Quarterly* 45, no. 2: pp. 179–89.

Hasker, W. 1998. "Religious Doctrine of Creation and Conservation." In *Routledge Encyclopedia of Philosophy*, vol. 2, edited by E. Craig. London and New York: Routledge, pp. 695–700.

Heidegger, M. 1953. Einführung in die Metaphysik. Tübingen: Max Niemeyer.

Leibniz, G. W. 1956. *Vernunftprinzipien der Natur und der Gnade; Monadologie-Principes de la Nature et de la Grace fondés en Raison; Monadologie*. Hamburg: Felix Meiner.

————. 1973a. "On the Ultimate Origination of Things." In *Leibniz: Philosophical Writings*, edited by G.H. R. Parkinson and translated by G. H. R. Parkinson and M. Morris. 1697; London: J. M. Dent & Sons, pp. 136–44.

————. 1973b. "Principles of Nature and of Grace Founded on Reason." In *Leibniz: Philosophical Writings*, edited by G. H. R. Parkinson and translated by G. H. R. Parkinson and M. Morris. 1714; London: J. M. Dent & Sons, pp. 195–204.

Leslie, J. 1978. "Efforts to Explain All Existence." *Mind* 87, no. 346: 181–94.

Loveley, E. 1967. "Creation: 1. In the Bible." In *New Catholic Encyclopedia*, vol. 4, edited by the Editorial Staff at Catholic University of America. New York: McGraw-Hill, pp. 417–19.

Lütkehaus, L. 1999. *Nichts: Abschied vom Sein, Ende der Angst*. Zürich: Haffmans Verlag.

May, G. 1994. *Schöpfung aus dem Nichts: Die Enstehung der Lehre von der Creatio Ex Nihilo*, translated by A. S. Worrall. Berlin: de Gruyter, 1978; Edinburgh: T & T Clark.

Nozick, R. 1981. *Philosophical Explanations*. Cambridge, MA: Harvard University Press.

Parfit, D. 1998a. "The Puzzle of Reality: Why Does the Universe Exist?" In *Metaphysics: The Big Questions*, edited by P. van Inwagen and D. Zimmerman. Malden, MA: Blackwell, pp. 418–27. Originally published in *Times Literary Supplement*, July 3, 1992.

———. 1998b. "Why Anything? Why This?" pt. 1, *London Review of Books* 20, no. 2 (January 22): 24–27.

Quinn, P. L. 1993. "Creation, Conservation, and the Big Bang." In *Philosophical Problems of the Internal and External Worlds: Essays on the Philosophy of Adolf Grünbaum*, edited by J. Earman, A. I. Janis, G. J. Massey, and N. Rescher. Pittsburgh/Konstanz: University of Pittsburgh Press/University of Konstanz Press, pp. 589–612.

———. 2003. "Cosmological Contingency and Theistic Explanation." *Faith and Philosophy* 20, special issue no. 5: 583–84.

Rescher, N. 1984. "On Explaining Existence." In *The Riddle of Existence: An Essay in Idealistic Metaphysics*. Lanham, MD: University Press of America, pp. 1–36.

———. 2003. "Contingentia Mundi: Leibniz on the World's Contingency." In *On Leibniz*. Pittsburgh: University of Pittsburgh Press, pp. 45–67.

Salmon, W. 1971. "Statistical Explanation." In *Statistical Explanation and Statistical Relevance*. Pittsburgh: University of Pittsburgh Press, pp. 29–87.

———. 2001. "Explanation and Confirmation: A Bayesian Critique of Inference to the Best Explanation." In *Explanation: Theoretical Approaches and Applications*, edited by G. Hon and S. Rackover. Dordrecht: Kluwer.

Samuelson, N. 2000. "Judaic Theories of Cosmology." In *The Encyclopedia of Judaism*, vol. 1, edited by J. Neusner, A. Avery-Peck, and W. Green. Leiden & Boston: Brill, pp. 126–36.

Schopenhauer, A. 1958, 1966. *The World as Will and Representation*, vols. 1 and 2, translated by E. F. J. Payne. New York/Indian Hills, CO: Dover/Falcon Wing Press. I am indebted to Charles Larmore for these references.

Swinburne, R. 1991. *The Existence of God*, rev. ed. Oxford/New York: Clarendon Press/Oxford University Press.

———. 1996. *Is There a God?* Oxford/New York: Oxford University Press.

———. 1998. "Response to Derek Parfit." In *Metaphysics: The Big Questions*, edited by P. Van Inwagen and D. W. Zimmerman. Oxford/New York: Oxford University Press.

———. 2000. "Reply to Grünbaum's 'A New Critique of Theological Interpretations of Physical Cosmology.'" *British Journal for the Philosophy of Science* 51: 481–85.

———. 2001. *Epistemic Justification*. New York: Oxford University Press.

Unger, P. 1984. "Minimizing Arbitrariness: Toward a Metaphysics of Infinitely Many Isolated Concrete Worlds." *Midwest Studies in Philosophy* 9: 29–51.

van Fraassen, B. C. 1980. *The Scientific Image*. Oxford/New York: Clarendon Press/Oxford University Press.

von Hayek, F. A. 1952. *The Counter-Revolution of Science: Studies on the Abuse of Reason*. Glencoe, IL: Free Press.

Wald, R. M. 1984. *General Relativity*. Chicago: University of Chicago Press.

Weinberg, S. 1993. *Dreams of a Final Theory*. New York: Vintage Books.

2.

ADOLF GRÜNBAUM ON RELIGION, COSMOLOGY, AND MORALS

An Appreciation and Critique

Nancey Murphy

1. Introduction

My assigned task is to respond to Professor Grünbaum's work on religion. I want to begin by thanking him for paying theists such as myself the compliment of disagreeing with us. Some of you might think that's a strange thing to be grateful for. If so, you are not aware of the low standing religion has in most philosophical circles. To illustrate: I was just finishing a doctorate in philosophy at Berkeley and told one of my fellow students that I'd decided to get a second doctorate in theology. "Well, okay," he replied, "but if you ever want a job in philosophy you shouldn't put that degree on your curriculum vitae." Well, how will I explain what I was doing all those years? He thought for a second or two and said: "Tell them you were in prison; that would be better."

Professor Grünbaum has paid the *compliment* of his attention to a number of issues that have a bearing on religion. Most of his writings fall into two categories: first, a critique of ways in which theists have made use of recent developments in physical cosmology; and, second, criticism of the claim that theism is needed as a foundation for morality. I shall describe and comment on some of Professor Grün-

235

baum's work in each of these two areas. Another relevant topic, which I shall not have time to address, is his commentary on Sigmund Freud's theory of religion.[1]

In the first part of my essay, then, I shall report on Grünbaum's criticism of theistic arguments. Despite the brilliance displayed in these moves and countermoves, I shall argue that the only likely outcome from such exchanges is to make clearer the interconnections among the assumptions of each side. To be more precise, I shall try to show that theism and secular humanism belong to contrasting *categorial frameworks*, neither of whose proponents will be convinced by arguments based on assumptions of the rival framework.

After a summary of Grünbaum's work on the relations between theism and morality, I shall present an overview of current work on this issue by some of the most significant philosophical ethicists of our day. In the process, I shall introduce the writings of Alasdair MacIntyre, and will then use his account of rational adjudication among competing traditions to suggest how the arguments between theists and secular humanists might become more productive. This will put me in position to express a final appreciation for the value of Grünbaum's writings on religion.

2. ON THEOLOGICAL INTERPRETATIONS OF CURRENT PHYSICAL COSMOLOGY

I shall not be able to do justice to all that Professor Grünbaum has written on theological interpretations of cosmology. Here is a list of some of the issues he has addressed; I shall explain some of these below:

1. A critique of authors who equate big bang cosmology's concept of the origin of the universe with the theological concept of creation. A related issue is the equation of the appearance of the physical universe in quantum cosmology with the theological doctrine of creation out of nothing.

2. Criticisms of arguments for the existence of God based on big bang cosmology; an important target here are the writings of William Lane Craig.
3. A critique of Philip Quinn's use of the law of conservation of matter and energy to argue for a concept of divine sustenance according to which the constituents of the universe need to be recreated by God from moment to moment or else they would cease to exist.[2]
4. A critique of Richard Swinburne's use of probability calculus to argue that the existence of God is more probable than not.[3]
5. A critique of a variety of arguments for God's existence that assume that something's existing calls for explanation whereas no explanation would be needed if nothing existed.

Before I take up any of these issues I shall present some background information on the cosmological theories and theological concepts involved. As is well known, in the 1920s cosmologists had to give up the idea that the universe, on a large scale, is static. Its observable expansion is the basis for the big bang theory. If we project back in time we come to a point, approximately twelve billion years ago, from which the expansion must have begun. This conclusion has been taken by some—theists and atheists alike—to have religious implications. Cosmologist Frederick Hoyle worked on an alternative "steady-state" model precisely because he saw the big bang to be too easily associated with "theological biases." Pope Pius XII welcomed the big bang theory as evidence for creation of the universe by God.

Some cosmologists have argued that with the big bang we reach the limits of science: the big bang is a "singularity" that cannot be encompassed under any scientific laws, and so its cause cannot be known (or perhaps even meaningfully discussed). More recently, though, some cosmologists have attempted to explain the big bang itself. Most notable here is the work of Stephen Hawking and James Hartle. Their work (as I understand it) depends on recognition that very early in the history of the universe there was a time when it would have been compressed enough in size for quantum effects to be significant. Because quantum events occur without causes in the classical sense, this raises the question of whether the origin of the universe can be explained without cause, that is, as the result of fluctuations in a quantum vacuum.

In addition, at this scale, the fluctuations would affect space-time itself. Hawking argues that before 10^{-47} seconds into the universe's existence, space and time would not have been distinguishable as they are now.[4] Physicist Paul Davies says, "one might say that time emerges gradually from space," so there is "no actual 'first moment' of time, no absolute beginning at a singular origin." Nevertheless, this does not mean that the universe is infinitely old; time is limited in the past but has no boundary.[5]

A number of authors have commented on possible theological implications of Hawking's work, including Hawking himself. "So long as the universe had a beginning," he says, "we could suppose it had a creator. But if the universe is completely self-contained, having no boundary or edge, it would have neither beginning nor end: it would simply be."[6]

I have been careful to say all along that only *some* theologians have seen these developments as relevant to the doctrine of creation. I need to explain why this is so. (I shall be speaking here of Christian theology, due to my lack of knowledge of other traditions, but I believe that much of what I say will have parallels in Jewish thought and to some extent in Islamic thought as well.) From the first few centuries of Christianity up through the Protestant Reformation in the sixteenth century, there was a consensus among theologians that the universe was created by God out of nothing and that creation involves temporal origin. However, during the Middle Ages, Aristotle's theory that the universe was eternal was discussed. Thomas Aquinas argued that the eternity of the universe would not contradict the doctrine of creation per se because creation is most basically about the dependence of all that exists on God; that is, it is the Christian answer to the question of why there is something rather than nothing. Be this as it may, Aquinas took the scriptures to be saying that the universe did in fact have a beginning.[7]

However, many modern theologians have argued that the doctrine of creation is not *at all* about temporal origin; it is *entirely* about dependence. On this view, an eternal, static universe would be no less dependent on God than one developing temporally. In general, it has been common among those indebted to the tradition of nineteenth-century liberal Protestantism, as well as to the Neo-orthodox movement of the mid-twentieth century, to understand the nature of religious faith and theology in such a way that scientific findings are almost by definition irrelevant to theology. These understandings can be traced ultimately to Immanuel Kant's distinction between the "phenomenal" world known to the senses and the "noumenal" world of things-in-themselves. Science pertains to the phenomenal world perceivable by the senses; but God, morality, and immortality pertain to the noumenal world. More recent versions of this strict separation distinguish between science and religion on the basis of the questions they answer—science answers "how" questions and religion answers "why" questions; or science tells us facts, and religion tells us about meaning and value.

So the question of what a Christian ought to make of the big bang theory is a difficult one. One thing can be said for sure: Grünbaum is correct to point out that the concept of the origination of the universe in cosmology cannot be simply *equated* with the Christian concept of creation. He is also correct to point out that the quantum vacuum out of which Hawking's universe emerges is not the "nothing" of Christian doctrine.

I shall now summarize some of Grünbaum's work by presenting his objections to increasingly sophisticated theistic arguments. A typical creationist argument of the least sophisticated type is the following: the big bang followed upon a state of nothingness, and the transition from nothing to something required an external cause, namely, God.[8] This formulation is problematic because, according to the theory of general relativity, time does not "exist" independently of matter. So the meaning of a time interval prior to creation is undefined. This means that (if we only consider the meaning of time in general relativity) there can be no *earlier* cause of the big bang.

This argument raises an important issue regarding the nature of philosophical concepts, but one I cannot pursue here. Is it the case that there is a single concept of time and that its use in general relativity exhausts its meaning? I believe that there is a rather large cluster of concepts of time, only partly overlapping. If it is the latter, then it is correct to say that the phrase "before the big bang" is undefined *in general relativity theory*, but we still have ordinary concepts of *before* and *after*, according to which we *can* meaningfully ask what happened before the big bang, just as cosmologists have done in asking whether or not "our" big bang was preceded by a series of big bangs and "big crunches." Thus, I would argue that theists are entitled to speak of God's existing before (our) big bang. However, I find most theological discussions of God and time to be flawed by a comparable insensitivity to the lack of a univocal meaning for the word "time."[9]

Grünbaum makes the more subtle point that $t = 0$ cannot be understood to represent any first *event*: "general relativity theory does not countenance t_0 as a bonafide instant of the Big Bang space-time."[10] The concept of an event in general relativity only makes sense if it can be located *in* time and in three-dimensional space. "Instead," Grünbaum says, "the Big Bang is a so-called 'singularity' in *at least* the sense that, as we approach it from ever earlier moments, the space-time metric of the GTR becomes degenerate, and the scalar curvature as well as the density become infinite."[11] But if $t = 0$ does not represent an event, it cannot be the effect of a cause. Thus, the whole question of what caused the big bang is undefined; it is a pseudoquestion.

A more sophisticated theistic argument has been proposed by William Lane Craig. It is an updated version of the "kalam" cosmological argument, first devised by Muslim scholars in the Middle Ages. ("Kalam" means speculative argument in Arabic.) In schematic form the argument can be stated in three steps:

1. What begins to exist is caused to exist by something else.
2. The universe began to exist.
3. Therefore the universe was caused to exist; and the cause of its existence is God.[12]

In the original version the second premise was supported by arguments that an actual infinity is not possible—the assumption that an infinite number of seconds has already elapsed, or that there have already existed an infinite number of beings runs into paradoxes. Craig amplifies the argument by claiming that big bang cosmology provides empirical support for the universe having a beginning. He sidesteps the problem of attempting to speak of a temporally prior cause by claiming that God's creative act was simultaneous with the origin of the universe.

Craig's argument still runs afoul of Grünbaum's denial of event status to the big bang and his denial that it is appropriate to look for its cause. Craig insists that the singularity must have a cause because it is impossible that it come out of absolutely nothing. Grünbaum replies in two ways: first, the phrase "it comes out of nothing" illegitimately smuggles in temporality—a prior state of "potentiality" out of which it comes; and second, the "metaphysical intuition" that something cannot come out of nothing is unjustified.

This intuition has been shared by Thomas Aquinas, Descartes, Berkeley, Leibniz, Locke, and currently by Philip Quinn and Richard Swinburne. All of these have taken the question "Why is there anything at all rather than just nothing?" to be a significant question in need of an answer. Grünbaum turns the tables: Why should one assume what he calls "the spontaneity of nothingness"? Why should it be assumed that the existence of anything needs explanation while nonexistence does not? Grünbaum has taken issue with arguments by both Leibniz and Swinburne for the spontaneity of nothingness; he contests their claims that the state of nothingness is *simpler* than the existence of any conceivable thing and therefore not in need of explanation.[13]

So back and forth the arguments go, becoming more complex and sophisticated. Does it matter whether the ball ends up in Grünbaum's court or that of the theists? I mean no disrespect to the contestants, but I want to argue that it does *not* matter—*if* the issue is one of converting opponents to one's own view. That is, I am entirely certain that neither Craig nor Swinburne is going to become an atheist after reading Grünbaum's next article, nor Grünbaum a theist. So what *is* the point? And how is it possible for equally intelligent, well-read, and well-intentioned people to come to diametrically opposed conclusions? I hope to provide some conceptual resources for understanding what is going on.

One helpful resource is Stephan Körner's concept of a categorial framework. Specifying a categorial framework involves listing all of the maximal categories—that is, the major classifications—that are needed for construing reality, along with their subordinate genera, criteria for membership in those categories, and the logical assumptions to be employed in reasoning. Körner's categorial framework is but a more formal account of what other philosophers would call a "lexicon" or "idiom" together with a "grammar." What is important for our purposes is Körner's claim that a categorial framework necessarily involves claims about the nature of reality.

> The manner in which a person classifies the objects of his experience into highest classes or categories, the standards of intelligibility which he applies, and the metaphysical beliefs which he holds are intimately related. To give an obvious example, the employment of the category of causally determined events, the demand that all or some explanations be causal, and the belief that nature is at least partly a deterministic system so involve each other that they are either all present in a person's thinking or else all absent from it. Groups of persons, societies, and whole civilizations exhibit, in so far as they can be said to think, a similar correlation between their categories, standards of intelligibility and metaphysical beliefs.[14]

So Körner's point is that to possess a category, a lexical item, such as "cause," commits a community to certain propositions about reality; for example, that nature is at least partly a determinate system.

This is due simply to the "grammar" governing the relations among the concepts, logic, and the rules for what count as instances of the concepts.

Körner describes such metaphysical claims as "internally incorrigible," that is, necessarily true so long as the categorial framework is not changed. We can produce an example from a Christian categorial framework by redescribing Anselm's ontological argument in Körner's terminology: "God" is a maximal category of the Christian categorial framework. The criterion for membership in this category is being "that than which none greater can be conceived." Objects that exist independently of thought are greater than those that are merely objects of thought (this is a part of the grammar of words such as "object," "reality," "existence," in Anselm's day). Therefore, the statement "God exists" is internally incorrigible. However, such claims are externally corrigible—they may appear unintelligible, questionable, or obviously false from the point of view of a different categorial framework.

While I do not believe that rival conceptual schemes are as neat and tidy as Körner's categorial frameworks (as I shall explain below), I do believe that his account of the interplay among categories, "grammar," and logic, resulting in internally incorrigible statements, is valuable for understanding disagreements between Grünbaum and the theists. These internally incorrigible statements show up as different and irreconcilable philosophical intuitions. For example, Grünbaum says, "I am rather at a loss to understand why a galaxy of philosophers thought that the mere logical or empirical contingency of the existence of any given particulars can support the spontaneity of utter nothingness."[15] However, this thesis is internally incorrigible for the theists in question. It has become common to argue that Thomas's five ways are not attempts to convince atheists of the existence of God, but rather are "grammatical remarks" in Ludwig Wittgenstein's sense. We learn in Thomas's third way that "contingency" *means* "would not exist without a creator." And being the one whose existence is not contingent is part of what is meant by "God."

So Grünbaum and his opponents inhabit different categorial

244 ADOLF GRÜNBAUM ON RELIGION, COSMOLOGY, AND MORALS

frameworks. One framework incorporates the category "God." In the other framework the universe itself is the ultimate existent. Since for Grünbaum the natural world is the whole of reality, metaphysical questions can be answered by looking to science. For example, if you want to know about causation, see what science tells us. In contrast, the theist has first to ask whether God is the chief exemplar of categories that apply to the physical universe, or whether God is radically different. For the theists whom Grünbaum criticizes, God is radically other. Therefore they will see some of Grünbaum's criticisms as misguided. For example, Grünbaum criticizes Quinn and Swinburne for their view that God's volition brings the world into existence; Grünbaum says that "in all our ordinary and scientific reasoning, it would be regarded as magical thinking to suppose that any *mere thought* could bring about the actual existence of the thought object, let alone out of nothing."[16] Swinburne and Quinn would reply that this criticism is beside the point; God's creative action is *supposed* to be different from any causal processes in the natural world.

So both the theist's and the naturalist's arguments are valuable in that they explore the connections within the two categorial frameworks, but neither side is going to be convinced by arguments that use as premises propositions that are at home only in the opponent's framework.

I mentioned above that Körner's concept of a categorial framework does not take account of the messiness of actual systems of beliefs. One way of seeing why this is so is to consider the work of George Lakoff and Mark Johnson on the role of metaphors in both philosophy and our everyday thinking. Most of our basic concepts such as "time," "causation," and "action" have very little content apart from metaphorical extension. The metaphors we use in reasoning with these concepts are based largely on our experience as embodied agents in the world of medium-sized physical objects. Our concept of causation has very little literal content—only something like the following: a cause is a determining factor for a situation. The concept of time, in fact, has *no* literal content. All of our understandings of time are relative to other concepts such as motion, space, and events.[17] We have a number of basic metaphors that structure our thinking about time. In one the observer is located in time as in a container; the past is behind and the future is in front. In another metaphor, the observer is stationary and time moves past. So most of our understanding of time is a metaphorical version of our understanding of motion in space.[18]

Because concepts such as "time" and "causation" arise within our everyday experience, I am suspicious of their extension to cases far outside of the everyday world, whether this be the denial that temporal language can be applied beyond the "beginning" of the universe or complex theories about God's relation to time. My position here reflects my stand on a controversy that has marked the whole of modern (Anglo-American) philosophy between those (such as Rudolf Carnap) who hope that science will bring order and univocity to our language and those (such as J. L. Austin) who are skeptical of the benefits of such attempts.

3. ON THEISM AND MORALITY

I turn now to report on a second topic that Grünbaum has addressed. On the one hand, this is a criticism of claims that theistic creeds are necessary as grounding for morality or ethics, and, on the other, a defense of the ethic of secular humanism.

Much of Grünbaum's motivation for writing on this theme comes, I take it, from the fact that "nowadays, theistic moral advocacy is readily turned into political intimidation, designed to browbeat into conformity or silence" those who do not believe in seeking religious remedies for society's ills. "Such coercive attempts are being made in our society by both Christians and Jews."[19] Secular humanism is said not to produce good citizens because a theistic belief is necessary *both logically and motivationally* to support morality.

So Grünbaum's goal is to show that morality does not require a theistic basis. His arguments are of three sorts. For one, he states that there simply is no evidence for claims that the behavior of religious believers is more moral than that of others, and in fact it is the reverse: religious believers are responsible for most of the evil done in the history of the West. Second, he claims that secular humanism is no less able to provide ethical guidance than theism, despite its denial of the existence of God.[20] Third (and this makes up the greatest part of his argument), he sets out to show that theism in fact is useless for grounding a moral code. He sums up his position as follows:

> Neither theism nor atheism as such permit the logical deduction of any judgments of moral value or of any ethical rules of conduct. Moral codes turn out to be logically extraneous to each of these competing philosophical theories. And if such a code is to be integrated with either of them in a wider system, the ethical component must be imported from elsewhere.
>
> In the case of theism, it will emerge that neither the attribution of omnibenevolence to God nor the invocation of divine commandments enables its theology to give a cogent justification for any particular actionable moral code.[21]

Revelation cannot serve as a source of moral guidance because there is no way to decide which scriptures to follow, the Koran, the Hebrew scriptures, or the Christian Bible. Even if this problem could be solved, he says, "all apologists, whether Christian or Jewish, for the divine inspiration of the Bible end up justifying . . . actions that in ordinary moral discourse we should regard as wicked or evil."[22] He mentions genocide in Deuteronomy and slavery and the subordination of women in the New Testament. If one replies that it is necessary to be selective in taking moral guidance from scripture, then, he asks, is it not on the basis of reason and intelligence that one makes such judgments and if so, why not simply rely on reason in the first place?[23]

In addition, the *fact* of there being so many different interpretations of what scripture teaches calls the sufficiency of scriptural norms into question. Grünbaum is most enthusiastic in citing conflicting "authorities" on matters of morality: Michael Novak, Jerry Falwell, Rabbi

Kahane; the pope (on birth control); Presidents Bush (George H. W.) and Carter; a number of Orthodox rabbis who argue that the Holocaust was God's punishment for Zionism; and Ayatollah Khomeini.

Grünbaum argues that the attempt to derive moral codes from the fact of God's *omnibenevolence* flounders on the problem of evil. There is no satisfactory account of how God's justice or benevolence can be reconciled with the fact of the Holocaust or the immense amount of suffering that is not caused by human choices. This is the point of citing the rabbis who argue that the Holocaust was punishment; he cites other authors who resign themselves to saying that God's ways are inscrutable.

Grünbaum has recalled and refurbished some powerful arguments against theism. In some cases he has expressed them with remarkable rhetorical force. In other cases their power would have been greater with a little less rhetorical flourish and overstatement. For example, if one wants to give evidence for discrepancies among the authoritative teachers of God's will, would Jesus and Muhammad not make more convincing examples than Jerry Falwell and the ayatollah? I suppose it is an important question for theists to ask themselves what it means, if anything, that such extremist voices can be found. Perhaps it means that the saner majority ought to police its ranks better, or perhaps we should be happy that heretics are no longer burnt.

A number of fine works have been written since David Hume's day addressing the problem of evil and the meaning of religious plurality. I shall address myself here only to the central issue Grünbaum has raised: the connection, if any, between theism and morality. Notice, though, that the question should not be whether the bare assertion of God's existence (or of God's existence and goodness) is an adequate basis for morality. It is rather whether the teachings of Jesus would provide an adequate basis if we had reason to believe that Jesus' God existed—or the teaching of Muhammad given the existence of Allah. This refinement of the question does not alone solve either the problem of religious plurality or that of varying interpretations of texts, but it does make the problems more manageable. I shall suggest

below (following Alasdair MacIntyre) that it is only large-scale traditions of moral reasoning that can profitably be compared as to their
rational acceptability.

Let us first consider the issue of religion as a *motivation* for moral
conduct. A recent tome on altruism reports on empirical research into
people's motives for altruistic behavior. The results are mixed
regarding the role of religion. For example, one study with college students in artificial circumstances found no role for religious commitment.[24] However, in interviews with people who had taken risks to
save Jews during the Holocaust, religion did appear as a significant
factor, especially if it instilled a sense of one's relation to the whole of
humankind.[25] Given the scarcity of research, and the need to make distinctions on the basis of level of commitment, content of the religious
teachings, and so forth, I believe we should say that the jury is still out
on the *motivational* connection between theism and behavior. A
problem with Grünbaum's attributing most of the evils in the past to
believers is that most *people* up to the present era have been believers;
thus we do not know whether the evil is a result of their religiosity, or
of their not being religious enough, or of extraneous factors such as
religion being used by the powers of empire and nation-state to pro-
mote their own interests.

So what, then, is the *logical* relation of theism to morality? Do
moral codes need to be backed by belief in a higher being? This issue
has been debated throughout all of Western history. One of the features
of modernity has been its quest for ethics and politics without religion.
Philosophers Stephen Toulmin and Jeffrey Stout show how this "flight
from authority" was motivated by the problem of too many religious
authorities after the Reformation, and especially motivated by the entanglement of religion with national interests.[26] Historian Wallace Matson
notes that the seventeenth-century political theorist Thomas Hobbes was
the first philosopher since Aristotle to argue for the rights of the
monarch without religious grounds.[27] Three major traditions have dominated ethical thought since then: There were further developments of
Hobbes's notion of the social contract, particularly influential among the

Founding Fathers of the United States. Second, there was Kant's deon-tological approach, according to which one could know one's duty by means of reason—the basic principle being not to make oneself an exception to moral rules. Third is the utilitarian tradition based on the motto: so act as to bring about the greatest good for the greatest number.

Several philosophical ethicists have recently called into question the Enlightenment's quest for ethics without religion. For this reason it is no overstatement to claim that a revolution has taken place in ethics within the past half generation. In the 1970s the most influen-tial book in philosophical ethics was John Rawls's *A Theory of Jus-tice*,[28] a brilliant synthesis of themes from the three most significant attempts to found ethics on pure reason: Kant's deontological pro-gram, utilitarianism, and social contract theory.

In 1981, however, Alasdair MacIntyre published *After Virtue*, in which he argued that the Enlightenment project of basing morals on pure reason was a costly mistake; an ethical system that was not based on a tradition (such as the Christian tradition) was bound to fail.[29] It has been estimated that *After Virtue* is now the most widely read book on philosophical ethics.[30]

MacIntyre's argument for the inevitability of the failure of the modern project, leading to emotivist theories among professional ethi-cists and extreme moral relativism on the streets, is as follows: modern philosophical ethicists were surprisingly traditional in their views of the *content* of morality; what they gave up in rejecting the authority of the Christian tradition was a concept of the human person that was necessary for justifying those traditional moral precepts.[31] Medieval Christianity (as well as its two sources, the Hebraic tradition and Greek philosophy) provided answers to the question *What is a human being for?* Or, in MacIntyre's words, *What is the end or purpose of human life?* (the Greek word "telos" captures both ideas). Without the traditional answer(s) to this question, traditional accounts of morality turned out not to be justifiable. This failure, however, has not been seen by moderns as their own particular historical predicament, but rather as a universal feature of morality itself.[32]

Less well known is a thin volume by Bernard Williams with the modest title *Morality*. Williams provides a devastating critique of both subjectivism and moral relativism, and then goes on to survey the major positive approaches to ethics from antiquity to the present. He finds most of them defective in that they are not capable of answering the question *Why be moral* (at all)? However, there is also a sort of theory, he says,

> that . . . seeks to provide, in terms of the transcendental framework, something that man is for: if he understands properly his role in the basic scheme of things, he will see that there are some particular sorts of ends which are properly his and which he ought to realize. One archetypal form of such a view is the belief that man was created by a God who also has certain expectations of him.[33]

Williams notes that it has been a philosopher's platitude that even if God did exist, this would make no difference to the situation of morality. But Williams believes this platitude to be based on mistaken reasoning: "If God existed," he says, "there might well be special, and acceptable, reasons for subscribing to morality."[34]

It is interesting to note that MacIntyre's own career as a philosophical atheist came to an end at about the same time that he recognized the need for some tradition in order to make sense of morality, and published a book arguing that the synthesis of the Aristotelian tradition of the virtues with Augustinian Christianity in the work of Thomas Aquinas is the most intellectually compelling of all available options.[35]

4. SO HOW SHALL WE REASON TOGETHER?

The *manner* in which MacIntyre argued for the intellectual superiority of his chosen tradition will be instructive for us. Earlier in this essay I pointed out the difficulties in attempting to defend or defeat theism or secular humanism by means of brief arguments. A major problem—

perhaps *the* major problem—is that these two positions are enmeshed in large-scale categorial frameworks. Arguments for each position or against its competitor seem inevitably to involve the use of assumptions that are at home in only one of the categorial frameworks.

Philosophers of science in the 1960s and 1970s noticed a similar problem in science. Thomas Kuhn showed how proponents of different major theories in science tend to talk past one another, and pointed out that it is impossible to argue for the new theory on the basis of the old theory's concepts. Such theories are embedded in paradigms or disciplinary matrices that provide assumptions about the nature of reality, of proper scientific method, and of relevant scientific problems.[36]

In response to the perceived relativism of Kuhn's account of science, Imre Lakatos proposed a means of adjudicating between competing theories on the basis of how they change over time. Does the increase in empirical content keep pace with the theoretical development or do the theoretical modifications amount to ad hoc maneuvers to save the theory from falsification?[37] Paul Feyerabend replied that Lakatos's proposal had no teeth, because the theory that was degenerating at one point in its history might be made progressive in the future if some bright new theorist came along.[38]

MacIntyre's approach to rationality builds on Lakatos's insights in a way that allows him to answer Feyerabend. He is interested in the bigger question of how we can argue for the rational superiority of a large-scale tradition, such as the Enlightenment tradition versus an updated Thomism, or, I would add, for the humanist tradition versus some particular theistic tradition.[39] First, he disputes the relativist's assumption that the proponents of each tradition will inevitably see their own tradition to be justified according to its own standards, while the competitors will always appear defective. This simply is not borne out by historical investigation. Sometimes major traditions fail—and are seen to fail—*by their own lights.*

Second, talented scholars are capable of learning to work in two different traditions and are thus capable of understanding each on its own terms. When this is the case, it is sometimes possible to recognize from the point of view of one tradition that the other tradition was *bound to fail* at just the point it did, and why. Such a judgment is the best evidence we could have for claiming that the surviving tradition has a more adequate grasp of the nature of reality.

So what can be said in such short space about the contest between secular humanism and theism? First, I acknowledge that the problems of religious plurality and evil are serious threats. We might even say that the religions in the West are in a state of intellectual crisis. However, if MacIntyre is correct, the humanist project is also in a state of crisis in that it has not found—in its three-hundred-year history— adequate resources for settling moral disputes. Furthermore, if religion is not to be accounted for on the basis of divine initiative, then it is necessary to give an account of the prevalence and persistence of religion, despite a number of predictions that it would soon disappear. Grünbaum's work on the psychoanalytic theory of religion is relevant here. So the question is whether the delusion hypothesis or any other is an adequate explanation.

Clearly we are in no position to settle these issues here. What can be said, though, is that traditions are tested in dialectical questioning by rival traditions. No tradition can claim to be worth its salt unless it has been severely tested by critical questions from outside. For his participation in this necessary questioning, theists of all stripes are to be grateful to Professor Grünbaum.

NOTES

1. See, for instance, A. Grünbaum, "Psychoanalysis and Theism," *Monist* 70, no. 2 (April 1987): 152–92.

2. See A. Grünbaum, "A New Critique of Theological Interpretations of Physical Cosmology," *British Journal for the Philosophy of Science* 51 (2000): 1–43, sec. 3.

3. Ibid., pp. 33–39.

4. See C. J. Isham, "Quantum Theories of the Creation of the Universe,"

in *Quantum Cosmology and the Laws of Nature: Scientific Perspectives on Divine Action*, ed. R. J. Russell, N. Murphy, and C. J. Isham (Vatican City State and Berkeley, CA: Vatican Observatory and Center for Theology and the Natural Sciences, 1993), pp. 49–89.

5. P. Davies, *The Mind of God: The Scientific Basis for a Rational World* (New York: Simon & Schuster, 1992), p. 67.

6. S. W. Hawking, *A Brief History of Time* (London: Bantam Books, 1988), p. 141.

7. For an overview see F. Copleston, *A History of Philosophy*, 11 vols. (Westminster, MD: Newman Press, 1950), vol. 2; chap. 36.

8. A. Grünbaum, "The Pseudo-problem of Creation in Physical Cosmology," chap. 7 in *Physical Cosmology and Philosophy*, ed. J. Leslie (New York and London: Macmillan, 1990).

9. See my remarks on philosophical concepts at the end of this section.

10. A. Grünbaum, "Pseudo-creation of the Big Bang," *Nature* 344, no. 6269 (April 1990): 821–22.

11. A. Grünbaum, "Some Comments on William Craig's 'Creation and Big Bang Cosmology,'" *Philosophia Naturalis* 31, no. 2 (1994): 226.

12. A. Plantinga, "Arguments for the Existence of God," in *Routledge Encyclopedia of Philosophy*, ed. Edward Craig, 10 vols. (London and New York: Routledge, 1998), 4: 86–87.

13. Grünbaum, "A New Critique of Theological Interpretations of Physical Cosmology," pp. 3–14.

14. S. Körner, *Categorial Frameworks* (Oxford: Basil Blackwell, 1970), p. ix.

15. Grünbaum, "Some Comments on William Craig's 'Creation and Big Bang Cosmology,'" p. 230.

16. Grünbaum, "New Critique of Theological Interpretations of Physical Cosmology," p. 23.

17. G. Lakoff and M. Johnson, *Philosophy in the Flesh: The Embodied Mind and Its Challenge to Western Thought* (New York: Basic Books, 1999), p. 137.

18. Ibid., p. 139.

19. A. Grünbaum, "In Defense of Secular Humanism," *Free Inquiry* 12, no. 4 (1992): 30.

20. A. Grünbaum, "The Poverty of Theistic Morality," in *Science, Mind and Art: Essays on Science and the Humanistic Understanding in Art, Epis-*

254 ADOLF GRÜNBAUM ON RELIGION, COSMOLOGY, AND MORALS

temology, Religion and Ethics, in Honor of Robert S. Cohen, ed. K.
Gavroglu, J. Stachel, and M. W. Wartofsky, vol. 165 of the Boston Studies in
the Philosophy of Science, 3 vols. (Dordrecht: Kluwer Academic, 1995), 3:
230.

21. Grünbaum, "In Defense of Secular Humanism," p. 31.

22. Ibid., p. 31, quoting S. Hook.

23. Ibid., p. 36.

24. C. D. Batson, "Addressing the Altruism Question Experimentally,"
in *Altruism and Altruistic Love: Science, Philosophy, and Religion in Dialogue*, ed. S. G. Post, L. G. Underwood, J. P. Schloss, and W. B. Hurlbut
(Oxford: Oxford University Press, 2000), pp. 89–105.

25. K. R. Monroe, "Explicating Altruism," in Post et al., eds., *Altruism
and Altruistic Love*, pp. 106–22.

26. S. Toulmin, *Cosmopolis: The Hidden Agenda of Modernity*
(Chicago: University of Chicago Press, 1990), chap. 2; J. Stout, *The Flight
from Authority: Religion, Morality, and the Quest for Autonomy* (South Bend,
IN: University of Notre Dame Press, 1981), chap. 2.

27. W. I. Matson, *A New History of Philosophy*, 2 vols. (San Diego: Har-
court, Brace, Jovanovich, 1987), 2: 292.

28. J. Rawls, *A Theory of Justice* (Cambridge, MA: Harvard University
Press, 1971).

29. A. MacIntyre, *After Virtue*, 2nd ed. (South Bend, IN: University of
Notre Dame Press, 1984).

30. "What Is Virtue?" *Newsweek*, June 13, 1994, pp. 38–39.

31. This raises the question of where the ideal of humanism—as
opposed to tribalism on the one hand and the claim that humanism is
"speciesism" on the other—comes from. To what extent are the ideals of secular humanism historically dependent on the theistic traditions of the West?
See by comparison MacIntyre's argument that Augustine's notion of the "city
of God" was needed to supplement Aristotle's moral system since Aristotle
could not give any justification for moral obligation to those outside of one's
own polis. See *Whose Justice? Which Rationality?* (South Bend, IN: University of Notre Dame Press, 1988), p. 146.

32. See MacIntyre, *After Virtue*.

33. B. Williams, *Morality*, 9th ed. (Cambridge: Cambridge University
Press, 1993), p. 63.

34. Ibid., p. 72.

35. MacIntyre, *Whose Justice?*; and again in MacIntyre, *Three Rival Versions of Moral Enquiry: Encyclopaedia, Genealogy, and Tradition* (South Bend, IN: University of Notre Dame Press, 1989).

36. T. Kuhn, *The Structure of Scientific Revolutions*, 2nd ed. (Chicago: University of Chicago Press, 1970).

37. I. Lakatos, "Falsification and the Methodology of Scientific Research Programmes," in *Criticism and the Growth of Knowledge*, ed. I. Lakatos and A. Musgrave (Cambridge: Cambridge University Press, 1970), pp. 91–196.

38. P. K. Feyerabend, *Against Method* (London: New Left Books, 1975).

39. To see the connection between the debate in philosophy of science and MacIntyre's work on practical rationality, see his essay "Epistemological Crises, Dramatic Narrative, and the Philosophy of Science," *Monist* 60, no. 4 (October 1977): 453–72.

PART TWO

TENSE, REALITY, AND INTERPRETATION

3.
ADOLF GRÜNBAUM
ON SPACE AND TIME

Michael Tooley

1. Introduction

In this paper, I shall be focusing upon two aspects of Adolf Grün-baum's discussion of space and time. First, I shall be considering Grünbaum's rejection of intrinsic metrics in favor of extrinsic metrics, both in the case of space and in the case of time. Here I shall be arguing for the following three claims:

(1) The idea of an extrinsic metric is open to very strong objections, both in the case of space and in the case of time;

(2) By contrast, there is no serious objection to the idea of an intrinsic metric, either in the case of space or in the case of time;

(3) There are good reasons for postulating an intrinsic metric, both in the case of space and in the case of time.

Second, I shall also be considering Grünbaum's account of the nature of time, and here I shall be arguing for the following theses:

(1) The argument that Grünbaum offers against accounts of the nature of time that postulate mind-independent tensed facts does not show that such accounts are unsound;

(2) Grünbaum's claim that all tensed facts are mind-dependent is open to a serious objection;

(3) Grünbaum's proposal that conceptualized awareness is an essential constituent of mind-dependent tensed facts is also very problematic.

2. GRÜNBAUM VERSUS NEWTON ON INTRINSIC METRICS FOR SPACE AND TIME

2.1 Newton's views

In formulating his physics, Newton made use of the notions of absolute time, absolute space, absolute place, and absolute motion. These he characterized in the "Scholium" to the definitions that he offered at the very beginning of his *Mathematical Principles of Natural Philosophy*,[1] and Adolf Grünbaum, very early in his own book, *Philosophical Problems of Space and Time*,[2] quotes at length the parts that are concerned with the ideas of absolute time and absolute space. Here are two crucial, and very famous quotes:

> Absolute, true, and mathematical time, of itself, and from its own nature, flows equably without relation to anything external, and by another name is called duration: relative, apparent, and common time, is some sensible and external (whether accurate or unequable) measure of duration by the means of motion, which is commonly used instead of true time; such as an hour, a day, a month, a year.
>
> Absolute space, in its own nature, without relation to anything external, remains always similar and immovable. Relative space is some movable dimension or measure of the absolute spaces; which our senses determine by its positions to bodies; and which is commonly taken for immovable space.

In these passages Newton is embracing, first of all, a realist conception of space and time, according to which both space and time would exist even if space were empty at all times, and second, the idea that both space and time contain intrinsic metrics, so that in the case of space, for example, not only can there be qualitative spatial relations —such as a betweenness relation involving three locations, *A, B,* and *C*—that obtain independently of any relations to any object or event in space, there can also be quantitative spatial relations that also obtain independently of any relation to any body—such as, for example, the relation of equidistance that obtains among four locations *A, B, C,* and *D* when the distance from *A* to *B* is equal to that from *C* to *D*. Similarly, in the case of time, not only can temporal locations stand in the qualitative relations of temporal betweenness, simultaneity, and temporal priority, independently of any relation to anything in space and time, they can also stand in quantitative temporal relations independently of any relation to anything in space and time, so that, for example, the temporal distance between times *A* and *B* may be twice that between times *C* and *D*.

2.2 Grünbaum's views

Grünbaum believes that Newton's views concerning the nature of space and time are unsatisfactory in at least two ways. First, there is the idea that there is a "flow" of time, understood as involving the existence of mind-independent tensed facts. Second, there is the idea that both space and time involve intrinsic metrics. Thus Grünbaum, in commenting on the very famous first sentence of the above quote from Newton—"Absolute, true, and mathematical time, of itself, and from its own nature, flows equably without relation to anything external . . ."—says in a footnote:

> It is Newton's conception of the attributes of "equable," (i.e. congruent) time-intervals which will be subjected to critical examination and found untenable in this chapter. But in Chapter Ten below, we shall give reasons for likewise rejecting Newton's view that the

concept of "flow" has relevance to the time of physics, as distinct from the time of psychology.[3]

What is Grünbaum's reason for rejecting the idea of an intrinsic metric, both in the case of time and in the case of space? A bit later in chapter 1, after saying that if space or time were discrete, rather than dense, then any finite spatial or temporal interval would contain an intrinsic metric—a metric that would be given simply by the number of spatial locations, or the number of instants, present in the interval in question—Grünbaum goes on to comment as follows on the situation where space and time are continuous:

> By contrast, upon confronting the extended continuous manifolds of physical space and time—their continuity being postulated in modern physical theories apart from programs of space and time quantization—we see that neither the cardinality of intervals nor any of their other topological properties provide a basis for an *intrinsically* defined metric.... Accordingly, the continuity we postulate for physical space and time furnishes a sufficient condition for their intrinsic metric amorphousness. And in this sense then metric geometry is concerned not with space itself but with the relations between bodies.[4]

This passage makes it sound as if Grünbaum is claiming that if the *topological* properties of space and time do not provide any basis for intrinsically defined metrics, then space and time do not involve any intrinsic metrics, since he says that "the continuity we postulate for physical space and time furnishes a sufficient condition for their intrinsic metric amorphousness." A bit later in this chapter, however, Grünbaum makes it clear that this is not what he is claiming:

> To justify and clarify this indictment, we shall now see that continuity cannot be held with Riemann to furnish a sufficient condition for the intrinsic metric amorphousness of any manifold *independently of the character of its elements*. For as Russell saw correctly, there are continuous manifolds, such as that of colors (in the physi-

cist's sense of spectral frequencies), in which the individual elements differ qualitatively from one another and have inherent magnitude, thus allowing for metrical comparison of the elements themselves. By contrast, in the continuous manifolds of *space* and of *time*, neither points nor instants have any inherent magnitude allowing an individual metrical comparison between them, since all points are alike, and similarly for instants. Hence in these manifolds metrical comparisons can be effected only among the *intervals* between the elements, not among the homogeneous elements themselves. And the continuity of these manifolds then assures the non-intrinsicality of the metric for their intervals.[5]

At this point, then, it appears that Grünbaum is advancing the following thesis:

If one has a continuous manifold, and if all the elements in that manifold are constitutionally homogeneous, then the manifold cannot contain an intrinsic metric.

Grünbaum, however, goes on to point out that there is always the possibility of an intrinsic metric that is based upon cardinality, so that each point gets assigned the measure *zero*, and all continuous intervals are assigned the measure *infinity*. Such an intrinsic metric is, of course, obviously unsatisfactory, since it entails that all intervals have precisely the same length. So let us refer to such obviously unsatisfactory measures as "trivial." What Grünbaum is claiming, then, appears to be this:

If one has a continuous manifold, and if all the elements in that manifold are constitutionally homogeneous, then the manifold cannot contain a *nontrivial*, intrinsic metric.

Later in his book *Philosophical Problems of Space and Time*, however, Grünbaum indicates that he is not advancing this thesis. For there he sets out a thesis that he refers to as "Riemann's Metrical Hypothesis" or "RMH":

> Be it topological or nontopological, there exists no kind of
> *implicit* (intrinsic) basis for nontrivial metric relations in contin-
> uous *P*-space.[6]

Grünbaum then goes on to say, with regard to Riemann's metrical
hypothesis:

> Being aware of the existence of intrinsic yet *non*topological proper-
> ties in the case of the real number continuum, I was only inductively
> inferring RMH as an empirical hypothesis (regarding both *P*-space
> and *T*-space) but *not deducing* it, when I wrote: "Accordingly, the
> continuity we postulate for physical space and time furnishes a *suf-
> ficient* condition for their intrinsic metric amorphousness." ... In
> particular, I was *not* deducing RMH from topological considera-
> tions, coupled with the claim that the elements of continuous *P*-
> space (and *T*-space) are of homogeneous constitution. And by the
> same token, I was *not deducing* any claim of conventionality from
> these premisses.[7]

In the end, then, it appears that Grünbaum wished to advance only the
following, rather more modest thesis:

> If one has a continuous manifold, and if all the elements in that man-
> ifold are constitutionally homogeneous, then it is *unlikely* that the
> manifold contains a *nontrivial*, intrinsic metric.

3. Extrinsic Metrics: A Fundamental Problem in the Case of Space

In this section, I shall argue that, in the case of space, there is a very
strong objection to the idea of an extrinsic metric, and that, as a consequence, there is no real alternative to accepting the idea of an
intrinsic spatial metric.

Consider an ordinary macroscopic object, such as a table. What makes it the case that a given table is, say, twice as long as it is wide? If there is an intrinsic spatial metric, the question poses no problem.

But what if there is *no* intrinsic metric? Then what makes it the case that the table is twice as long as it is wide? One idea is to appeal to an extrinsic metric that is given by a certain measuring rod, or by wavelengths of light produced by a certain isotope, such as Krypton 86. But how does reference to such extrinsic standards make it the case that the table is twice as long as it is wide? If there were a number of measuring rods of the relevant type actually on top of the table, then there would be an actual state of affairs that would serve as the truthmaker for the claim in question. But it will only rarely be the case that there is, for example, any measuring rod of the relevant sort in contact with the table, let alone in contact with it at the requisite number of locations. So what makes it the case that the table is twice as long as it is wide when the table is not in contact with the relevant extrinsic standards in the relevant ways?

The natural response is to appeal to the results of *possible* measurements, to what *would* be the case if one *were* to make appropriate measurements, and thus to analyze statements about length—such as the statement about the dimensions of the table—in terms of counterfactual statements.

One point worth noting is that some of those counterfactual statements may be very complex, since it might require a large number of operations to determine the length of something. Indeed, if the length happens to be *irrational* relative to the unit of length in question, an infinite number of operations will be required to determine the length.

This, however, is not the crucial point. The fundamental objection here is that if one is going to make use of subjunctive conditional statements, those subjunctive conditionals need to be such that it is possible for there to be truthmakers. But what, then, are the truthmakers for the claim, say, that if one were to measure a given table, one would find that it was (approximately) four feet long? A natural answer is that the truthmaker involves the fact that one is reasonably competent at measuring things, together with the fact that the table is about four feet long. But the latter fact cannot be part of the truthmaker in the present context, on pain of circularity.

The objection here is precisely the same as one of the crucial objections to classical phenomenalism. Thus, the phenomenalist claims that truths about physical objects logically supervene upon truths about the experiences that people actually have, together with truths about the experiences that they would have if they had certain other experiences. The problem is that there are no satisfactory truth-makers available to the phenomenalist for the *counterfactual* statements about experiences. Similarly, if one holds that truths about lengths logically supervene on actual comparisons involving some standard, together with the results that one *would* obtain if certain comparisons *were* made, the problem is that, given *that* approach to length, there are no truthmakers for those counterfactual statements concerning the results of possible measurements.

In the case of space, then, there appears to be a decisive objection to the idea of an extrinsic standard. What, then, is the alternative? In thinking about this issue, it is useful, I suggest, to begin by considering the case of *qualitative* spatial relations. In particular, consider the relation of *spatial betweenness*. Is there any problem in the idea that spatial betweenness is a relation that is intrinsic to a spatial manifold, so that it is simply a basic fact, say, that locations A, B, and C are such that B is between A and C? It is not easy to see why there should be a problem here. But if there can be a basic, three-termed relation of betweenness that is intrinsic to spatial manifolds, why can there not also be either a basic, dyadic relation of being equally long, that can hold between spatial intervals, or a basic quaternary relation that obtains among ordered quadruples of spatial points when the distance between the first pair of elements is equal to that between the second pair?

4. THE CASE OF TIME:
INTRINSIC VERSUS EXTRINSIC METRICS

How does the situation compare in the case of time? Is it simply a matter of arguing along parallel lines?

In the case of spatial distance or length, it was clear that distance or length cannot logically supervene upon *actual* comparisons or measurements, and if one chooses any sort of macroscopic object to measure time, the point will be the same: such objects do not exist at all times, and so duration cannot logically supervene upon actual processes involving such objects. Then, if one invoked subjunctive conditionals concerning the processes that would have taken place if such a macroscopic clock had been present, the problem, once again, would be that there are no truthmakers for such counterfactuals.

Perhaps, however, one can find a type of submicroscopic object that, first, is an accurate clock, and that, second, is present at absolutely every time? Atomic clocks will not do, since there were times when no atoms were present in the universe. But perhaps there are submicroscopic objects, or processes, that are accurate clocks and that are present at absolutely every time—past, present, and future? If so, then the problem of there not being truthmakers for crucial counterfactuals could be avoided, since all temporal distances would supervene on actual states of affairs.

Imagine, then, that there is some type of periodic process—A, B, C, A, B, C, A, B, C, A, B, C ... and so forth—that is always taking place. But not just any process that repeats itself in an orderly fashion can function as a clock. This can be seen as follows. Imagine a world where, given a pendulum, if the bob is pulled to one side and then released, the path that it traverses is the same as in our world, so that, just as in our world, there is a gradual reduction in the amplitude of successive swings, until the bob finally comes to rest. So far, a world just like our own. But now the differences. When two pendulums, of identical construction, have their bobs raised the same amount, and then released simultaneously, it turns out that, although they go through the same states, in the same order, they do not go through them in step. Thus, by the time that the amplitude of the one pendulum has decreased by 10 percent, the amplitude of the other may have decreased by only 7 percent. Later, however, this situation may reverse itself, so that the amplitude of the first pendulum may, overall, have decreased by less than the second. And so the process may continue, with first one pendulum, and then the other, taking the lead in the progression from a state of enjoying swings of wide amplitude to a state of rest.

In such a world, pendulums would obey causal laws, but those laws would be what one might call laws of *pure succession*. Such laws would involve relations that fixed the *order* of events, and the direction of time, but not the temporal distance between events. Consequently, in such a world, pendulums could not function as clocks.

If, then, a periodically repeating process A, B, C, is to function as a clock, the appropriate correlations must exist. In particular, it must be the case that when two or more A, B, C, processes start off at the same time, they always march completely in step, and so are always in the same state at any given time.

But what ensures that this is so? One could hold that there was some *basic* law to the effect that if two states of type A were simultaneous, they would always give rise to two simultaneous states of type B, which would then give rise to two simultaneous states of type C, which in turn would give rise to two simultaneous states of type A, and so on. But if one thinks, for example, of a Newtonian world, is it plausible to hold that there would be *basic* laws of such a form? Is it not more plausible that any laws governing periodic processes derive from laws—such as the laws of motion, the law of gravity, the law of electrostatic attraction, and so on—that are in no way restricted to *periodic* processes? But if this is right, then given that time enters into the law that $F = ma$, if time were defined in terms of the A, B, C, process, one would have a situation where one was explaining why certain correlations held by appealing to laws, such as $F = ma$, that involve a concept of time that in turn presupposes that the correlations in question do hold, and this seems circular in an unacceptable way.

In addition, however, doesn't one need a periodic process that could not possibly have been absent? For if the process A, B, C, is such that, had the initial conditions been different in a certain way, there would have been no processes of that type, then we have a situation where, if temporal distance is defined in terms of the process in question, it will be true that, had initial conditions been different, there would have been no quantitative temporal relations in the world. But surely this is not acceptable: a change in the initial conditions should not have the consequence that the world could not start out from the changed initial conditions and then proceed along in accordance with the relevant basic laws, such as $F = ma$.

The upshot is that although the issue is somewhat more complicated in the case of time than in the case of space, the idea of an extrinsic standard of temporal distance appears very problematic. In the first place, it is not clear that there is any ever-present periodic process that can serve as the measure of temporal distance. In the second place, and more seriously, it seems very unlikely that any physical process is such that it would have been present even if initial conditions had been different. But if the process does not meet this constraint, it would seem that it could not be the basis of the definition of temporal distance that enters into the relevant basic laws.

What, then, are the alternatives? One possibility would be to parallel the intrinsic alternative that was suggested in the case of space. That is to say, one could hold that, just as there can be a basic, dyadic relation of equality of length in the case of spatial intervals, so there can be a precisely parallel, basic, dyadic relation of equality of temporal intervals—a relation that is intrinsic to temporal manifolds.

I am inclined to think, however, that the postulation of such a primitive, quantitative temporal relation does not do everything that needs to be done. To see why, consider two types of causal processes, A and B, and four times, t_1, t_2, t_3, and t_4. Suppose, further, that there is a process of type A that begins at time t_1 and ends at time t_2, and that there is a process of type B that begins at t_3 and ends at t_4. Suppose, finally that the temporal interval starting at t_1 and ending at t_2, and the temporal interval starting at t_3 and ending at t_4 stand in the quantitative relation of equality of temporal length. Then if the quantitative relation of equality with respect to temporal length enters into the laws that underlie causal processes of types A and B, the desired correlations, both with regard to different instances of processes of a given type and with regard to processes of different types, will follow. Thus it will follow, for example, both that if there were another process of type A that started at time t_1, it would also finish at time t_2, and that if there were another B-type process that began at time t_1, it would also conclude at time t_2.

So far, so good. There is, however, a question that arises in the case of time for which no parallel question arises in the case of space. This question emerges, perhaps, most clearly if one imagines that, up until some time t_1, there had been no processes of type A. Then, at t_1, two A-type processes begin. If one has laws that, rather than involving some temporal unit, incorporate only the relation that obtains between two temporal intervals if and only if they are of equal length, then the laws will ensure that the two processes of type A will end at the same moment, but they will not determine what that instant will be. So if the two causal processes end at time t_2, this will not be something that was determined by the initial conditions plus the laws: the two processes could equally well have ended at some earlier time, or at some later time.

But is there any alternative of the intrinsic-metric variety that can avoid this problem? I want to suggest that there is. It emerges if one considers the thesis mentioned earlier, and which Grünbaum appears to have advanced, namely:

> If one has a continuous manifold, and if all the elements in that manifold are constitutionally homogeneous, then it is *unlikely* that the manifold contains a *nontrivial*, intrinsic metric.

Grünbaum's idea was that the proposition that a manifold is continuous, together with the proposition that all of the elements of the manifold are constitutionally homogeneous, provides inductive support for the conclusion that the manifold contains a *nontrivial*, intrinsic metric. Now, precisely why the former provides inductive support for the latter is not, perhaps, all that clear, though it may be that the idea is simply that both continuity and constitutional homogeneity of the elements rule out some ways—and some quite natural ways—in which the manifold might contain an intrinsic metric, and so, by eliminating those possibilities, the probability of their being an intrinsic metric must be lowered.

What I am interested in here, however, is not the force of this argument, but, rather, the plausibility of one of the premises involved, namely, what might be called the homogeneity postulate:

> All of the elements of a temporal manifold, or of a spatial manifold, are constitutionally homogeneous—that is, they do not differ with respect to their intrinsic properties.

Why should one think that this proposition is true? In the case of the purely mathematical points in geometry, of course, there are presumably good reasons for holding that they are constitutionally homogeneous, since one can maintain that their nature is fully given by the relevant axioms. But when one shifts to moments of time, or to spatial locations, or to space-time points—that is, to things in the actual world—why should one carry over this assumption? For one thing, the idea that there can be actual entities that stand in relations to other things, but that have *no intrinsic properties themselves*, seems very problematic. One might try to argue, of course, that points do have an intrinsic nature—namely, that of having no extension. But this rather looks like a negative property, and so if one holds, as many metaphysicians do, that there are no negative universals, not having any extension is not going to count as an intrinsic property.

In addition, there is good reason to suppose that many things—including very small things—have natures that, as of yet, fall outside our ken. If so, what reason is there for holding that this is not likely to be true in the case of, for example, space-time points?

In short, I cannot see that there is any real reason—let alone a compelling one—to accept the homogeneity postulate. On the contrary, it seems to me that if one is considering the possibility of an intrinsic temporal metric, it is very natural to reject the homogeneity postulate. For if there is be a temporal unit, it seems plausible that this requires the existence of processes that involve change, and so if there is to be an intrinsic metric that is embedded in space-time itself, it would seem that one must hold that different times themselves have different intrinsic properties.

Given this line of thought, one picture that naturally suggests itself is, in brief, as follows. Suppose, first, that properties C and D are intrinsic properties of space-time points; second, that all space-time points have either property C or property D, but not both; third, that if two space-time points x and y are simultaneous, then either both have property C, or both have property D; fourth, that spatial locations change from having property C to having property D, then go back to having property C, and so on; fifth, that there is a basic quantitative relation of equality with respect to temporal length, and that locations always have property C for the same length of time, and similarly for property D. Then the idea is that by postulating this *one* type of law-governed, causal sequence—that is, the C, D, C, D, etc., sequence—one avoids the problem that the D-type events that are caused by simultaneous C-type events might not occur at the same time. Moreover, given that the causal law in question concerns properties of space-time points themselves, there is no possibility, given such a space-time, for there not to be such a causal sequence present in the world, regardless of the initial conditions involving physical objects and events in space-time. In addition, such sequences will be present at every time and every place, and so will be occurring wherever any other type of causal process is occurring.

The conclusion, accordingly, is that, given these facts, there is no problem concerning the unit of temporal length that plays a role in all causal laws, since it can be defined, for example, as the temporal distance from the acquisition by a location of property C to the time that that location next acquires property D. This unit, together with the relation of equality with respect to temporal length, will then serve to define all possible lengths of temporal intervals.

To sum up this section: Though the situation is, as we have seen, more complicated in the case of time than in the case of space, the idea of an extrinsic temporal metric also appears deeply problematic. By contrast, there does not seem to be any objection to an intrinsic temporal metric, and it seems that an especially satisfying type of intrinsic metric can be arrived at by rejecting the homogeneity postulate.

5. GRÜNBAUM'S CLAIM THAT TENSED FACTS ARE MIND-DEPENDENT

5.1 Grünbaum's rejection of mind-independent tensed facts

In his paper "The Meaning of Time," Grünbaum criticizes, generally in a very effective and persuasive fashion, a number of arguments that have been offered in support of dynamic or tensed views of the nature of time—such as arguments that appeal to causal indeterminism. By contrast, however, both Grünbaum's rejection of the view that there are *mind-independent* tensed facts, and his postulation of the existence of *mind-dependent* tensed facts, seem much less satisfactory, since in neither case does he really seem to offer a detailed and carefully developed argument.

Thus, as regards the former, what Grünbaum says is as follows:

Clearly, an account of becoming which provides answers to these questions is *not* an analysis of what the common-sense man actually *means* when he says that a physical event belongs to the present, past, or future; instead, such an account sets forth how these ascriptions ought to be construed within the framework of a theory which would supplant the scientifically untutored view of common sense. That the common-sense view is indeed scientifically untutored is evident from the fact that *at a time t*, both of the following physical events qualify as occurring "now" or "belonging to the present" according to that view: (i) a stellar explosion that occurred several million years before time *t* but which is first seen on earth at time *t*, (ii) a lightning flash originating only a fraction of a second before *t* and observed at time *t*. If it be objected that present-day common-sense beliefs have *begun* to allow for the finitude of the speed of light, then I reply that they err at least to the extent of associating absolute simultaneity with the now.[8]

As regards Grünbaum's rejection of mind-independent tensed facts, then, the argument appears to be contained in the final sentence, and it can, I think, be expanded as follows:

(1) In view of the special theory of relativity, it is reasonable to hold that the world does not contain a relation of absolute simultaneity.

(2) There cannot be mind-independent tensed facts unless there is a mind-independent now.

(3) There cannot be a mind-independent now unless events stand in the relation of absolute simultaneity.

(4) Therefore, it is reasonable to conclude that there are no mind-independent tensed facts.

I am not sure that this argument is Grünbaum's only reason for rejecting the idea of mind-independent tensed facts. But if it is, then that rejection is not on firm ground. For, in the first place, there are views that postulate mind-independent tensed facts but that reject the claim that there cannot be a mind-independent now unless events stand in the relation of absolute simultaneity. Here I am thinking, for example, of the tensed views of time advanced by Howard Stein in his article "On Einstein-Minkowski Space-Time,"[9] and by Storrs McCall in his book *A Model of the Universe*.[10]

In the second place, the view that the special theory of relativity provides good grounds for concluding that no events in our world stand in the relation of absolute simultaneity is by no means unproblematic, for at least two reasons. First of all, it can be argued that there is another, very well-confirmed physical theory—namely, quantum mechanics—that provides support for the claim that events in our world do stand in relations of absolute simultaneity. Thus one can appeal, on the one hand, to the very general, and absolutely central idea, of the collapse of a wave packet—an idea that seems very clearly to involve the notion of absolute simultaneity. On the other hand, this line of argument can also be reinforced by a consideration of Bell's Theorem, and of the experiments that showed, contrary to the thrust of the Einstein-Podolsky-Rosen thought experiment, that certain correlations cannot be explained in terms of local hidden variables. (I have set out both of these lines of argument, and especially the second, in a detailed way in *Time, Tense, and Causation*, pages 357–62.)[11]

Second, it is also possible to construct alternatives to the special theory of relativity that do entail that events in our world stand in relations of absolute simultaneity. If done in an ad hoc way, that would not, of course, be of great interest. But what I have argued elsewhere is, first, that if one postulates an absolute space-time, one is postulating something whose existence at future times is extremely unlikely unless there is some causal law relating the existence of later space-time points to earlier ones, and, second, that the simplest and most plausible conservation principle for space-time is one that, because it involves *non-branching* causal chains connecting space-time points, entails the existence of privileged frames of reference, and, thereby, the existence of a relation of absolute simultaneity that holds between events in the world. (This argument is developed at length in *Time, Tense, and Causation*, pages 342–46.)

5.2 Grünbaum's proposal concerning tensed sentences: the existence of mind-dependent tensed facts

Grünbaum, in "The Meaning of Time," proposes a certain account of the meaning of tensed sentences. As we have just seen, however, he emphasizes that what he is offering is not an analysis of tensed sentences as ordinarily used. For, as we have just seen, he views the ordinary understanding of tensed sentences as involving a mistaken tensed view—namely, one that entails that physical events stand in relations of absolute simultaneity. What Grünbaum is doing, accordingly, is putting forward a different way of interpreting tensed sentences—an interpretation that, on the one hand, will enable one to use tensed sentences in just the contexts that one normally uses them, but that, on the other hand, will not incorporate any false claims about the world.

The interpretation that Grünbaum proposes reflects his view that becoming, rather than being a feature of the world outside of consciousness, is, like the secondary qualities—such as color properties—a *mind-dependent* feature of things. Thus he says, in "The Meaning of Time":

> It is apparent that the becoming of physical events in our temporal awareness does not itself guarantee that becoming has a mind-independent physical status. Commonsense color attributes, for example, surely *appear* to be properties of physical objects independently of our awareness of them and are held to be such by common sense. And yet scientific theory tells us that they are mind-dependent qualities like sweet and sour are. Of course, if physical theory claims that, contrary to common sense, becoming is not a feature of the temporal order of physical events with respect to earlier and later, then a more comprehensive scientific and philosophical theory must take suitable cognizance of becoming as a conspicuous characteristic of our *temporal awareness* of both physical and mental events.[12]

But what exactly is involved in temporal awareness, and what is the "becoming" that Grünbaum says is a "conspicuous characteristic of our temporal awareness"? Part of the answer emerges in the following passage:

> What qualifies a physical event at a time t as belonging to the present or as now is *not* some physical attribute of the event or some relation it sustains to other *purely physical* events. Instead what is *necessary*

so to qualify the event is that at the time t at least one human or other *mind-possessing* organism M is conceptually aware of experiencing at that time either the event itself or another event simultaneous with it in M's reference frame.[13]

This passage brings out two important aspects of Grünbaum's approach. First, in a world without any experiences, nothing would possess the property of presentness. So tensed properties are mind-dependent. But second, mere experience is not enough: there must also be an appropriate *conceptual awareness*.

What is involved in the latter? Here Grünbaum's answer is as follows:

What then is the content of M's conceptual awareness at time t that he *is experiencing* a certain event *at that time*? M's experience of the event at time t is coupled with an awareness of the temporal coincidence of his experience of the event with a state of *knowing* that he has that experience at all. In other words, M experiences the event at t *and* knows that he is experiencing it. Thus, presentness or nowness of an event requires conceptual awareness of the presentational immediacy of either the experience of the event or, if the event is itself *unperceived*, of the *experience* of another event simultaneous with it.[14]

So far, then, the account that Grünbaum is offering of what it is for an event E to lie in the present appears to be as follows:

Event E is present if and only if there is some mind M and some event F such that, first, F is either identical with E or simultaneous with E; secondly, M is in a first-order mental state of experiencing F; and, thirdly, M is in a second-order mental state of being aware, or knowing, that he is experiencing F.

One crucial question remains, however, and that is whether Grünbaum, in setting out this analysis, is offering a tenseless account of tensed sentences. The answer is that this is not what Grünbaum is doing:

My characterization of *present* happening or occurring *now* is
intended to *deny* that belonging to the present is a physical attribute
of a physical event E which is *independent* of any *judgmental awareness* of the occurrence of either E itself or of another event simultaneous with it. But I am *not* offering any kind of *definition* of the
adverbial attribute now, which belongs to the conceptual framework
of tensed discourse, solely in terms of attributes and relations drawn
from the tenseless (Minkowskian) framework of temporal discourse
familiar from physics. In particular, I avowedly invoked the present
tense when I made the nowness of an event E at time t dependent on
someone's knowing at t that he *is experiencing E*. And this is tantamount to someone's judging at t: I am experiencing E *now*. But this
formulation is *non*viciously circular. For it serves to articulate the
mind-dependence of nowness, *not* to claim erroneously that nowness
has been eliminated by explicit definition in favor of tenseless temporal attributes or relations.[15]

The upshot is that Grünbaum is offering a very distinctive and unusual
view of tensed facts. On the one hand, he agrees with advocates of
tenseless approaches to time that the external, physical world itself
involves no tensed properties: there is nothing beyond the properties
and relations that enter into the fundamental theories in physics. On
the other hand, however, he agrees with advocates of tensed
approaches to time that reality does contain tensed properties. But he
holds that these are only to be found in certain sorts of mental states.

If we make all of this explicit, the analysis being proposed by
Grünbaum can, I think, be put as follows:

Event E is present if and only if there is some mind M and some
event F such that, first, F is either identical with E or simultaneous
with E; secondly, M is in a first-order mental state of having an experience of F, where this experience involves the tensed property of
presentness; and, thirdly, M is in a second-order mental state of
being aware, or knowing, that he is having such an experience.

5.3 An evaluation of Grünbaum's proposal

In evaluating Grünbaum's proposal, it will be useful to separate the two most crucial elements: first, the claim that there are mind-dependent tensed facts; second, the claim that those mind-dependent tensed facts involve the existence of second-order mental states that incorporate a conceptual awareness of a first-order mental state.

5.3.1 The first claim: tensed facts are mind-dependent

There is an important division within tensed approaches to time that is helpful, I think, in considering this first element of Grünbaum's proposal. On the one hand, some tensed views postulate a special, *intrinsic* property of presentness that every event has, but only for a moment. Sometimes intrinsic properties of pastness and futurity are also postulated, but that is not crucial in the present context; what matters is that presentness is viewed as an intrinsic property of events.

By contrast, some other tensed views are formulated instead in terms of the idea that past events, present events, and future events are not all equally real. So, for example, it has often been claimed that while past and present events are real, future events are not. Given such a view, there is no need to postulate any intrinsic tensed properties: one can say, for example, that an event is present at a given time if it is actual as of that time, and there are no later events that are actual as of that time. Presentness is then not an intrinsic property of events, but a *relational* property.

How does Grünbaum regard presentness? I am not sure that Grünbaum's discussion in "The Meaning of Time" enables one to advance a confident answer to this question. Consider, however, a sentence quoted above:

> Thus, presentness or nowness of an event requires conceptual awareness of the presentational immediacy of either the experience of the event or, if the event is itself *un*perceived, of the *experience* of another event simultaneous with it.[16]

What is crucial here, I suggest, is the phrase "presentational imme-diacy." How is this phrase to be interpreted? One suggestion would be that to characterize an experience as involving presentational imme-diacy is simply to say that the experience is a state of consciousness. But that interpretation will not generate a tensed view, since someone who holds a *tenseless* view of time, and who is, say, a property dualist, will certainly want to say that experiences are states of consciousness, and that one is often conceptually aware of enjoying such states of consciousness.

If one is to have mind-dependent tensed facts, presentational immediacy must involve some special property, and one that is present when one is having the experience, but not present when the experi-ence lies in the past. Suppose, then, that the special property in ques-tion is an *intrinsic* property of experiences. Such an intrinsic property will not, of course, on Grünbaum's view, be identical with presentness, since the latter will be identical instead with the property of being a second-order experience in which one is conceptually aware of the presentational immediacy of a first-order experience. But that does not matter. The crucial point is simply that, on this interpretation, *part* of what is involved in Grünbaum's account of presentness is an intrinsic property that experiences have—but only for a moment.

If this is Grünbaum's view, the question arises as to how an experi-ence can have an intrinsic property at one time, and then not have it at later times. In the case of persisting entities, of course, this sort of ques-tion does not pose any problem, since there is at least the perdurantist answer—namely, that a persisting object may, for example, be red at one time and not red at some other time because an object's being red at a time is a matter of the relevant *temporal part* of it at that time being red. So, for a persisting object to be red at one time and not at another is simply for there to be two distinct temporal parts of the object that exist at the relevant times, one of which is red, and the other not.

But such an explanation is not available in the case of experiences: if an experience's changing from being present to being past involves the loss of an intrinsic property, this cannot be analyzed in terms of the experience's having two distinct temporal parts, one of which has the intrinsic property, and the other of which lacks it.

In short, the problem is that there is an objection to any view that identifies presentness with some *intrinsic* property, and which also applies to Grünbaum's view if the presentational immediacy that is part of Grünbaum's account involves an experience's having any intrinsic property that it has only for a moment. (I have developed this objection in a more detailed way in *Time, Tense, and Causation*, pages 228–30.)

Let us consider, then, the other main alternative with respect to tensed accounts of the nature of presentness—the view, namely, that the present is the point at which events come into existence, so that presentness, rather than being an intrinsic property of events, is a relational property: an event possesses the property of presentness at a time t if and only if the event is among the things that are actual as of time t, but not among the things that are actual as of any time earlier than t.

Do we get a satisfactory account, then, if an experience's possessing presentational immediacy is construed in this way? It does not seem to me that we do, for at least two reasons. In the first place, consider some experience, E, that occurs at time t_1. Unless an epiphenomenalist view of experiences is true, E may very well be a cause of some physical event, P, that occurs at a later time t_2. Assume that this is so, and let t_0 be any time earlier than t_1. If presentational immediacy is construed in terms of an experience's coming into existence at the time of its occurrence, then experience E is not among the things that are actual as of time t_0. But what about physical event P? If it is not among the things that are actual as of time t_0, then there must be some later time—presumably t_2—at which P becomes actual. But then one has *mind-independent* tensed facts. Accordingly, if Grünbaum's view is to be right, it must be that physical event P is actual as of time t_0. But now one has an event—namely, P—that is actual as of a time when a cause of that event—namely, E—is not actual. But how can this be? How can an effect be real as of a time when its cause is not real?

Second, let Q be any physical event that is simultaneous with E. Is it not plausible that if two events occur at the same time, then any time as of which one is actual must also be a time as of which the other is actual? If so, then Q cannot be actual as of any time as of which E is not actual. On the present view, however, E is not actual as of any time prior to the time when it occurs, and so physical event Q will not be actual as of any time prior to the time when E occurs, and so comes into existence. But then there are mind-independent tensed facts concerning physical event Q: it, too, comes into existence at a time, and it, too, is not part of the totality that is actual as of any earlier time.

To sum up: If, as Grünbaum holds, presentational immediacy is not to be reduced to tenseless facts, the question arises as to *what sort* of tensed facts are involved, and here there seem to be essentially two possibilities. According to the first, there is some intrinsic tensed property that an experience has only at the moment when it occurs. This gives rise to the problem of how an experience can have an intrinsic property at one moment, and yet lack it at other moments.

According to the other alternative, the tensed fact is a matter of an experience coming into existence, of there being a first moment as of which it is actual. Here the problem is that it may very well be that an experience is simultaneous with some physical events, and causally prior to others, and when either of these things is the case, there does not seem to be any satisfactory account of when the physical events in question are actual, *unless* one holds that not only experiences, but also physical events, come into being. But then one is embracing mind-independent tensed facts.

5.3.2 The second element: conceptual awareness of an experience

One way in which Grünbaum diverges from both standard tensed and tenseless approaches to time is in proposing an interpretation of tensed sentences according to which part of the state of affairs that serves to make it true that a certain experience lies in the present involves the existence of a certain sort of *conceptualized* awareness possessed by the person who is having the experience in question.

In evaluating any proposal concerning the meaning of tensed sentences, it is important to keep in mind that any account, if it is to be acceptable, must deal not just with the use of tensed sentences on their own, but also with the use of such sentences when they are embedded within various contexts. So, for example, a satisfactory account must work in the case of tensed sentences embedded within *propositional attitude* contexts. But Grünbaum's proposal, as I shall now argue, seems to face a serious problem in this regard.

Consider, for example, the following sentence, in which the tensed proposition that I am *now* in state *S* is the object of a certain propositional attitude:

"I wish that I were *now* in state *S*."

What is it that I am wishing for, according to Grünbaum's account of present-tense sentences, when I wish that I were now in state *S*? According to Grünbaum's account, in order for me *now* to be in state *S*, my being in that state would have to be identical with, or simultaneous with, some experience that I was now having, and of which I was conceptually aware.

But suppose, now, that state *S* is the state of being completely unconscious, so that we have the following sentence:

"I wish that I were now completely unconscious."

What is it that I am wishing for in this case, on Grünbaum's account? The answer is that, given his proposal concerning the meaning to be assigned to present-tense sentences, what I am wishing for is that it were now true that I was having some experience *E*, of which I was conceptually aware, and that was simultaneous with the state of my being completely unconscious. This, however, involves a contradiction.

The problem, in short, is that there are tensed propositions—such as the proposition that I am now completely unconscious—that, on the one hand, although they cannot be truthfully affirmed or thought, can perfectly well be coherent objects of certain propositional attitudes, but that, on the other hand, could not be coherent objects of those propositional attitudes if Grünbaum's account were correct.

6. SUMMING UP

I have focused upon two aspects of Adolf Grünbaum's discussion of space and time. First of all, I considered Grünbaum's rejection of intrinsic metrics in favor of extrinsic metrics, and there I argued, first, that the idea of an extrinsic metric is open to strong objections, both in the case of space and in the case of time; second, that there is no serious objection to the idea of an intrinsic metric, either in the case of space or in the case of time; and, third, that there are good reasons for postulating an intrinsic metric, both in the case of space and in the case of time.

I then considered Grünbaum's approach to tensed sentences, and there I argued, first, that Grünbaum has not provided us with a good reason to abandon tensed accounts of the nature of time that incorporate the idea that there are mind-independent tensed facts; second, that there does not seem to be any reason to accept the claim that tensed facts are mind-dependent; and, third, that the idea that conceptualized awareness is an essential constituent of mind-dependent tensed facts has the unwelcome consequence that some perfectly coherent objects of propositional attitudes turn out to involve a contradiction, given Grünbaum's proposal concerning the meaning of tensed sentences.

NOTES

1. Sir I. Newton, *Philosophiae Naturalis Principia Mathematica* (London: 1687). Quotations are from the 1934 translation by F. Cajori, *Sir Isaac Newton's* Mathematical Principles of Natural Philosophy and His System of the World (Berkeley: University of California Press, 1934), see pp. 6–7.

2. A. Grünbaum, *Philosophical Problems of Space and Time*, 2nd ed. (Dordrecht: Reidel Publishing, 1973), p. 5.

3. Ibid., p. 5.

4. Ibid., pp. 9–10.

5. Ibid., p. 16.

6. Ibid., p. 498.

7. Ibid., p. 499.

8. A. Grünbaum, "The Meaning of Time," in *Basic Issues in the Philosophy of Time*, edited by Eugene Freeman and Wilfrid Sellars (LaSalle, IL: Open Court, 1971), see p. 196.

9. H. Stein, "On Einstein-Minkowski Space-Time," *Journal of Philosophy* 65 (1968): 5–23.

10. S. McCall, *A Model of the Universe* (Oxford: Oxford University Press, 1994).

11. M. Tooley, *Time, Tense, and Causation* (Oxford: Oxford University Press, 1967).

12. Grünbaum, "The Meaning of Time," p. 196.

13. Ibid., pp. 206–207.

14. Ibid., p. 207.

15. Ibid., p. 209.

16. Ibid., p. 207.

4.

GENERAL COVARIANCE AND EINSTEIN'S "HOLE" ARGUMENT

Or, Once More into the Breach, Dear Friends

STEVEN HUMPHREY

I n the years before 1913, Einstein was convinced that the field equations of the general theory of relativity (GTR) he was seeking must be what is known as "generally covariant." But between 1913 and 1915, he was diverted from the principle of general covariance by an argument he called the "hole argument." After decades of neglect, philosophers and physicists began to examine this period in the development of the GTR, led by the research efforts of John Stachel and John Norton, among others. For the philosophers and historians of science, this episode represents an opportunity to investigate how Einstein came to his final version of his theory of gravitation. Some, like Earman, Norton, Cartwright, and others, have seen the hole argument as relevant to the questions of space-time substantivalism and determinism. Physicists, on the other hand, find this issue relevant to the search for a quantum theory of gravity, that is, a theory uniting the two major successes of twentieth-century physics. The principle of general covariance is seen as a significant stumbling block in the development of any theory unifying GTR and quantum field theory. As Julian Barbour says, "Many people attribute the difficulties inherent in the quantization of gravity to the general covariance of Einstein's general theory of relativity, so clarification of its true nature is important."[1] In this paper, I will explore some of these issues and end by arguing that the hole argument has been frequently and importantly misinterpreted, even by Einstein, and suggest that general covariance has implications that have been neglected or ignored.

There has been much dispute and disagreement as to the precise meaning of the principle of general covariance.[2] In searching for the field equations for the general theory of relativity, Einstein, at least until 1913, believed that those equations should be generally covariant, in the sense that they should hold in any arbitrary reference frame. Physically, his intuition amounts to this. "All our space-time verifications invariably amount to a determination of space-time coincidences. If, for example, events consisted merely in the motions of material points, then ultimately nothing would be observable but the meetings of two or more of these points. Moreover, the results of our measurings are nothing but verifications of such meetings of the material points of our measuring instruments with other material points, coincidences between the hands of a clock and points on the clock dial, and observed point-events happening at the same place at the same time. The introduction of a system of reference serves no other purpose than to facilitate the description of the totality of such coincidences."[3] And again, "The laws of nature are merely statements about space-time coincidences; they therefore find their only natural expression in generally covariant equations."[4] I take his point as being that physics is the study of the behavior and interactions of material bodies and fields. The framework from within which one observes, or measures, such interactions should be irrelevant to the dynamics of those interactions. Thus, the equations which govern that behavior should hold equally well in any reference frame or coordinate system. The principle of general covariance, as Einstein saw it, could be seen as an adequacy requirement placed on the formulation of the field equations of general relativity (as Kretschmann argued)[5] and would thus have no physical significance. The debate over the physical significance of the requirement of general covariance has been well documented.[6]

The principle is also understood as pertaining to any model of the field equations. Given any relativistic space-time consisting of a manifold together with some set of geometrical objects, in particular the stress-energy tensor field and metric tensor field, specified relative to some coordinate framework, the result of applying an arbitrary coordinate transformation to that model would also be a model of the theory. That is, to say that the field equations are generally covariant is to say that any space-time related by any arbitrary coordinate transformation to some solution of those equations will also be a solution. Nowadays, these arbitrary coordinate transformations are referred to as *passive diffeomorphisms*. A passive diffeomorphism is any one-to-one mapping of one coordinate system onto another that is both continuous and smooth, and whose inverse is both continuous and smooth. Thus, any space-time that is related diffeomorphically to some relativisitic space-time is itself a model of the field equations. We call an equivalence class of such space-times a class of *diffeomorphic spaces*.

Einstein saw the principle of general covariance as an obvious generalization of the principle of relativity that formed the basis of the special theory of relativity. Recall that principle stated the laws of physics should hold in every inertial reference frame, that is, the laws are invariant under Lorentz coordinate transformations. The covariance group for the special theory is the set of Lorentz transformations. Based upon the insight of the equivalence of inertial and gravitational mass, he thought he should be able to generalize the principle of relativity beyond inertial reference frames, to include *all* reference frames. That is, the laws should be invariant under *any* arbitrary coordinate transformation. The covariance group for the general theory is thus the set of diffeomorphisms.[7] There have been numerous familiar arguments presented to show that the principle of general covariance is not a generalization of the principle of relativity.[8]

For Einstein, however, based on his physical intuitions about gravitation and the dynamics of the interactions of material bodies, general covariance seemed like a necessary requirement for any adequate physical theory. But sometime in 1913 he developed his "hole argument," which caused him to abandon the goal of finding generally covariant field equations for almost two years.

Here is the "hole argument," as Einstein presented it in 1914.

We consider a finite portion Σ of the continuum, in which no material process occurs. What happens physically in Σ is then completely determined if the quantities $g_{\mu\nu}$ are given as functions of the coordinates x_ν with respect to a coordinate system K used for the description. The totality of these functions will be symbolically designated by $G(x)$.

Let a new coordinate system K' be introduced that, outside of Σ, coincides with K, within Σ, however, diverges from K in such a way that the $g'_{\mu\nu}$ referred to K', like the $g_{\mu\nu}$ (and their derivatives), are everywhere continuous. The totality of the $g'_{\mu\nu}$ will be symbolically designated by $G'(x')$. $G'(x')$ and $G(x)$ represent the same gravitational field. If we replace the coordinates x'_ν by the coordinates x_ν in the functions $g'_{\mu\nu}$, i.e., if we form $G'(x)$, then $G'(x)$ also describes a gravitational field with respect to K, which however does not correspond to the actual (i.e., originally given) gravitational field.

Now if we assume that the differential equations of the gravitational field are generally covariant, then they are satisfied by $G'(x')$ (with respect to K') if they are satisfied by $G(x)$ with respect to K. They are then satisfied with respect to K by $G'(x)$. Relative to K there then exist the solutions $G(x)$ and $G'(x)$, which are different from each other, in spite of the fact that at the boundary of the region both solutions coincide; i.e., *what happens in the gravitational field cannot be uniquely determined by generally covariant differential equations for the gravitational field.*

Therefore, if we demand that the course of events in the gravitational field be completely determined by the laws to be set up, then we are forced so to restrict the choice of coordinate systems that it is impossible, without abandonment of the restrictive conditions, to introduce a new coordinate system K' of the type previously characterized. The continuation of the coordinate system into the interior of a region Σ must not be arbitrary.[9]

Eventually, of course, Einstein realized that the hole argument did not imply that general covariance entailed a pernicious indeterminism, and resumed his search for generally covariant field equations, the final version of which appear in his famous paper of 1916. When asked, years later, why it took so long to come to this conclusion, he replied, "The main reason lies in the fact that it is not so easy to free oneself from the idea that co-ordinates must have an immediate metrical significance."[10] And in a letter written to Paul Ehrenfest at the end of 1915, he says, "The apparent compulsion of [the hole] argument disappears at once if one considers that 1) the reference system signifies nothing real, 2) that the (simultaneous) realization of two different g-systems (better said, of two different grav[itational] fields) in the same region of the continuum is impossible by the nature of the theory. . . . The physically real in what happens in the world (as opposed to what depends on the choice of the reference system) consists of *spatio-temporal coincidences*."[11] Thus, for Einstein at least, it would seem that the issue hinges on the physical significance (or lack thereof) of reference frames defined on the manifold, and that the mistake he made was in failing to realize that a coordinate transformation results only in the *renaming* of the points on the manifold.

John Stachel has argued that, even though Einstein worked exclusively in language of coordinate-dependent differential geometry and expressed the principle of general covariance in terms of coordinate transformations, his geometrical intuition was deep enough to allow us to infer that the point he was making somewhat more profound. Thus, Stachel rejects the view that Einstein failed to realize that "the transformation of the components of the metric tensor under a coordinate transformation results in a redescription of the *same* gravitational field in a *different* coordinate system."[12] And, in fact, Einstein explicitly states that "$G'(x')$ and $G(x)$ represent the same gravitational field." But then where does $G'(x)$ come from? Einstein says that it comes from replacing the coordinates x'_ν by the coordinates x_ν in the functions $g'_{\mu\nu}$, i.e., forming $G'(x)$. In the modern language of abstract, coordinate-independent differential geometry, it is difficult to translate Einstein's

292   General Covariance and Einstein's "Hole" Argument

claim into something coherent. At some point $\mathscr{P}$ on the manifold there is a metric tensor $g(\mathscr{P})$. The components of this tensor will of course be different in different coordinate systems. Let '$g_{\mu\nu}(\mathscr{P})$' represent the components of g in coordinate system x_ν, and let '$g'_{\mu\nu}(\mathscr{P})$' represent the components of g in coordinate system x'_ν. The functions $g_{\mu\nu}(\mathscr{P})$ and $g'_{\mu\nu}(\mathscr{P})$ are related by a passive diffeomorphism, or what Einstein calls an "arbitrary coordinate transformation." Seen in this way, there is no room at $\mathscr{P}$ for a third entity, $G'(x)$. There is a single metric tensor field, and two coordinate systems. For Einstein, however, the functions $g_{\mu\nu}$ and $g'_{\mu\nu}$ are functions of coordinate system values, rather than points on the manifold. That is, each of the sixteen components of $g'_{\mu\nu}$ is going to be a function of some coordinate system x'_ν. Now, replace all reference to the coordinates of x'_ν with coordinates x_ν. For example, if the line element of $g'_{\mu\nu}$ relative to x'_ν is given by

(1) $ds'^2 = -x'dt'^2 + t'dx'^2 + dy'^2 + dz'^2$

then the result of applying the coordinate transformation D to this yields

(2) $ds^2 = D(-x'dt'^2 + t'dx'^2 + dy'^2 + dz'^2) = -D(x')dt^2 + D(t')dx^2 + dy^2 + dz^2$.

But if we do as Einstein suggests, and replace all of the x'_ν by $x_\nu = D(x'_\nu)$, we get

(3) $ds'^2 = -D(x')dt'^2 + D(t')dx'^2 + dy'^2 + dz'^2$.

This is certainly different from (1) and (2). Einstein claims that, whereas (1) and (2) represent the same gravitational field, (3) does not. If $x' \neq D(x')$, the values of the line elements will be different. That is, the space-time interval between two events will be the same, according to (1) and (2), but different according to (3). But (3) does not result from a *coordinate* transformation on (1). We might say that we have moved the tensor $g'_{\mu\nu}$ from (t', x') to $(D(t'), D(x'))$. That is, we have created the new field $G'(x)$ by transforming (i.e., moving) $G'(x')$ to a new point on the manifold. What Einstein has done in creating $G'(x)$ is performed a point transformation on the manifold. Stachel argues that what Einstein was asserting was that in a region where nongravitational forces are absent, the *dragged-along* metric tensor field under a point transformation results in a physically distinct gravitational field.

What does this mean? First, we have to distinguish passive from *active* diffeomorphisms. As described above, a passive diffeomorphism is a coordinate transformation, which maps the coordinate values of one coordinate system onto another. Insofar as a coordinate system itself is a mapping from the points of the manifold to (in this case) R^4, the operation of a diffeomorphism upon a coordinate system simply changes each $x_\nu(\mathcal{P})$ to $x'_\nu(\mathcal{P})$, at some point $\mathcal{P}$. That is, a passive diffeomorphism *reinterprets* the geometrical objects on the manifold, describing them first in the language of one coordinate system, and then in the language of another. The numerical and functional values of the components of these geometrical objects, relative to the original coordinate system, will be changed by the transformation.

What about an *active diffeomorphism*? What are its effects? An active diffeomorphism is an actual *point* transformation in which the points of the manifold themselves are "moved." The effect of an active diffeomorphism would be to map each point $\mathcal{P}$ in some region of the manifold onto some (not necessarily different) point $\mathcal{Q}$. Over the years, the interpretation of the principle of general covariance has shifted from a statement about coordinate transformations to one about point transformations. A modern version of the principle would read ". . . if a tensor field $\mathcal{X}$ on the manifold M is a solution of the set of field equations, then the pushed-forward tensor field $\varphi_*\mathcal{X}$ of $\mathcal{X}$ is also a solution of the same set of equations for any *active* diffeomorphism $\varphi : \mathcal{M} \rightarrow \mathcal{M}$ of the manifold onto itself."[13] Or "Let M be a 4-D differentiable manifold, g be a metric tensor defined on $\mathcal{M}$, and $\mathcal{T}$ be the stress-energy tensor defined on $\mathcal{M}$, giving the material contents of space-time. The given a model $<\mathcal{M}, g, \mathcal{T}>$ of GTR, the space-time model generated by applying an arbitrary (smooth, 'one-to-one') transformation (diffeomorphism) to the points of $\mathcal{M}$ is also a model of GTR."[14]

There seems to be some question about exactly what changes under an active diffeomorphism. In the case of passive diffeomorphisms, the case is clear: Only the names of the points on the manifold change, which then changes the component values of the geometrical objects at those points. But in the case of active diffeomorphisms, for example, Stachel says, "If there are two geometrical object fields on a manifold, one may be dragged along with a point transformation, while the other is left undragged, or left behind, as we shall say. Then we may refer to one field as dragged along relative to the other."[15] Later he says, "Suppose a set of geometrical object fields in some region of the manifold completely describes the physical situation in that region. Now drag along *all* of these fields with some point transformation restricted to this region."[16] Well, which is it? We might say that in a purely mathematical sense, it is possible to define a diffeomorphism that will do just about anything you like. But what kind of diffeomorphism is intended by the principle of general covariance, interpreted as a principle governing the formulation of a physical theory? That is, which geometrical objects are pushed forward or dragged along by the active diffeomorphism on a manifold intended to represent some region of physical space-time? All of them? Some of them? Or do we get to choose? It seems there are three alternatives.

1. A diffeomorphism transforms only the points, leaving the objects unaffected (equivalently, it moves all of the objects, leaving the points unaffected). That is, it disconnects the geometrical objects from the points at which they are defined. If this is possible, then the symmetry group of a relativistic space-time becomes simply the set of active diffeomorphisms, i.e., the symmetry group of relativistic space-time becomes identical to the covariance group of GTR. But this clearly violates the spirit

of general relativity, in that space-time itself would be non-dynamical. It is the metric tensor field and its derivatives that define the "shape" of space-time. If we are allowed to separate the metric from the manifold, then we have created a background space, a prior geometry with a Minkowski metric, which itself is nondynamical but on which various "physical" metrics may be defined. Few would regard this as compatible with GTR and general covariance. To paraphrase Stachel, space-time and the gravitational field in the GTR are like love and marriage in the old song. Mathematically, we would say that the same holds for the manifold and the metric tensor field.

2. The active diffeomorphism of general covariance transforms some of the objects, relative to the others. There doesn't seem to be any principled way of deciding which objects to move and which to leave alone. And which sets of geometrical objects might go along for the ride without violating the field equations? The field equations of general relativity would not be generally covariant, since we would be able to move the metric tensor field of a relativistic space-time, while leaving the stress-energy tensor field unaffected, in which case the field equations would hold in one space and be violated in another, diffeomorphically related to the first.

3. An active diffeomorphism transforms the points of the manifold, and all of the geometrical objects defined at those points, including the coordinate systems. This seems most consistent with Einstein's creation of $G'(x)$, above, and the only option consistent with the spirit of general covariance.

Now, what does the hole argument look like when described in terms of active diffeomorphisms? Let's look at the space-times described in the hole argument. Both contain nonzero mass-energy distributions surrounding regions containing no nongravitational fields. Outside the hole, the spaces are identical, in that one results from an identity transformation performed on the points of the other.

Inside the hole, it is claimed, the spaces might differ as much as you like. The diffeomorphism will come to smoothly differ within the hole. The problem that Einstein thought he recognized was that this would imply a form of indeterminism, in that the boundary conditions would not uniquely determine the characteristics of the gravitational field inside the hole. Stachel has argued that there is no physical way of distinguishing these spaces, that they are purely mathematically different spaces, all of which are physically indistinguishable. Hawking and Ellis seem to be adopting some version of this view when they describe a model for the field equations of general relativity as an equivalence class of space-times, related by active diffeomorophisms. I would like to suggest something stronger, namely, that there is also no mathematical way of distinguishing them. In particular, I will argue that no two diffeomorphic space-times, satisfying the field equations, will be mathematically distinct, and that the assumption that they are is based upon a fairly simple misconception.

Consider two putatively distinct relativistic space-times that embody the situation described in the hole argument. They are identical outside the hole but different inside the hole, though related by a diffeomorphism. But in what way might they be different inside the hole? Insofar as the geometric objects residing at each point inside the hole are "dragged-along" with the points by the point transformation, and since all of the topological and geometrical properties of the space are determined by those geometric objects, and since those properties exhaust the mathematical properties of the spaces, there can be no difference between the spaces.

Consider, for example, a relativistic space-time containing the sort of hole described above. Let S be a curve on that space-time, traversing the hole. Without loss of generality, let the curve be a geodesic of the space-time. Interpreting the principle of general covariance in terms of active diffeomorphisms, there is another space-time, identical with the first except that the set of manifold points that constitute S are mapped onto a different set of points, which constitute the curve S'. That is, the diffeomorphism changes the path of the curve. Inasmuch as geodesic curves are intended to represent the trajectories of freely falling test bodies in physical space-time, the fact that both of these models are solutions of the field equations and share all boundary conditions would seem to imply an essential indeterminism as to the trajectory of a test body through the hole. There are an infinite number of answers to the question, what trajectory would a test body follow through the hole, all of which are consistent with the field equations and the boundary conditions. It is examples like this that I suspect many writers on this subject have in mind when they discuss the indeterminacy that seems to accompany general covariance. But in what ways are S and S' thought to be distinct? If S is a geodesic, could the diffeomorphism map S onto a curve, S', which was not? No. The equation for a geodesic on a semi-Riemannian manifold is

$$\nabla_u u = 0$$

where ∇ is the covariant derivative operator and u is the tangent vector to the curve. A geodesic is a curve that parallel transports its tangent vector along itself. Now, remember that a curve is a set of points $\mathscr{P}(\lambda)$, where λ is the "affine parameter" of the curve, and a tangent vector at a point on the curve is equivalent to a directional derivative at that point. Thus, even if the active diffeomorphism "moves" part of a geodesic curve, it drags the covariant derivative operator and tangent vector field along with it. (The geodesic, of course, will remain smooth and continuous, due to the nature of a diffeomorphism.) These geometrical objects are defined at particular points; if you move the points, the objects go along for the ride, and retain their particular values.

But what about the metric tensor field itself inside the hole? Won't it be warped or changed in some significant way by an arbitrary active diffeomorphism? No. Again, the geometrical object that is the metric tensor field is a mapping that assigns to each point on the manifold a particular metric tensor, whose component values might differ from one coordinate system to another but which is the same object in each of those systems. Each metric tensor is tightly bound to the point to which the field assigns it. As I argued above, moving the points without dragging along the resident metric tensors violates the fundamental principle of general relativity, which holds that the gravitational field just is the geometrical structure of space-time. Space-time itself is a dynamical object in GTR. Its properties are determined by the distribution of mass-energy throughout space-time. The manifold without the metric cannot represent physical space-time.

All of the topological and geometrical relations that hold between points on the manifold are defined by the geometrical objects residing at those points. The distance between two points is strictly a function of the metric tensors at those points. The properties of the curves on the manifold are determined by the values of the covariant derivative operator along those curves. The curvature of some region of the manifold is identical with the behavior of the Riemann and Ricci tensors over that region. But all these objects are defined at particular points, and when an active diffeomorphism "moves" the points, it moves the objects as well. Further, the objects do not change under this point transformation. They are still the same objects, residing at the same points, having the same values. The properties of the region of the manifold inside the hole are exhausted by specification of the values of these various geometrical objects residing at the points inside the hole. (All of these objects are defined in terms of the metric tensor, from the covariant derivative operator to the curvature tensors.) There are no mathematical properties that differ from one diffeomorphic space to another. Stachel has argued that these diffeomorphic spaces are physically indistinguishable, in the absence of what he calls "individuating fields." I am simply arguing that they are mathematically indistinguishable, as well.[17]

My position, and the arguments for it, are actually pretty simple, and their roots can be traced to the Leibniz-Clarke correspondence. How can one move point $\mathscr{P}$ to point $\mathscr{Q}$? Movement involves a change in position, moving from one location to another. But how can $\mathscr{P}$ move or change location? $\mathscr{P}$ will always be at $\mathscr{P}$. If we introduce a background space, with our manifold as a foreground space, then one can imagine moving a point of the foreground space relative to the background space. But what happens when we don't have a background space—when all we have is our manifold?

Years after Einstein developed the GTR, after the development of the abstract, coordinate-free approach to differential geometry (due primarily to Elie Cartan), the principle of general covariance has been interpreted as the requirement that the equations of general relativity admit of a coordinate-free formulation. Some have argued that at least one other consequence of the principle is a demand that there be "no prior geometry." Misner, Thorne, and Wheeler claim that "mathematics was not sufficiently refined in 1917 to cleave apart the demands for 'no prior geometry' and for a 'geometric, coordinate-independent formulation of physics.' Einstein described both demands by a single phrase, 'general covariance.' The 'no-prior-geometry' demand actually fathered general relativity, but by doing so anonymously, disguised as 'general covariance,' it also fathered half a century of confusion."[18] They then cite Kretschman's 1917 paper as an example of such confusion and discuss Nordstrom's theory of gravitation as an example of a theory that violates the principle of general covariance, thus implying that the principle does indeed have physical significance. What does the "no-prior-geometry" requirement amount to? Essentially, that there are no nondynamical geometrical objects defined on the space-time manifold, including the manifold itself. This has implications for the "hole argument." The identity of these diffeomorphic spaces implies the "no prior geometry" requirement of MTW, for if there were a "prior geometry," a "background space" endowed with its own topological and geometrical properties, then the diffeomorphic spaces would be easily distinguishable, by noting the change in the positions of the points, under the active diffeomorphism, relative to this background space. Such a background space might serve as Stachel's individuating field.[19] The failure to recognize the necessity of a prior geometry for distinguishing diffeomorphic spaces has led some to mischaracterize both the principle of general covariance and the hole argument. For example, John Norton writes, regarding the consequences of the hole argument, "If we recall that the metric determines the inertial trajectories of the space-time, then we can see just how disastrous is this result. Given the fullest specification of the space-time outside the hole, the theory will be unable to determine the trajectory along which a particle in free fall will traverse the hole, even though its trajectory before and after the hole is known exactly. . . . This is an extremely awkward form of indeterminism, for the hole might be both of very small spatial size and temporal duration. Even given a full specification of the fields in its future, past and everywhere else in space, the theory is still unable to specify what happens inside the hole."[20] Of course, this is exactly what I am denying.

The diffeomorphism certainly does change the inertial trajectories. But the page is serving as a background, "prior" geometry. Without that, and restricted to the manifold itself, and its resident geometrical objects, there is no difference between the trajectories before and after the action of the diffeomorphism.

Recently, some have argued that the indeterminism engendered by general covariance and the hole argument is related to the question of substantivalism about space-time points. As Hoefer and Cartwright describe it, the argument goes like this: "If points of the space-time manifold were real, that would imply that any generally covariant theory, such as the general theory of relativity, would be indeterministic; but the issue of determinism versus indeterminism is an empirical question, and should not be settled by metaphysical assumptions."[21] How would the substantial reality of space-time points lead to indeterminism? Presumably, it is because it is thought that if the points are real, an active diffeomorphism "really" moves them. But if my arguments above are correct, and the diffeomorphic spaces that constitute models for GTR are, in fact, mathematically identical, then the reality of the points is irrelevant to questions of indeterminacy. Whether the points are real or not, the models are deterministic, so long as there is no background space with a prior geometry, which is disallowed by the GTR. The question of the "awkward indeterminacy" and its implications for substantivalism arises only if one regards space-time models for the field equations, related by active diffeomorphism, as different, in some way. Stachel has argued that they are physically indistinguishable. I have argued that they are mathematically indistinguishable.

Notes

1. Julian Barbour, "On General Covariance and Best Matching," in *Physics Meets Philosophy at the Plank Scale*, ed. C. Callender and N. Huggett (Cambridge: Cambridge University Press, 2001), p. 199.

2. See, e.g., J. Norton, "Philosophy of Space and Time," in *Introduction to the Philosophy of Science*, ed. M. Salmon et al. (Englewood Cliffs, NJ: Prentice-Hall, 1992); J. J. Stachel, "Einstein's Search for General Covariance, 1912–1915," in *Einstein Studies*, vol. 1, *Einstein and the History of General Relativity*, ed. D. Howard and J. Stachel (Basel: Birkhäuser, 1989); J. Earman and J. Norton, "What Price Spacetime Substantivalism? The Hole Story," *British Journal for the Philosophy of Science* 38 (1987): 515–25.

3. A. Einstein, "Die Grundlage der allgemeinen Relativitätstheorie," *Annelen der Physik* 49 (1919): 769–822; English translation in H. A. Lorentz et al., *The Principle of Relativity* (London: Methuen, 1923; reprint, New York: Dover, 1952), pp. 109–64. This passage is translated in Stachel, "Einstein's Search for General Covariance, 1912–1915," p. 87, and also appears in Barbour, "On General Covariance and Best Matching," p. 202.

4. A. Einstein, "Prinzipielles zur allgemeinen Relativitätstheorie," *Annelen der Physik* 55 (1918): 241–44. A translation appears in Barbour, "On General Covariance and Best Matching," p. 202.

5. See E. Kretschmann, "Über die prinzipielle Bestimmbarkeit der berechtigten Bezugssystems beliebiger Relatixistätstheorien," *Annelen der Physik* 53 (Leipzig): 575–614.

6. See, for example, J. Christian, "Why the Quantum Must Yield to Gravity," in *Physics Meets Philosophy at the Planck Scale*, ed. C. Callender and N. Huggett (Cambridge: Cambridge University Press, 2001), pp. 305–38; R. Penrose, "On Gravity's Role in Quantum State Reduction," in *Physics Meets Philosophy at the Planck Scale*, ed. C. Callender and N. Huggett (Cambridge: Cambridge University Press, 2001), pp. 290–304; Barbour, "On General Covariance and Best Matching."

302 GENERAL COVARIANCE AND EINSTEIN'S "HOLE" ARGUMENT

7. Actually, this is not quite correct. In GTR, the laws are not invariant under diffeomorphisms. Performing the appropriate transformation upon an equation that is true in the absence of gravitation, i.e., making the equation generally covariant, produces new ingredients in the coordinate-dependent equations, in particular, the metric tensor and the affine connection. In contrast with the principle of special relativity, which forbids the introduction of new elements, the principle of general covariance does not require that these drop out, and in fact, we exploit the presence of these geometrical objects to represent gravitational fields. Thus, general covariance is not an invariance principle. See S. Weinberg, *Gravitation and Cosmology* (New York: John Wiley & Sons, 1972), p. 92.

8. M. Friedman, *Foundations of Space-Time Theories* (Princeton, NJ: Princeton University Press, 1983).

9. A. Einstein, "Die formale Grundlage der allgemeinen Relativitätstreorie," *Königlich Preussische Akademie der Wissenschaften (Berlin)*, *Sitzungsberichte* (1914): 1066–77. A translation appears in Stachel, "Einstein's Search for General Covariance, 1912–1915," pp. 72–73.

10. A. Einstein, Autobiographical Notes, in *Albert Einstein: Philosopher-Scientist*, ed. P. A. Schilpp (LaSalle, IL: Open Court, 1949), p. 67.

11. As quoted in Stachel, "Einstein's Search for General Covariance, 1912–1915," p. 86.

12. Ibid., p. 73.

13. Christian, "Why the Quantum Must Yield to Gravity," p. 311.

14. C. Hoefer and N. Cartwright, "Substantivalism and the Hole Argument," in *Philosophical Problems of the Internal and External Worlds*, ed. J. Earman et al. (Pittsburgh: University of Pittsburgh Press, 1993), p. 25.

15. Stachel, "Einstein's Search for General Covariance, 1912–1915," p. 74.

16. Ibid., p. 75.

17. It is tempting to consider "eggbeater diffeomorphisms," which scramble the manifold inside the hole dramatically and which it would be hard to believe are identical with the original space. But the restrictions on diffeomorphisms (i.e., that they are smooth and continuous) entail that they preserve the topological properties of a space. Spatial orders, for example, relations of "next to," "in the neighborhood," and "between," are all preserved by a diffeomorphism.

18. C. W. Misner, K. S. Thorne, and J. A. Wheeler, *Gravitation* (New York: W. H. Freeman and Company, 1973), p. 431.

19. Interestingly, in their discussion of the Cauchy problem for GTR, Hawking and Ellis say, "In order to obtain a definite member of the equivalence class of metrics which represent a space-time, one introduces a fixed 'background' metric and imposes four 'gauge conditions' on the covariant derivatives of the physical metric with respect to the background metric. These conditions remove the four degrees of freedom to make diffeomorphisms and lead to a unique solution for the metric components" (S. W. Hawking and G. F. R. Ellis, *The Large Scale Structure of Space-Time* [Cambridge: Cambridge University Press, 1973], p. 227). This would be akin to defining a privileged coordinate system in which the components of the metric can be uniquely specified.

20. Norton, "Philosophy of Space and Time," p. 229.

21. Hoefer and Cartwright, "Substantivalism and the Hole Argument," p. 23.

5.

THE REPRESENTATION OF NATURE IN PHYSICS

A Reflection on Adolf Grünbaum's Early Writings on the Quantum Theory

BAS C. VAN FRAASSEN

Before I turn to my main topic, which I shall explore here in the light of Adolf Grünbaum's early work in the philosophy of quantum mechanics, I want to say something to express my own debt to Professor Grünbaum, both for his guidance and for his work.

Our symposium included the opportunity to hear his autobiographical remarks, both touching and disturbing to those of us who have shared some of the last century's painful history. The remarkable outcome, a true sign of hope in my view, is that Adolf Grünbaum himself could have emerged from such a childhood and youth, all but lost to the oppression and alienation of that era, as the example of social conscience and personal charity for which I thank and admire him. For I learned as much from his personal engagement—with us students on a personal level, with social and moral issues, and with philosophy itself, where he never aimed to turn a student into a disciple—as from his scholarly achievements, which provided the initial basis for my own reflections on the character of physical theory.

After graduation my first position was at Yale University, where I was fortunate to know and learn from one of Grünbaum's own teachers, Henry Margenau. It was with some curiosity, of course, that at this point I read two of Grünbaum's earliest publications, both critical responses to Margenau's philosophy of quantum mechanics. The first, "Realism and Neo-Kantianism in Professor Margenau's Philosophy of Quantum Mechanics," includes a strong defense of a scientific realist position with respect to quantum mechanics, but mainly as against neo-Kantian and idealist themes then current in attempts to interpret the theory.[1] I will concentrate here on the second, "Complementarity in Quantum Physics and Its Philosophical Generalization," together with some of Grünbaum's early articles on the special theory of relativity (henceforth "STR").[2]

Although Margenau's interpretation of quantum mechanics and the Copenhagen interpretation were not at all the same, they shared an antirealist orientation. The idea of complementarity was generalized quite far beyond the use it had seen in illuminating the appearance of incompatible observables, but those generalizations were then drawn on to suggest support for that orientation in ways that cried out for the philosophical critique that Grünbaum supplied. But the paper also engages actively with the question of how we should or can understand the world described by quantum theory, acknowledging that theory's radical departures from classical physics. This will also be my focus here. While I will not disagree nearly as much with Grünbaum's views as Grünbaum disagreed with Margenau's, I will defend one aspect of the Copenhagen interpretation of quantum theory that I see as radicalizing our understanding of physical theory.

1. COMPLETENESS CRITERIA FOR SCIENCE

Before we turn specifically to quantum mechanics and the Copenhagen interpretation, I need to draw on another philosopher whose writings were a paradigm and a guide for Adolf Grünbaum as well as for myself: Hans Reichenbach. One lesson that I took away from those writings was that science is to be understood as an enterprise with a distinctive cognitive aim and with loyalty to a distinctive empirical methodology. Reichenbach himself was seriously concerned with the requirement such an understanding places on us to display the criteria of success that an enterprise thus understood must aim to satisfy—if only as an ideal, even if the perfect success, that would consist in completely meeting those criteria, is beyond reach of such finite beings with finite resources as ourselves. His early writings on quantum mechanics included a strong defense of relinquishing certain earlier criteria, which had been held sacrosanct in modern science but which rested, as he argued, on empirically vulnerable presuppositions.

As I learned from Reichenbach and Grünbaum, the second great scientific revolution in modern times took place around the previous turn of the century, with the coming first of all of relativity, and second, of the quantum. Inherited from modern science was the claim that all phenomena in nature derive from underlying deterministic mechanics and the philosophical conviction that a scientific account is complete only if it is deterministic. Supporting that conviction was the philosophical creed, current among neo-Kantians, that the very intelligibility of nature and the very coherence of experience require their possibility of being conceivable as set in a rigidly deterministic causal order. That is precisely the completeness criterion challenged most saliently in Reichenbach's early writings. The probabilistic resources of classical statistical mechanics were newly adapted in such a way that, as it seemed then, no grounding in an underlying deterministic mechanics was possible. Reichenbach sided with a vocal part of the physics community that explicitly rejected the task of finding or postulating hidden mechanisms behind such apparently stochastic processes as radioactive decay. Nature is indeterministic, or at least it can be or may be—and if that is so, determinism is a *mistaken* completeness criterion for theory.

Now Reichenbach, who did much to provide a rationale for this rejection of determinism, introduced an apparently weaker but still substantive new completeness criterion: the *common cause principle*.[3] This principle is satisfied by the causal models of general use in social sciences and for many purposes in the natural sciences as well. They are models in which all pervasive correlations derive from common causes (in a technical, probabilistically definable sense). But the demonstration in the 1960s and later that quantum mechanics violates Bell's inequalities shows that even this third criterion was rejected, in effect, by the new physics.[4]

Determinism without causality?
Grünbaum on Bohm's mechanics

As Grünbaum discusses (1957, note on pp. 715–16) both the denial of determinism and the analysis of causality connected with the Copenhagen interpretation of quantum mechanics was challenged by Bohmian mechanics—proposed alternatively as an interpretation of quantum mechanics and as a rival theory. Bohm describes a world of particles that have position as their sole physically significant attribute, and whose motion is strictly deterministic. But that motion derives from the total quantum state, which implies that there are non-local correlations among these particle motions that cannot be traced back to any "synchronizing" process in their common past. So there are clear violations there too of the common cause principle, and the Bell inequalities are of course violated. This example shows both that a deterministic interpretation is possible and that satisfaction of the common cause criterion is not logically implied by determinism. Bohm's mechanics surprisingly presents us with a picture of *determinism without causality* so to speak. Citing Heisenberg and Reichenbach, Grünbaum emphasizes that on Bohm's reformulation of quantum mechanics "the behavior exhibited by the waves [the governing wave function for the multiparticle system] would be, classically speaking, every bit as strange as that of a "particle" whose motion depends on the presence of a slit through which it could not have passed." Since Bohm's mechanics has been given sophisticated new formulations in recent decades, we know now that it can accommodate the phenomena. But, as Grünbaum in effect pointed out at the time, Bohm's purport to give us a world picture close to the classical one is certainly not borne out.[5] In any case, both the lesson that a science can be indeterministic and that the common cause criterion may be violated by physical theory, so that these cannot be defining criteria of success in science, remain.

However that may be, I'll now turn to a further completeness criterion that seems compatible with these rejections and appears to be quite generally accepted at least among philosophers and by the general public.

2. APPEARANCE VERSUS REALITY AS A SCIENTIFIC PROBLEM

The completeness criteria of determinism and causality involve demands for explanation. The sort of explanation demanded there from science is not just a supply of missing information needed for a simple, systematic account of the phenomena, but requires connections deeper than brute or factual regularity. Thus as an aspirant empiricist I tend to see those demands as placing a burden of unwanted metaphysics on the sciences. But now I want to suggest that there has in fact been a still deeper-going demand (criterion of success) upon modern science, which also came to be challenged precisely in the development of the new quantum theory. That is the demand that, however different the appearances (to us) may be from the reality depicted in theory, theory must *derive* the appearances from that reality. The sense of "derive" is strong, and once more connected with substantive notions of explanation. This demand can also be read as again a criterion of completeness: science is asserted to be incomplete until and unless it meets that demand. I need to explain this further, but let us at once give it a name: the *appearance from reality criterion*.[6]

Before asking for the precise sense of "derive" involved here, let us take some familiar examples in which science does offer us such a derivation. We credit science with adequate and satisfactory explanations of how many familiar phenomena are produced: how ash is produced when we burn a cigarette or some logs, how methane is naturally produced in a swamp, and how a flame is turned yellow when a sodium sample is inserted. Copernicus also explained the planets' retrograde motion, and the theory of sound as waves in air explains the Doppler shift. In all these cases the familiar witnessed events are not just located in the theoretical world picture but shown to derive from the theoretically described reality, although such terms as "ash," "cigarette," "swamp," "yellow," and "retrograde" have at best a derivative status there and at worst need to be treated rather cavalierly in the theoretical context.

A distinction: phenomena and appearances

Let us introduce a distinction here that may at first blush seem to be a distinction without a difference for such examples. By the *phenomena* I mean the observable parts of the world, whether objects, events, or processes. These the sciences must save (in the ancient phrase)—but these admit of objective and indeed purely theoretical description, which does not link their reality to contexts of observation or to acts of measurement. By the *appearances* I mean the contents of observation and of measurement outcomes. The phrase "to save the phenomena" is often rendered more colloquially as "to save the appearances" and "appearances" is generally taken as synonymous with "phenomena," but of course these are terms with a past. On philosophical lips they are often loaded with connotations that link them to mind and thought. So I am introducing a new technical distinction and use here, specially adapting these terms to our present discussion, eschewing such connotations. In this new sense the planetary motions—whatever they are or are like—are phenomena, but planetary "retrograde" motions are, *pace* Copernicus, (mere) appearances.

Galileo famously promised in *The Assayer* that the colors, smells, and sounds in the experienced world would be fully explained by physics. Among whose descriptive parameters those qualities were not allowed. Descartes' posthumous work *The World, or Treatise on Light* purported to lay the foundation of a world picture entirely transparent to the human understanding—although the theory that was to provide us with that world picture was but a barely enriched kinematics. The colorful, tasty, smelly, and noisy *appearances* will be shown to be produced as (yes!) *interactional* events, in which the relata are in principle completely characterized in terms of primary qualities, respectively, quantities of spatial and temporal extension alone. These promises were not empty: there were solid achievements behind them. Those achievements accumulated into awesome riches by the late nineteenth century. Describing the success that Galileo had promised, and using our new terminological distinction, we can say: combustion of sodium samples is an observable process (phenomenon) that can be exhaustively described in physical terms, but this description can also be utilized to explain how that process produces a yellow appearance to the human eye, and of course, to a camera.

The appearance from reality criterion

Given all that success, it is not surprising if the appearance from reality criterion should begin to pervade the ideals set for science. Theory must *derive* the appearances from that reality. By "derive" I do not mean a bare logical deduction. I mean a connection of the order of explanation through necessity and/or causal mechanisms to be displayed. The derivation is required to show and make intelligible the structure of the appearances as being *produced* by the reality behind them to their (possible) observers. The demand for such a derivation is not met if science should simply issue successful predictions of measurement outcomes, by means of systematic rules of calculation, from the state of nature theoretically described. That sort of success does not ipso facto amount to an explanation of why and how the appearances must be the way they are. The stronger demand that we should be able to see science as providing that sort of derivation/explanation is a continuing theme in much scientific realist writing on the sciences. I'll quote from Jarrett Leplin's *A Novel Defense of Scientific Realism*:

> A theory is not simply an empirical law or generalization to the
> effect that certain observable phenomena occur, but an explanation
> of their occurrence that provides some mechanism to produce them,
> or some deeper principles to which their production is reducible.[7]

On the other hand, I am not contrasting successful explanation with
indeterministic or stochastic accounts.[8] An adequate explanation of
how an event is produced need not be deterministic. For example, if
we were to say that statistical thermodynamics explains how burning
a cigarette produces ash, and add that there are random small fluctuations that modify the underlying mechanics, that would still leave us
with a derivation (in the appropriate sense) of the phenomenon. Note
well, though, that we can't turn this around and conclude that just any
indeterministic account of how things happen then counts as such an
explanatory derivation!

I am also not contrasting the appearance from reality criterion with
instrumentalism. For the instrumentalist denies that a theory is in any
sense a story about what the world is like. That is implausible, for a
story is usually precisely what a theory seems to be, though what it
describes the world to be like is very unlike how it appears to us. At
precisely that point the criterion applies: there is then a felt gap in the
story that must be filled so that the appearances clearly derive from
what is really going on. But on the other hand, again, we cannot turn
this around. The assertion that the appearances are produced in some
specific way, without displaying that way, does not by itself provide a
satisfactory explanation. Indeed, the idealist and (quasi-) instrumentalist accounts, which Grünbaum depicted as prevalent forms of easy
antirealism among scientists, seem to me to fall under this heading.[9]
For it is easy enough, but entirely uninformative, to wave a hand at
some relation of the theorizing and measuring agent to the aspects of
nature that are measured and represented, and claim that this accounts
for how the appearances differ from what science ostensibly describes.
If left as a mere claim that does not suffice.

Suppose then, for a moment, that we have a science that does not manage to derive the appearance from the reality it describes. What are the philosophical alternatives? Those who accept the theory may well still believe that it correctly describes what things are really like. On the other hand, they have the alternative of *rejecting* the criterion. Then after this rejection they have still two further alternatives. For they can either deny that there is a gap "in nature," so to speak (that turns out to be the Copenhagen alternative),[10] or they can adhere to a certain metaphysical doctrine. That doctrine would have the form: there is a sense of "derive" (presumably related to notions of causality or necessity in nature) in which the appearances do derive from the theoretically described reality, but that connection in nature is beyond the resources of theory to make explicit. To avoid the emptiness of such antirealist reactions as those critically discussed by Grünbaum, that metaphysical doctrine then needs to be given some content. This game is not over; there are examples still today in the literature on the interpretation of quantum mechanics, which attempt to do so. But we will concentrate here on what may be of value in the Copenhagen approach, however much it enmeshed itself in some of those philosophical tangles.

Apparent rejection of the criterion

The new quantum mechanics developed in the 1920s was perceived even by some of the physicists most closely involved as not bridging the explanatory gap between reality and appearance. Some, as I will discuss, disagree that there is a gap there for science to fill. Some—at times the same as those who deny that there is a gap—strained to bridge it with philosophical as well as putatively physical explanations. Let us first note the bare facts of the matter. The vehicle for prediction in quantum mechanics is, at heart, the Born rule:

If observable A is measured on a system in quantum state ψ, the expectation value of the outcome is $< \psi, A \psi>$

I take it that measurement outcomes are prime examples of what we should classify as appearances. The quantum states are then the theoretically described reality. At this point in my story it is of course not excluded that those appearances are completely describable in terms of quantum states. But is that so in fact?

3. GRÜNBAUM'S CRITIQUE OF NONREALIST INTERPRETATIONS OF QUANTUM MECHANICS

Reportedly, the Copenhagen physicists were surprised by Einstein's resistance to their approach to atomic physics because they felt they had followed the example he set in constructing the theory of relativity. Hadn't Einstein in effect subjected classical concepts of spatial and temporal relations to an "operationalist" critique, and thereby shown that they needed to be replaced by a radically different representation of nature? Einstein's critique had still left intact the possibility of ascribing simultaneous positions and velocities to given bodies within any given frame of reference. Now a similar operationalist critique had removed the warrant for such simultaneous ascription, and accordingly shown that the classical representation in phase space must be replaced by a radically different representation of a material body's physical state.

Grünbaum (1957) begins by pointing out very clearly that in Heisenberg's and Margenau's accounts we can find consistent operational prescriptions for the simultaneous ascription of position and velocity. There is no logical contradiction, for example, in taking the data of a sequence of position and time measurements on particles emitted from a given source and ascribing a velocity during the relevant interval by means of the classical formula, dividing distance covered by time elapsed. But such ascriptions do not have the value that assertions about physical magnitudes are meant to have. The prediction of a future position on the basis of such an ascription of a current position and velocity will, *according to the quantum theory itself*, not be verified. In fact, neither operational incompatibility nor operational compatibility offers us any logically sufficient or necessary clue to theoretical significance.

What is a theoretically significant quantity?

What relations do such data then really have to future events? From the data we can, via the theory, infer *backward* to something about the quantum mechanical state of the emitted particles, and then on that basis infer *forward* to probabilities of various outcomes in further measurements to be made. What shall we conclude about the notion of a physical magnitude under these conditions? The notion of a simultaneous position-cum-velocity, while "operationally definable," has something wrong with it from a theoretical point of view. Just what is that?

Within elementary quantum mechanics we can offer the following criterion.[11] (I will restrict this discussion to discrete observables, thus shying away from position and velocity, which are continuous. Continuous quantities can be treated in terms of families of discrete observables, but this is not the moment to go into technical details.)

> **Criterion:** There is a theoretically significant physical quantity A with possible values a(1), a(2), . . . , a(k), . . . if and only if there is for each index k a possible state Ψ(k) such that if A is measured on a body in state Ψ(k) then the value found will be a(k), with probability 1.

To fix terminology: in that case Ψ(k) is called an *eigenstate* of A corresponding to *eigenvalue* k. In general, let us write P(Ψ, A, a) = r to say that the probability equals r that an A-measurement on a system in state Ψ will have value a as outcome.

Complementarity now appears as follows: by this criterion it may well be that both A and B are theoretically significant physical quantities, but there is no such quantity that would correspond to their simultaneous ascription. That is:

Quantities A and B are complementary if and only if they have no eigenstates in common.

This is the extreme case: A and B might share some eigenstates, while not having all of them in common. In neither case is there a theoretically significant "conjunctive" quantity (by the above criterion), though in the less extreme case some relevant conjunctions do make sense.

Technical sense(s) of complementarity

Thinking of it this way we have of course no operational criterion, but rather (as Grünbaum points out) a theoretical one, since the availability of eigenstates is a matter on which the theory pronounces. But note also that this criterion takes for granted that we have an independent specification of what it is to measure (putative) quantity A. We can illustrate this perhaps most clearly with an example of a well-known measurement of a discrete observable, which allows for the description of another putative observable, one that has an operational recipe for its (equally putative) measurement, but one that fails the criterion.

Imagine then a Stern-Gerlach apparatus with two exits, intuitively separating particles with spin Up and spin Down along its vertical axis. Let the first exit be the entry to a second such apparatus, rotated 45 degrees with respect to the former. The first we can certainly regard as measuring spin along the vertical axis, and it is possible to prepare particles in a state so that they are guaranteed to emerge from the first exit, or respectively from the second exit. But thinking of the two apparatuses as constituting a single compound measurement of a new observable A does not work. There is no state such that, if the particle is in that state and subjected to this compound measurement, it is guaranteed to emerge from the top exit of the second apparatus. So this composite setup, while it has the looks of a measurement setup for a physical quantity characterizing the particle state, does not satisfy our criterion. In fact, there is no observable—theoretically significant physical quantity—that is being measured by this contraption.

Clearly there are further questions we could investigate about the relationships between operational procedures and theoretically represented physical quantities.[12] But this so far underlines Grünbaum's main conclusion concerning the actual status of complementarity in physics: "operational incompatibility of jointly significant sharp simultaneous values is [neither a necessary nor] *a sufficient condition* for their lack of joint theoretical meaning."[13]

What does a measurement reveal?

Given this negative conclusion, how shall we think of those physical quantities? If we could take it that the value found in a measurement were precisely the value that the quantity had when the object entered the measurement setup—that is, if the measurement *reveals* a possessed value—then it would seem that those complementary physical quantities would have simultaneous values after all. For it would seem then that if A is measured and value a(k) is found, we could add that if B had been measured, then one of its possible values b(j) would have been found—and since measurements reveal possessed values, then A had value a(k) while B had some value or other, simultaneously.

Logically speaking, there would be no contradiction between this conclusion and anything we have said above. But of course that conclusion would run counter to the Copenhagen view of the matter. At least as I read Bohr, to the above explication of what counts as a theoretically significant physical quantity, his view appears to add the tacit principle:

> no circumstance can be the case in nature unless it possible for nature to be such that a suitable measurement would be certain to reveal that circumstance.[14]

"Nothing can be real unless it can really so appear"—that might be the slogan to convey this. And of course that principle implies that quantities that are complementary in the above strong sense cannot, according to the theory, have simultaneous values ever.

If measurement outcomes do not reveal possessed values, what do they do? As I indicated before, they allow "backward" inference to features of the physical state. For example, if particles emitted by a certain source are all submitted to an A-measurement and values a(j) are found in proportion q(j) of the outcomes, that may be taken as evidence that the source prepares particles in a quantum state Ψ such that (in the notation introduced above) $P(\Psi, A, a(1)) = q$. (Equivalently, in the notation I used to express the Born rule, such that $\langle\psi, A\psi\rangle = \Sigma a(j)q(j)$.) But then of course A itself is not being mentioned as a feature of the physical state! How should we think of that state then? In what terms shall we describe it, once terms pertaining to familiar quantities such as position, momentum, spin, and so forth do not have the role of characterizing that state directly?

Relational properties and invariants

There are, as we now know, various possible replies to this, which provide the seed-kernels for various basically tenable interpretations of the quantum theory (even if none of them shall ever be beyond dispute). Grünbaum aligns himself with a two-pronged reply, in which he draws an explicit parallel to discussions of the theory of relativity, thus vindicating to some extent the Copenhagen contention to have followed Einstein's example. The first prong is to assign to the measured quantities a relational status: what is revealed by measurement is not a feature of the system that it had when entering the measurement setup, but a relationship between the system and the setup with which it is made to interact:

> Instead of being simultaneous "autonomous" attributes of the microphysical object, belonging to it *independently* of the *particular experimental arrangement* into which it enters, exact theoretical values of conjugate parameters are each only *interactional* properties of an atomic object that is coupled *indivisibly* to a particular kind of observational macro-setup. For the incompatibility of the circumstances allowing the theoretical ascription of a sharp value of one

conjugate parameter with those allowing the corresponding assignment for the other renders these attributes *theoretically disjunctive*
and interactional.[15]

(For the distinction between *relational* and *interactional*, see below.)
The second prong follows Born in looking to the invariants in the situation for the basic features that characterize the system and its physical state independently of any measurement setup. Indeed, Grünbaum
sees this as a necessary addition, given the above, if realism with
respect to the theory is to be tenable:

> To assert that the particular pairs of attributes which furnished the
> mechanical state descriptions in classical physics are neither
> autonomous nor theoretically conjunctive but rather interactional
> and theoretically disjunctive does *not* itself entail that there are also
> no *other* attributes of, say, an electron whose existence is *inde-*
> *pendent* of whether the electron is observed. If one denies the exis-
> tence of any such other attributes as well, then indeed there is no
> articulate sense in which the electron can be supposed to exist inde-
> pendently of being observed.[16]

Born and Rosenfeld identified these other attributes as the invariant
quantities—such as, in the case of elementary particles—the rest-
mass, charge, and total spin. They are invariant in two senses: they do
not vary as the state evolves in time, and if measured, they do not
allow of alternative possible values as outcomes.

The precise connection with realism about the individual particles
is subject to further questioning. All particles of a given type (e.g., all
electrons) are exactly alike with respect to those features. So it would
seem that terms are also needed to describe differences—for example,
to say that in a given atom there are electrons at several different
energy levels, or to say that one emitted photon was absorbed and
another reflected. At first blush, at least, these are not descriptions of
relationships to measurement setups. But at least we have here a first
step in a description of the quantum state in terms that do not simply

relate it to measurement contexts. In that sense the basic requirements of realism with respect to the theory are served.

Relational properties: perspectival or interactional?

As Grünbaum points out, there is an interesting and instructive parallel here to the reconception of the attributes that "furnished the mechanical state descriptions in classical physics" in the transition from classical to relativity physics, but one that can all too easily be exaggerated. It is of course correct to say that the spatial distance between two events is a magnitude that relates those events to a given frame of reference, while the space-time interval between them, an invariant, is not a relational quantity in that sense. But spatial distance, though relational in that way, is not what Grünbaum calls "interactional": it is not as if those two events are spatially a certain distance apart only if they are involved in some interaction—let alone, involved in a measurement. We might say that, in the theory of relativity, spatial distance is "perspectival" (or perhaps more literally "frame-dependent") while position and velocity in quantum mechanics (on the interpretation we are presently discussing) are "interactional," thus dividing "relational" into several subcategories. Of course, this very remark also leads us to the question whether that interactional interpretation of the classically familiar mechanical attributes in the context of quantum mechanics was really forced upon us, or a matter of one interpretation among others. Indeed, it raises the question whether there might not be another interpretation according to which those classically familiar attributes are perspectival in a similar sense or a similar way.

To explore this question I will proceed in several stages, beginning with some points about perspective, frames of reference, and measurement. I am very conscious of the tempting fallacies that one certainly sees beckoning in even famous physicists' writings when the scientific description of nature is linked to measurement as opposed to measurement-independent fact. These fallacies belong precisely to the family of fallacies that Grünbaum exposes in his discussions both of

Margenau's interpretation of quantum theory and of Copenhagen-related writings about complementarity. But it may be possible to avoid those fallacies while still arriving at an understanding of the physical sciences that gives pride of place to the relationship between theory and what appears in perception and measurement.

4. PERSPECTIVES AND MEASUREMENT OUTCOMES

I mentioned above that the Copenhagen physicists, elaborating their interpretation of quantum mechanics, tried at various times, admittedly with not such great success, to import insights from relativity theory into the putative observer-relativity of measurement results. Putative philosophical insights that exploited relations between frames of reference and visual perspectives were the subject of Grünbaum's still earlier "The Clock Paradox in the Special Theory of Relativity" and his related writings on the perspectival character of time dilation and length contraction (see the list in footnote 2 above). These early articles were a mainstay of my own first few steps in the philosophy of physics; I'll draw here especially on "Logical and Philosophical Foundations of the Special Theory of Relativity."[17]

As in the critique of Margenau we see Grünbaum here insistent in his rejection of idealist, subjectivist, or homocentric interpretations. Length contraction is a good example to illustrate the fallacies and confusions that led various authors into such interpretations. The correct understanding of the STR identifies the limiting character, constancy, and source-independence of the speed of light (all features having nothing to do with limitations of measurement) as accounting for the relativity of simultaneity and for the frame-dependence of length in the STR.[18] Just to see how this point played out in the literature it suffices to cite a small passage from Herbert Dingle and Grünbaum's commentary. Dingle had written:

> Every relativist will admit that if two rods, A and B, of equal length
> when relatively at rest, are in relative motion along their common

direction, then A is longer or shorter than B, or equal to it, exactly as you please. It is therefore impossible to evade the conclusion that its length is not a property of either rod; and what is true of length is true of every other so-called physical property. Physics is therefore not the investigation of the nature of the external world.[19]

In response to this Grünbaum rightly remarks: "Far from having demonstrated that relativity physics is subjective, Professor Dingle has merely succeeded in exhibiting his unawareness of the fact that *relational* properties do not cease to be *bona fide* objective properties just because they involve relations between individuals rather than belong to individuals themselves."[20]

Yet Grünbaum finds a legitimate place for the introduction of measurement and perspective into a foundational discussion of the STR. When he explains the difference between the physical length contraction postulated (and explained) in Lorentz's theory on the one hand and the length contraction implied by the STR, he exhibits the latter as a perspectival effect that appears in measurements made under different conditions. This is perfectly consistent with the foregoing, but the nuances of the discussion will allow me to display the (for our discussion, crucial) distinction between the ("objective") observable *phenomena* (which admit of frame-independent description without loss) and the *appearances* in the outcomes of measurement operations:

the Lorentz-Fitzgerald contraction is measured in the very system in which the contracted arm is *at rest*, whereas the contraction that Einstein derived from the Lorentz transformations pertains to the length measured in a system relative to which the arm is *in motion*.

. . . Unlike the Lorentz-Fitzgerald contraction this "Einstein contraction" is a *symmetrical* relation between the measurements made in any two inertial systems and is a consequence of the inter-systemic relativity of simultaneity, because it relates lengths determined from *different* inertial perspectives of measurement. . . . What Einstein did explain, therefore, is this "metrogenic" contraction, . . . which poses no more logical difficulties than the differences in the angular sizes of bodies that are observed from different distances.[21]

We have therefore to look more carefully into this, and give explicit status to the distinctions on which the correct discussion of the relativity of certain physical quantities relies.

Differences between perspective and frame of reference

The length of a body is a frame-dependent quantity: its value is not invariant under a shift in its description from one frame of reference to another. If we think of frames of reference as belonging to specific inertial systems—bodies experiencing no acceleration—then length is a relational property: A has length b in relation to body B but length b' in relation to body B'. That, I take it, is Grünbaum's reply to Dingle. We can of course add to this that there is an invariant quantity in the immediate neighborhood, which Dingle appears to ignore and which can be illustrated Einstein-style by mentioning two lightning strikes at the two ends of body A. The space-time interval between those two events has a value that is the same in every inertial frame of reference, an "absolute" that, so to speak, takes over the role of the classically invariant length.

But there is something more to be said about this, which pertains to the relationship between measurements and frames of reference. Grünbaum is quite right to describe the content of the measurement outcome as perspectival, when he speaks here of "lengths determined from *different* inertial perspectives of measurement" and when he likens this to "differences in the angular sizes of bodies that are observed from different distances."

This is a very important point about all measurement in general: measurement is perspectival. Let us think ourselves briefly back into pre-STR days when the illustrative example of angular separation in visual perspective could be conceived of in a space independent of time. The most advanced scientific measurements of Galileo's day, for example, were those of astronomy applied in navigation. For a simple example, think of two navigators on different ships. They are, let us say, simultaneously sighting the same mountain peak as well as one of the circumpolar stars and they record respectively:

1. Peak: direction NNW, elevation α
2. Star: direction NNE, elevation β

1. Peak: direction N, elevation α'
2. Star: direction E, elevation β'

From these together with time measurements and earlier log entries they calculate their "objective" position on the ocean (latitude and longitude). But the initial measurement outcome reports are perspectival: *the "from here" accompanies every observation judgment*!

Thus the quite common use of the analogy to visual perspectives when we discuss frames of reference and measurement outcomes hides an important difference as well. Measurement outcomes are perspectival in one way; descriptions or representations within a frame of reference (coordinate dependent representations) are perspectival in a different way.

Let us for a moment take the content of a visual perspective as itself a paradigmatic example of the content of a measurement outcome. The visual perspective contains (in some sense) the spatial relations between bodies as they appear at a certain time from a certain point of view. This content can be "intercepted" on a plane surface, by means of a painting, drawing, or photo. The character of that interception is described in projective geometry rather than Euclidean geometry. In contrast, the spatial relations between bodies in a frame of reference are described in a coordinatized Euclidean space. In the visual perspective content we see systematic marginal distortion as we inspect the projection farther away from the center, and we find many parts of the seen (measured) bodies entirely unrepresented (due to occlusion or to their location relative to the "eye").

Can we perhaps think of the frame of reference as an idealizing abstraction from the contents of such measurements? It is true that we arrive at a Euclidean space if we think of the "eye" as moving farther and farther away from the objects: when it becomes a point at infinity, the projective space has become a Euclidean space.[22] But that hardly counts as assimilating frames of reference to visual perspectives. Think of Galileo's two observers, one stationary on the shore and one on board ship. If we try to think of the latter's frame of reference as the content of a visual perspective with "eye" at infinity, we must still have that "eye" move with the ship. In what sense is that "eye" connected with the point where the mast meets the deck, which is the spatial origin of that frame of reference? The visual perspective metaphor has been strained to breaking at this point: it is as if we are invited to think of the "eye" simultaneously at infinity and at a specific location on the ship, from where it can look into all directions at once.

Phenomena and appearances: the distinction continued

To undo our confusion here I suggest that we should mobilize the above introduced terminological distinction between *phenomena* and *appearances*. The observable processes, which can be described both in a coordinate-independent way and also relative to any frame of reference (i.e., within any suitable coordinate system) are the phenomena. The content of a measurement outcome of which the content of a visual perspective (whether personal or in the more hygienic form of such observations by the navigators I described above) can also show us those observable processes, but what it delivers are the appearances.[23]

Now we face a question with respect to the traditional demand that science has "to save the phenomena" until now expressed synonymously as "to save the appearances." In our terms, which is it? I want to insist firmly that it is the former. The observable processes must have a proper home in the models a science makes available for the representation of nature—that is the empirical criterion of adequacy. From the beginning of modern science there was also the claim that science was saving, or would save, the appearances: the entire smelly, colorful, noisy mess of them. And I do not think we do the traditional writers an injustice if we take them to have subscribed to the appearance from reality criterion of adequacy for the sciences in the sense that I have now given that criterion. But I submit that this is an additional criterion, going beyond the demand to save the phenomena.

Einstein and Minkowski

There was—in the modern science initiated by Galileo's ship or shore observers and Descartes coordinate systems, for the formal representation of what they saw—one central idealization that was removed in the transition to the STR. In a visual perspective the lines of projection are light rays. When representing the content of a perspective at a given moment, such as in a painting, we can think of the light ray connection as instantaneous. In fact, Descartes suggested that it was instantaneous. By the end of the seventeenth century it was already known to be a connection that takes time. The classical kinematics frame of reference, conceived of as a projective space with the "eyes" at infinity, seems to be oblivious of this point. The observer on Galileo's ship, describing a motion on the shore in his own frame, does not take this into account. Enter Einstein: in relativistic kinematics this was corrected and as a consequence not only speeds but lengths and time intervals are affected—they are not the same in moving frames of reference. By taking the speed of light into account we arrive at descriptions of the phenomena much closer to actual deliverances of visual observation. Einstein's thought experiments with moving trains, clocks, and lightning strikes shift us from painting or still-photo representation to motion pictures.

So far then the physics and geometry of light and moving bodies allow a perfect derivation of the structure of the kinematical phenomena—and indeed the appearances—from the underlying physical reality.

Such successes in the history of modern physics must surely be in large part responsible for the appearance from the reality criterion's grip on our imagination. It would have been very hard for anyone in the modern period, even extended to take in the special and general theories of relativity, to resist the conviction that in this derivation of the appearances science is doing precisely what all science is in principle required to do. Which is to say: to satisfy the appearance from reality completeness criterion.[24] For after all, nothing succeeds like success, and philosophers have never been very resistant to what we might call the Inference from Success to Design![25]

5. DOES COMPLEMENTARITY REALLY HAVE TO DO WITH FRAMES OF REFERENCE?

It has often been noticed that observables in quantum theory are associated with something like frames of reference or coordinate systems, and there is therefore a tempting parallel or analogy to exploit in connection with complementarity. In fact, it has been suggested that the same physical situation could, for example, be looked at in complementary ways in a "position frame of reference" and a "momentum frame of reference": in the one, positions are attributed to the bodies involved and, in the other, momenta. Then one could think that these are two mutually incompatible representations of the same part of nature with equal rights as to truth and reality. But that really does not make sense at all.

The point behind the analogy is of course that a pure state space in this theory is a separable Hilbert space, and the eigenvectors of a discrete observable furnish a base for that space. If A and B are two such observables, complementary in the strong sense I mentioned above, then a base of eigenvectors of A will have members inclined at some angle to every element of a base consisting of eigenvectors of B. Considered as geometry that is precisely also how we can think of two Cartesian coordinate systems in a Euclidean space or two Galilean frames in its kinematic generalization. Hence the pair of bases for the Hilbert space consisting of eigenvectors of two complementary observables is the formal counterpart of a pair of spatial or kinematic frames of reference with different orientations.

But purely geometric point obscures the relevant difference

between the two. In the case of physical space we can think of this pair of frames as containing all the bodies in that space described from two different points of view, associated with—as it were—two persons looking in different directions. In the kinematic case these might be Galileo's shorebound observer and mariner on a moving ship. So the descriptions of situations in those two frames of reference are *of the same situations in the same actual world*. The pure states represented by points in a Hilbert space, on the other hand, are alternative states that the represented system can have—alternative possible states. They are not the states of different systems in one world but *the states of one system in different possible worlds*, so to speak.

The analogy can be attempted in a more sophisticated way by looking then at a model of a system consisting of different parts. The Hilbert space whose vectors represent the pure states of the whole system is a tensor product of several Hilbert spaces. Both Hugh Everett and Simon Kochen introduced the idea of the state of one part relative to or witnessed by another part. That is much closer to the idea of a spatial or kinematic frame of reference. However, there is also a big gap in that analogy. Suppose the total system is X+Y. Then the notion really defined as a relative state is the state ψ of Y relative to state ψ of X, where ψ is a pure state that X can have. But there is nothing in the representation of X+Y that warrants associating ψ with X. If X+Y is, for example, in pure state Φ, then X is typically in a mixed state (a reduction of Φ), with various of its possible pure states as components. So the nearest we have here to a perspective would be something like "what Y would look like to X if, per impossibile, X were in possible pure state ψ."

Of course, Everett's idea was then transformed into "many-worlds" interpretations, on the one hand, and "modal" interpretations, on the other.[26] Below we will take a look at the latter, and we will see that there is indeed a way to think of measurement-outcome contents as analogous to the contents of visual perspectives, though not in the naive way which I sketched above.

6. Is the Appearance from Reality Criterion Abandoned in Quantum Mechanics?

This criterion appeared to be blatantly violated by the Born rule when that was offered as the sole and sufficient bridge from quantum reality to observable phenomenon. The theoretically described reality presents us with a fully deterministic evolution of the unvisualizable quantum state, while the observable phenomena display an irreducibly stochastic process. Born gives us *precisely*, but *nothing more than*, the rule to calculate the probabilities in the latter from the former.

Heisenberg was the most straightforward advocate for the view that this is enough and completes the task of physics. If we look at the story since then, it seems to me that the currently more or less acceptable interpretations offered fall into three classes: (i) the sort that purport to derive the appearance from the reality but fail, and (ii) those that do not purport to do this, but either (ii-1) merely pay lip service to the old ideal or (ii-2) more honestly content themselves to flesh out the Born interpretation in a way that precludes this third sort of completeness altogether.[27] If that is correct, of course, then it was right to reject (in effect) the appearance from reality criterion as imperative for the sciences.

Is there a "collapse"?

When we look at the philosophy of quantum mechanics, we must clearly distinguish between *changes* to the theory and *interpretations*. The former, even if proposed as interpretations, differ in that they change the empirical predictions. Von Neumann's version of the "collapse of the wave function" introduced a change, although that was not at once apparent. He proposed that in a measurement, the quantum state of the object is projected or "collapsed" into one of the eigenstates of the measured quantity. The immediate questions are: *What constitutes a measurement?* and *What explains this collapse?* An answer to the first question that remains within quantum theory itself offers no room for that collapse, let alone an explanation.

The two sorts of responses that attempt to maintain von Neumann's proposal were initially typified by Wigner, on the one hand, and by Groenewold and Margenau, on the other.

Wigner answered that a measurement is not an event completely describable in physics; it must include consciousness, a mind-body interaction. That was certainly a radical suggestion. Imagine Schrödinger's dismay—he wrote later: "it must have given to de Broglie the same shock and disappointment as it gave to me, when we learnt that a sort of transcendental, almost psychical interpretation of the wave phenomenon had been put forward, which was very soon hailed by the majority of leading theorists as the only one reconcilable with experiments, and which has now become the orthodox creed, accepted by almost everybody, with a few notable exceptions."[28]

Unfortunately Wigner's reply only looks like it answers the second question. We must insist here on the difference between providing an explanation and merely postulating that there is something that explains! To add the postulate that there is a mechanism of a certain sort, even if intelligible, that changes a certain pure state into a mixture—thus "collapsing the state"—is a far cry from having a science that derives the measurement outcomes from the quantum states in the relevant sense. In fact, Wigner provides no clue at all to *how* the appearances thus derive from the reality.

Groenewold and Margenau argued instead that von Neumann's added postulate was purely interpretative and did not really augment the Born rule. We can illustrate their argument with Schrödinger's famous cat:

> The Cat itself is a measuring instrument, with "dead" and "alive" as pointer states. So the collapse into one of these states may happen inside its closed box, consistently with Von Neumann's postulate. When the box is opened, a second measurement occurs, but no further collapse is needed. Alternatively we can suggests that the Cat lacks something unspecified so that it does not function as a measurement apparatus, and there is no collapse till the box is opened. Now the point is that the probability of finding a dead cat at the end is the same regardless of which scenario we assume.

The ostensibly correct conclusion is that von Neumann's postulate does not affect the empirical content of the theory. Of course, if that is so, its "derivation" of the appearances is still not by an empirically accessible mechanism. But actually, as David Albert has forcefully pointed out, that conclusion is not correct anyway. For there is a definable quantity pertaining to the system *as a whole* (box with cat etc. inside) for which measurement outcome probabilities are certainly different on the two scenarios.[29]

Let's admit that von Neumann's alteration of the quantum theory, together with Wigner's addition of a consciousness-matter interaction, implies that the phenomena do derive from the quantum-mechanically described reality. But recall the distinctions I made early on: this is a case where the appearance from reality criterion is nevertheless not satisfied because physics cannot *provide* the derivation.

What if we ignore Wigner as well as Margenau and Groenewold, and just propose that von Neumann's projection postulate explains what happens in measurement? The story of the world is that it is after all a stochastic process on the level of the quantum states themselves: these states develop deterministically except for abrupt "swerves" during a class of special interactions, the measurements. Very well; but then we run up against the question that Wigner wanted to answer along the way, so to say: "*When* is a measurement made?" If we try to describe it quantum mechanically, we can't easily distinguish the cat's interaction with the device, in the middle, from our interaction with the cat at the end.

It is probably for that reason that some physicists have insisted strongly that quantum mechanics is only a theory of *measured* systems. Wigner would of course hold that, but I think here of views that would allow also any macroscopic apparatus as truly measuring, without regard to consciousness, while ascribing quantum states only to the measured objects and not to the measuring set ups.[30] Quite obviously this sort of view implies that there will be no explanation in physics of how measurements can have determinate outcomes, for the latter belong then to the part specifically not modeled in terms of quantum states.

These early discussions are illuminating not only because they begin to chart our range of options but also because they were closely related to practice. Whatever the theoretical status of "collapse" the way the working physicist calculates does always assume that the appearances will be at least *as if* states thus collapse in measurement. The Born rule is not genuinely explained, but certainly most easily conveyed in practice by the assertion that upon measurement the object will be in one of the eigenstates of the measured observable, with given probability. *The appearances are as if von Neumann's projection postulate is true.*

Mismatches in space

I can't go far into the idea of complementarity for now but want to draw your attention to one of its most provocative claims: space-time description and quantum state description are complementary. That means at least that both are indispensable to our full appreciation of nature and that they cannot be meaningfully combined in any straightforward way.

How could the Copenhagen school say that? First of all, Schrödinger's quantum state is a function ψ (x , t) defined on (classical) space and time, with "x" the position parameter. Second, Heisenberg's famous uncertainty principle is an inequality relating the position and momenta parameters to the quantum state. So what is going on? Does this not mean that the quantum state description is fully integrated, or even based on, a spatiotemporal description of nature?

It does not. We can't think of ψ, despite its appearance, as a quantity pertaining to space points like a classical field.[31] And Heisenberg's principle is statistical, deduced from the Born rule, which relates the quantum state to probabilities of measurement outcomes of those parameters. The spatiotemporal description pertains to the phenomena and to the appearances, and is maintained in the way Bohr explained: the concepts of prequantum (classical and relativistic) physics are the ones we must keep using, suitably restricted, to describe nature as it appears to us.

This is a point that is generally obscured even in quite theoretical, philosophical discussions. As illustration I'll take the Aspect experiment, which is now the standard "Einstein-Podolsky-Rosen paradox" example.[32] Here is how it is typically described: A pair of photons is emitted with opposite momenta toward two polarization filters. With the filters properly oriented with respect to each other, one photon passes its filter if and only if the other does not. Notice that, in this description, two distinct directions of motion are attributed to the two photons. While we can't also specify positions in the same way, the time of emission is pretty definite, and the passing of a filter recorded with a click—so in that time interval the two photons are in clearly separate halves of the laboratory. Right?

Not right. For such a photon pair must have, taken as a whole, a symmetric state. If we derive from this total state, by reduction, the states for the individual photons, each receives the *same* mixture of the two pure states. Based on the quantum state *alone*, we cannot ascribe different directions of motion, nor different spatial regions, nor any other differentiating features to these photons. Yet everything appears to happen *as if*!

Now photons may be special, but we could repeat this entire discussion, *mutatis mutandis*, for heavy particle pairs. The same point applies. One reaction is to interpret the mixed states of the photons as "ignorance mixtures." The ignorance interpretation of mixtures says that to be in a mixture of *this* and *that* is to be really in one or the other, with no further information. That is, like "collapse," a staple of working physics problems, but it is untenable theoretically precisely in cases like this. However, practice is not wrong as practice. *The appearances are as if the ignorance interpretation is correct.*

In the case of photons we are not so emotionally involved. We can give up on the idea that these are individual photons, and regard that way of speaking as but an asses' bridge toward a quantum field description. But if we think of that as a general solution, we had better take the consequences for all physical objects. In that case there aren't any, at least not in the sense of objects that are really in one half of the room and not in the other half—as you and I and these chairs and tables appear to be.

Eddington gave a famous example contrasting the manifest image with the scientific image: the table as science describes it is utterly unlike the way we describe the observed table. But his tables are not all that unlike: they are both precisely located in space, and the contrast is not so different from that between a cloud of locusts seen from afar and from close by. In quantum mechanics his example has returned with a vengeance: everything we observe, even in Aspect's laboratory, appears to have a determinate place, and if it moves, it does not shock us with discontinuities. But the theoretical description of that laboratory assigns no determinate locations or movements to its parts—it is not too far-fetched to say that that laboratory is only partly manifest in the spatiotemporal realm at all!

I think it is time to look at more recent interpretations, and for me to make the case that these vindicate what I take to be the Copenhagen insight that rejected the appearance from reality criterion.

7. THE APPEARANCES YOKED UNTO A FORBEARING REALITY

So I turn finally to the class of interpretations that seem to me to endorse—implicitly, explicitly, or cryptically—the rejection of the appearance from reality criterion. In his recent book *Interpreting the Quantum World*, Jeffrey Bub displays a very large class of interpretations under the same heading: *modal interpretations* in a general sense.[33] On all of them, an observable (that is, a physical quantity) can have a determinate value even if the quantum state does not make it so.[34] Thus no collapse is needed for measurement outcomes—or indeed any other sort of event—to be characterized by a definite position, or definite velocity, or definite charge, or definite death-or-life.

While in a quantum state which does not imply that at all, the object is *as if* it is in an eigenstate of the pertinent observable. So we have here a clear reality (the quantum state) contrasted with appearance (the "value state" or "property state").

Note well: in aligning these two aspects of a system (under such an interpretation) with the reality/appearance dichotomy, I am departing from how modal interpretations have been presented so far (including by myself). The "value state" or "property state" is introduced to validate the assertion that physical measurement processes do have "definite" outcomes. The pointer is really at the "17," the cat is really dead inside the box—although the quantum state does not make it so. But now I want to say: that is not a separate aspect of the real situation; it is not the case that a system has two states. Rather what is called the value state or property state is the content of a perspective on the system—of a perspective of a (possible) measurer or viewer or measurement setup.

I am not suggesting either that there is a measurement apparatus located at every point, or that the description of the world is restricted to what happens in actual measurements. In the case of visual perspectives as treated in projective geometry, and equally in classical kinematics, we think of every point and orientation determining a perspective, regardless of whether there is a thus oriented measurement apparatus or viewer present at that point. The mechanics plus optics does allow us to derive the contents of all those visual perspectives. While omitting this conviction that the appearances can be thus derived, adding only that they can be predicted probabilistically, think of it here in the same way. The appearances are the contents of possible as well as actual measurement outcomes.[35]

The measurement outcomes, these are the appearances to be saved! They are saved in that the interpretation makes room for them in the theoretical world picture. But they are saved in a way that explicitly rejects their derivability from the quantum state. (In fact, they preclude *even supervenience on* the quantum state: for two systems in the same quantum state may have different value states.) As I use the terms now, on the other hand, the phenomena are those observable processes, objects, and events that can be described without loss entirely in terms of the quantum states and their evolution in time. On the view we are presently exploring, all real processes can be thus represented in quantum mechanics, though this representation does not determine what appears in the possible measurement setups in our actual world—let alone show how those specific appearances are produced.

What are the appearances like, on such an interpretation? We do see quite some variation there.[36] Bub's interpretation implies that the actual state of the world is characterized by the definiteness of a single "privileged" observable. It need not be position. We can think of his world as follows: it has a quantum state and in addition there is an observable that has a definite value, *just as if that observable was just measured on the world, with a collapse precipitated by that measurement.* Note well that this is a matter of appearance only: the quantum state is *not* collapsed.

In my own favored interpretation, the Copenhagen Variant of the Modal Interpretation (CVMI), there is no simple privileged observable.[37] But it is *as if the ignorance interpretation of mixtures is correct,* for every object in the world has a "value state" that is pure. These value states are related to the quantum states and to measurement processes (quantum mechanically defined) so that in consequence it is also *as if the projection postulate (postulate of collapse of the wave function) is true.* Again the "as if" describes the appearances, that is, the value states (which include the measurement outcomes) but not the quantum state.

On this interpretation we cannot say: it is just as if the "collapse" idea is right and the world looks as if it has just been subjected to great single comprehensive measurement. Rather every object, including every part of an object, "looks" as if it has just been subjected to a collapsing measurement of some observable (though not necessarily an observable with a familiar classical counterpart). That is not compatible with the idea that the appearances of the individual objects are all precipitated by an (imagined) single measurement carried out on the whole. The reason for this incompatibility lies in the holism of the quantum theory. If the state of a whole, compound system is projected into a given pure state, the result will in general not have pure states as its reductions to specific parts. Yet it is possible for both the whole and the parts to be definite in their apparent characteristics. So it is as if each part is seen individually from some measuring vantage point.

8. THE STRUCTURE OF APPEARANCE

On this view, what is the world like? We restrict ourselves here to elementary quantum theory. The world consists of things that, however, as Bohr said, resist description consonant with the older ideas of causality and locality:

> the renunciation of the ideal of causality in atomic physics which has been forced on us is founded logically only on our not being any longer in a position to speak of the autonomous behavior of a physical object.[38]

But these objects each have a quantum state (dynamical state); they are often compound, and then their parts all have quantum states too (derivable by "reduction of the density matrix"). The quantum state of an isolated system develops in time in accordance with the Schrödinger equation, that is, deterministically. All of this applies equally well to those cases in which one part of a system is a measuring apparatus in appropriate interaction with another part.

But besides these physical states that are the subject of dynamics, there are the appearances of these very things in possible determinate measurement setups. These appearances are described in the same language as the dynamical states. They can be described as value states or property states, represented by vectors in the same Hilbert space, which represent the pure quantum states. In actual measurement, these value states are what appear in the measurement outcomes.

Appearances systematically unlike the postulated reality

I can't emphasize strongly enough that *what is "seen" in a measurement outcome is never what the objects are really like.* What is "seen" is what the object "looks like" to the apparatus, viewer, or setup. One should say, "That is what it is like *from here.*" So in ordinary life, a still photo displays the shape of the object projected on a plane, in a one-point linear perspective.[39] The content of the photo is the content of a measurement outcome. Similarly, a video or motion picture displays the objects moving at determinate speeds—these are the speeds in its frame of reference. Those speeds are not part of the "objective," frame-independent quantities to be found in a more advanced classical mechanics model, whose basic quantities are the invariants of the motion.

Perhaps these analogies are going against one aspect of the perspectival version of the modal interpretation that I am presently offering. For I say that the appearances are described in the same language as the physical states—by attributions of value states represented in precisely the same way as pure quantum states. But in saying that I am taking a shortcut. The outcome of a measurement, for example, a record of a track in a cloud chamber, would be recorded in frame-dependent language too: for example, the speed and direction relative to a frame connected with the apparatus. That is, the value state would be described both partially and relatively in the particular case. It would still be *as if* the alpha particle had a determinate momentum. Moreover, the relevant measurement outcomes here are typically not of individual events but of mean values, to be compared with the predicted expectation values. The insistence on representing the value states in the way I do implies that the appearances are never *as if* the objects are simultaneously characterized by determinate values of incompatible observables.

There is an ambiguity here too, as is well brought out by Grünbaum's and Margenau's discussions of the possibility of attributing both precise locations and precise momenta to individual particles for a certain time in a measurement. This can be done by, for example, two position measurements and a time of flight calculation. But, as mentioned above, the data thus gathered cannot be used to infer a real quantum state in which those quantities are determinate, they have absolutely no empirical predictive value, and they have no analogue on a statistical (as opposed to individual) level. For these reasons, the addition of a time of flight calculation to position measurement outcomes could be greeted with "so the appearances are as if the particle had simultaneous determinate location and momentum," but that would have to be countered by "the appearances displayed in the aggregate destroy that initial impression."[40]

When I say that the appearances are describable in the same language as the quantum states, I am honoring one of the principles that became entrenched as the working Copenhagen ("orthodox") interpretation: *no two things can be true together unless they can have probability 1 together.* I add to this that the world of appearances is at all times as fully determined as it can be, compatible with this constraint. From this it follows that it is the language of pure quantum states that has the right structure to describe the appearances. And thus it is, as far as what appears in actual or possible measurement, *just as if* both the ignorance interpretation of mixtures and the projection postulate are true.

Relative states and the invariants

The value states are what appear in actual and possible measurement outcomes, in perfect analogy with how certain spatial aspects of bodies appear in the contents of visual perspectives on those bodies (that is, their projections on certain planes from certain vantage points, whether on photo or motion picture). But the appearances in different such measurement setups are systematically related to each other. In the classical case in which spatial shapes are registered, that is fully explained by means of the three-dimensional shape of the object and the straight line propagation of light, as described in geometric optics. The appearances are *relative* quantities, such as "shape as seen from" and, in mechanics, speed relative to a frame of reference, and they derive from *invariant* quantities that are the same from all visual vantage points (the cross-ratio in projective geometry), respectively in all frames of reference (such as acceleration). What about the quantum case, as we now wish to interpret it?

There are many different possible measurements to make on systems in a given quantum state. Because of the statistical character of Born's rule, individual outcomes have little or no significance (I'll say more about that below), but averages do. The recorded averages in different sorts of measurements all follow from the quantum state via the Born rule—"follow" of course in the minimal calculational, and not the explanatory, sense. The quantum state is characterized in terms of invariants; it is only very partially revealed in any measurement setup, but how it is there revealed follows (in that sense) from that invariant character of the system. This was strongly urged as crucial to any interpretation by Born in his Nobel Prize lecture.[41] But we must be careful not to assimilate this too closely to the relation between spatiotemporal invariants in mechanics to the spatial and temporal frame-relative quantities there. As Adolf Grünbaum comments:

> Now that we know the independently existing attributes of atomic entities defined in Born's sense by the quantum mechanical invariants, it is perfectly clear that they do *not* constitute attributes of *individual* events *in space and time* which are the values of a set of state variables linked by deterministic laws. This result was to be expected from any philosophical interpretation compatible with complementarity, since, as Bohr has explained, "the renunciation of the ideal of causality in atomic physics which has been forced on us is founded logically only on our not being any longer in a position to speak of the autonomous behavior of a physical object."[42]

In the interpretation I am now proposing, the quantum state contributes the invariants displayed in the relationships between the statistics of outcomes in different measurement setups. To take the simplest sort of example, presented naively: if many particles are prepared in the same quantum state, and position measurements are made on one subaggregate while momentum measurements are made on another subaggregate, then the statistics will bear out Heisenberg's uncertainty relations.

Appearance "kinematics"

How do those appearances change over time? That process is indeterministic, but it is strongly constrained by the quantum states of the objects involved. The value state must always be *possible with respect to* the quantum state.[43] And at those moments which mark the end of a measurement process (as identified by purely quantum mechanical criteria on the quantum states of object and apparatus), the Born rule gives the probabilities of the various possible value states.[44]

The Copenhagen school consisted of physicists, and we must all agree that their expositions often muddled themselves with half-baked bits of philosophy. But it would be silly of us to extrapolate from those, and not from the cases in which their dicta and practice agreed with each other. Thus Heisenberg's idealist and quasi-Kantian sentiments can be ignored, it seems to me, while such passages as the following are consonant with his actual work:

> the introduction of the observer must not be misunderstood to imply that some kind of subjective features are to be brought into the description of Nature. The observer has only the function of registering decisions, i.e., processes in space and time, and it does not matter whether the observer is an apparatus or a human being. . . . The Copenhagen interpretation regards things and processes which are describable in terms of classical concepts, i.e., the actual, as the foundation of any physical interpretation.[45]

So what happens now to our naive idea of objects that have specific and quite definite locations in space, remain there or move continuously from one place to another, with definite velocities?

This classical conception was always a vast idealization, and what appears to us in experience was compatible with it only under restricted conditions. But while a dynamic state governed by

Schrödinger's equation cannot have its spatial support restricted to a finite region for more than a moment, the value state can remain localized in that way for some time. The "rigid" connections between bodies which are reflected in the quantum state through correlations ("entanglements") of the states of the parts will keep the value states of the parts connected as well. The spatiotemporal description of the process, as it appears, corresponds approximately but not mistakenly to the value state description.

The appearances do not supervene

On this view, the appearances do not even supervene on the quantum states, let alone be explicable from them by mechanisms of perception. That is not due simply to the indeterminism in this theory. It is easy enough to imagine classical stochastic processes, observed or measured at different times, but with the measurement outcomes perfectly derivable from the instantaneous state of the process, the state of the measuring apparatus, and the character of their interaction. That is the picture of the Lucretian universe, and there is no reason at all to see it as violating either the common cause or appearance from reality criteria. The nonclassical indeterminism of quantum theory breaks the mold.

Note here well, however, the difference between the perspectival version of the modal interpretation that I am now presenting, and the original "empirically superfluous hidden variable" version. For if we classify the value state and the quantum state as together representing the complete physical state of the object, then how the object appears in the measurement outcome *does* supervene on its (combined) physical state. But supervenience fails too if the value state is simply classified as the content displayed (partially and relatively) in the given measurement outcome, and the physical state consists of the quantum state alone. Without stretching our reading of Bohr too far, I would say that is also precisely how it was when Born had introduced his rule and the Copenhagen physicists refused to add hidden variables, whether empirically contentful or superfluous.

So there is one striking difference between the original "empirically superfluous hidden variable" version of the modal interpretation
and the present "perspectival" version. For a given quantum state there
may be many value states possible relative to it. In the original version,
one of these will be the real one, the actual one, presenting the definite
properties the system actually has, in addition to its quantum state. But
in the perspectival version, all those relatively possible value states are
on a par: they are simply how the system "looks" from one possible
vantage point or another. If I (or a robot voice mechanism on a measurement apparatus) enunciate the content of an observation (of a measurement outcome) it may take the form, "The iron bar has negative
charge," but it is tacitly indexical, accompanied by the tacit "from
here"—so it does not contradict the statement that the quantum state
of the bar is a superposition or mixture of positive and negative charge.

The final challenge

The details of quantum theory interpretation are fascinating, challenging, and frustrating, and its problems are by no means all settled.
But my main aim in this paper is not to defend a specific interpretation—let alone its details in one form or another! Rather, what I mean
to do is to argue that this actual part of recent history of science should
convince us that it is perfectly scientific, and scientifically acceptable,
to reject the completeness criteria for science that I outlined. That is,
a thesis concerning the aim and methodology of science, directed
against at least certain traditional themes in "realist" philosophies of
science.

If my view of it is right, and if in addition the Copenhagen physicists were acting in a way that counts as real physics when they introduced and developed quite explicitly a theory and an interpretation
incompatible with the appearance from reality completeness criterion,
then that criterion is *not* a constraint on the sciences. It is, in that case,
just another of those philosophically or metaphysically motivated
imperatives that could hamper science if they were obeyed, and

receive much lip service, but are anyway quickly flouted when that hampering is felt.

NOTES

1. A. Grünbaum, "Realism and Neo-Kantianism in Professor Margenau's Philosophy of Quantum Mechanics," *Philosophy of Science* 17 (1950): 26–34.

2. A. Grünbaum, "Complementarity in Quantum Physics and Its Philosophical Generalization," *Journal of Philosophy* 54 (1957): 713–27; A. Grünbaum, "The Clock Paradox in the Special Theory of Relativity," *Philosophy of Science* 21 (1954): 249–53; A. Grünbaum, "Reply to Dr. Tornebohm's Comments on My Article," *Philosophy of Science* 22 (1955): 233; A. Grünbaum, "Reply to Dr. Leaf," *Philosophy of Science* 22 (1955): 53; A. Grünbaum, "Logical and Philosophical Foundations of the Special Theory of Relativity," in *Philosophy of Science*, ed. A. Danto and S. Morgenbesser (New York: Meridian Books, 1960), pp. 399–434; and A. Grünbaum, "The Relevance of Philosophy to the History of the Special Theory of Relativity," *Journal of Philosophy* 59 (1962): 561–74.

3. See for comparison my "Rational Belief and the Common Cause Principle," *What? Where? When? Why? Essays in honor of Wesley Salmon*, ed. R. McLaughlin (Dordrecht: Reidel, 1982), pp. 193–209, and references therein. Under certain conditions this criterion actually demands determinism, as I show there.

4. See my "The Charybdis of Realism: Epistemological Implications of Bell's Inequality," *Synthése* 5 (1982): 25–38, reprinted in J. Cushing and E. McMullen, eds., *The Philosophical Consequences of Quantum Mechanics* (Notre Dame, IN: University of Notre Dame Press, 1989).

5. See K. Bedard, "Material Objects in Bohm's Interpretation," *Philosophy of Science* 66 (1999): 221–42; B. van Fraassen, "Interpretation of QM: Parallels and Choices," in *The Interpretation of Quantum Theory: Where Do We Stand?* ed. L. Accardi (Rome: Istituto della Enciclopedia Italiana; New York: Fordham University Press, 1994), pp. 7–14.

6. See my "Science as Representation: Flouting the Criteria," *Philosophy of Science* 71 (2004): 794–804, for a related exploration of this theme.

7. Jarrett Leplin, *A Novel Defense of Scientific Realism* (New York: Oxford University Press, 1997), p. 15.

8. Attempts to provide such accounts that could perhaps provide such explanations even today include work by Stephen L. Adler and his colleagues on generalized quantum dynamics, as well as earlier work by, for example, E. Nelson, *Dynamical Theories of Brownian Motion* (Princeton, NJ: Princeton University Press, 1967). Such accounts, if successful, can satisfy the appearance from reality criterion no less than deterministic theories.

9. Grünbaum, "Complementarity in Quantum Physics and Its Philosophical Generalization," pp. 717–19.

10. In view of Don Howard's illuminating historical studies we must be very careful not to read too much into the term "Copenhagen"; I intend to keep its connotations minimal, and even so realize that my reading of Bohr may be contentious. See D. Howard, "Bohr's Philosophy of Quantum Theory: A New Look—'Who Invented the Copenhagen Interpretation?' A Study in Mythology," *Philosophy of Science* 71, no. 5 (2004): 669–82 (Supplement).

11. I am staying here in the historical context in which observables are represented by Hermitean operators; the discussion would have to take a different form in more recent contexts in which the concept of observable is generalized to representation by positive operator valued measures.

12. In fact, the compound measurement setup just described can be used to motivate a generalized concept of observable, not restricted to representation by Hermitean operators.

13. Grünbaum, "Complementarity in Quantum Physics and Its Philosophical Generalization," p. 716.

14. Notice that I have made this weak enough to serve the present purpose without implying the so-called eigenstate-eigenvalue link, that is, the stronger principle that an observable actually has a certain value if and only if the system is in a corresponding eigenstate of that observable. I have not ruled out that A has value a(k) in a given state, provided only there is some state (perhaps another one) that is an eigenstate of A corresponding to that value.

15. Grünbaum, "Complementarity in Quantum Physics and Its Philosophical Generalization," p. 717.

16. Ibid.

17. That was the first paper by Grünbaum that I read, when still an undergraduate, and which made me want to study with him—lo these many years ago.

18. Grünbaum, "Logical and Philosophical Foundations of the Special Theory of Relativity," pp. 411–12.

19. Cited in ibid., p. 433, n. 29.

20. Ibid.

21. Ibid., pp. 419–20

22. I am oversimplifying, of course, but not so much as to obscure the main point. For a more detailed story of projective, affine, and Euclidean spaces along these lines, see, for example, B. E. Meserve, *Fundamental Concepts of Geometry* (New York: Dover, 1983), chaps. 4–6.

23. I argue elsewhere that the content of a measurement outcome is to be conceived of as an indexical proposition. The description of a process in the language of physics itself, whether coordinate free or coordinate dependent, however, expresses a proposition that is not indexical. We need of course to distinguish between the attribution of a relational property on the one hand and the indexical attribution of some such property on the other. To say of a body that it has length b in the frame of the fixed stars is an example of the former—to say that its length is b, full stop, is to be understood as the indexical assertion that it has length b in the speaker's frame of reference.

24. That the structure of the phenomena as observed within a given frame of reference can be derived from the invariant features of the situation in both classical and relativistic physics "serve[s] to explain, though *not* to justify," I agree with Adolf Grünbaum, Einstein's rejection of the Copenhagen line, despite, as Grünbaum says, charges that his reasoning was here akin to that of early opponents of relativity. Grünbaum, "Complementarity in Quantum Physics and Its Philosophical Generalization," p. 720.

25. Nor are ordinary people, of course: if someone succeeds in the stock market, s/he is automatically thought of as very clever, for example.

26. Because of their current interest in the field, it would certainly be appropriate to go into the question whether the appearances are saved on a many-world interpretation. But that will have to be a later project.

27. I will actually only look at some sorts of interpretations, and realize that both the range I inspect and my assessment of what are currently more or less acceptable interpretations are controvertible. With respect to the Bohmian option, I'll again avoid a direct confrontation, but I place it in the first class.

28. E. Schrödinger, "The Meaning of Wave Mechanics," in *Louis de*

Broglie Physicien et Penseur, ed. A. George (Paris: Editions Albin Michel, 1953), p. 16.

29. "Recombination" experiments furnish today the most psychologically compelling support for rejecting collapse, but in my view David Albert's point is the most solid reason. Note of course that Albert's point does not give a reason to reject collapse theories—there is no a priori reason to expect the predictions of a no-collapse theory to be vindicated, as opposed to those of a collapse theory. His point serves only to reject the Groenewold-Margenau contention that the collapse adds no empirical import.

30. This sort of view is to be contrasted with one congenial especially to cosmologists, and I think most discussants from the side of philosophy, to the effect we are to think of quantum mechanics as potentially applying as well to the universe as a whole. The choice between these two views was clearly and explicitly laid out by J. Wheeler's commentary on Everett's original paper, "Assessment of Everett's 'Relative State' Formulation of Quantum Theory," *Reviews of Modern Physics* 29 (1957): 463–65.

31. This becomes very clear as soon as we look at a many-particle state when, as Schrödinger rapidly appreciated, we cannot regard ψ as denoting a wave in physical space but rather in a many-dimensional configuration space. It is possible of course to regard what ψ denotes as physically real, and what happens in physical space as an aspect thereof, manifested on that level, or instead guiding and constraining what happens on that level, that is not being denied here. See further Bradley Monton, "Wave Function Ontology," *Synthése* 130, no. 2 (February 2002): 265–77.

32. With thanks to Soazig LeBihan for a discussion of the experiment with respect to this point.

33. J. Bub, *Interpreting the Quantum World* (Cambridge: Cambridge University Press, 1997).

34. Ibid., p. 178. This class of interpretations include Bohm's interpretation, Bub's own, versions of Bohr, Kochen, and many others, though it does not in fact include all modal interpretations—see my review of his book. The Copenhagen variant of the modal interpretation, which I shall discuss below, is not included, but shares the features I am outlining here.

35. This must be read very carefully. All those measurement outcome contents must cohere together in a certain way, so that they can be thought of as all perspectives on a single world in some specific quantum state. In just the same way, the entire set of contents of visual perspectives, with origins in

both possible and actual viewers, in a given room, for example, must cohere so that they can be regarded as being "of" the same room. In the case of the modal interpretations I am discussing, the delineation of what the joint value states can be of the parts of a compound system, given a quantum state for the whole, is directed to this point.

36. Although Bub lists it as one of the interpretations covered in his framework, I am not going to take up Bohmian mechanics here. Bohm allows only one parameter to have a definite value—always the same one, always definite—namely, position. This world is one of particles that are always somewhere—and larger objects "made up" of those particles, always in a precise spatial region. Their motions are continuous in time. This view may have been inspired by the extreme operationalist idea, going back to Mach, that in the last analysis every measurement is a length measurement. (Not very plausible: could you describe even a length measurement operation using only predicates denoting lengths?) Or perhaps it derives even further back from Descartes' dream of a world whose only objective properties are attributes of extension. That the phenomena are saved in a weak sense only and that there is still an appearance/reality gap here is argued in my "Interpretation of QM: Parallels and Choices," as well as in papers by K. Bedard (see references in note 5) and A. D. Stone, "Does the Bohm Theory Solve the Measurement Problem?" *Philosophy of Science* 61 (1994): 250–66.

37. See my review of *Interpreting the Quantum World*, by Jeffrey Bub, *Foundations of Physics* 28 (1998): 683–89, for an explanation of how the CVMI is related to, but does not fall into, the class described in his book.

38. N. Bohr, "Causality and Complementarity," *Philosophy of Science* 4 (1937): 293; cited in Grünbaum, "Complementarity in Quantum Physics and Its Philosophical Generalization," p. 722.

39. This is actually only correct for the pinhole camera. Its photos don't look nearly as lifelike as the ones made with cameras that have lenses, another demonstration that, to create "realistic" appearances, distortion is precisely what is in order.

40. See by comparison Grünbaum, "Complementarity in Quantum Physics and Its Philosophical Generalization," pp. 713–15; H. Margenau, *The Nature of Physical Reality* (New York: McGraw-Hill, 1950), pp. 376–77; and H. Reichenbach, *Philosophic Foundations of Quantum Mechanics* (Los Angeles: University of California Press, 1944; New York: Dover, 1998), p. 119.

41. M. Born, "Statistical Interpretation of Quantum Mechanics," *Science* 122 (1955); see also his "Physical Reality," *Philosophical Quarterly* 3 (1953), reprinted in his *Physics in My Generation* (New York: Pergamon Press, 1956), pp. 151–63. See the discussion by Grünbaum "Complementarity in Quantum Physics and Its Philosophical Generalization," pp. 720–22, who points out that L. Rosenfeld made the same point in "Strife about Complementarity," *Science Progress* 41 (1953): 405–406.

42. Grünbaum, "Complementarity in Quantum Physics and Its Philosophical Generalization," p. 722; italics and quotation marks in original.

43. That is, one represented by a vector that is not orthogonal to the vector or statistical operator which represents the quantum state.

44. In the case of modal interpretations, it has been strongly suggested that they must be supplemented by a "value state dynamics." This may derive from not wanting to give up on the appearance from reality criterion. However, as Bradley Monton pointed out to me, just adding a value state dynamics would in any case not suffice to satisfy that criterion.

45. W. Heisenberg, "The Development of the Interpretation of Quantum Theory," in *Niels Bohr and the Development of Physics*, ed. W. Pauli (London: Pergamon Press, 1955), p. 22; cited in Grünbaum, "Complementarity in Quantum Physics and Its Philosophical Generalization," p. 719.

PART THREE

CAUSATION AND PSYCHOTHERAPY

6.
PSYCHOANALYTIC RESEARCH METHODS

EDWARD ERWIN

The discipline of psychoanalysis has expanded considerably since it was created by Sigmund Freud. Orthodox Freudian theory must now compete with newer psychoanalytic theories including ego psychology, Kleinian theory, self-psychology, and object relations theory. There are variants of each of these, and each theory has inspired a separate psychoanalytic therapy. One thing that has not changed to any great degree, however, is the reliance on analytic case studies for the empirical confirmation of psychoanalytic hypotheses. If one examines for the past five years the *International Journal of Psychoanalysis* and *JAPA*, the *Journal of the American Psychoanalytic Association,* one will find very few experimental studies; with relatively few exceptions, the research reports are either case reports or theoretical discussions that cite no studies except case reports. Many analysts continue to believe that case study research is useful for validating psychoanalytic hypotheses. As one well-known analyst says, "Psychoanalysis is a clinical and not an experimental discipline; all the knowledge that we have comes from clinical observation. If our clinical observations cannot be used inductively to test, correct and validate our theories, these theories can be no more than a personal credo based on personal experience" (Hanly 1992, 293).

Some leading analysts, however, are now expressing serious doubts about the epistemic value of analytic case studies. One recently complained of the persisting belief among analysts that "conducting psychoanalysis in one's consulting room is a form of research, an attitude still reflected in the bulk of articles published in psychoanalytic journals" (Compton 1998), and an editorial in *JAPA* (Tassman 1998, 670) has noted that "to be credible contributors to present-day research literature, we must move beyond the historical emphasis on the single case study approach."

Disagreements about the value of clinical case studies have been around a long time, but they took a radically different turn in 1984 when a philosopher, Adolf Grünbaum, published the most detailed and most powerful critique ever published of the Freudian clinical evidence and the epistemic role of analytic cases. Thus began what became known in the psychoanalytic literature as "the Grünbaum debate." There have been many replies to Grünbaum's 1984 book, but virtually all of them have been answered in the published literature (see, for example, Erwin 1993, 1996; Eagle 1993; Grünbaum 2002). There is another type of reply, however, which concedes that Grünbaum is right about the profound weaknesses of the existing Freudian clinical evidence, but disagrees with him about what to do next. Grünbaum argues that the needed epistemic cure is to do well-controlled experimental studies or perhaps epidemiological studies. Some analysts agree with this point too, but others argue that new methodologies of a nonexperimental form will do the job. Thus, a number of relatively new methodologies have been proposed and some are now being widely used in studying some of the newer psychoanalytic theories or therapies.

PART 1: STANDARDS

To what degree the newer psychoanalytic research methods are an improvement over the traditional case study method has not yet been determined. Indeed, investigators have plunged ahead with their research with very little discussion of the validity of their methods. Although empirical details about these methods are crucially important, other issues are important too, and on some of them, philosophers can make a potentially useful contribution to a debate that has yet to take place. One issue concerns the appropriate weights and measures for evaluating the newer methods. Any proposed standards will no doubt be subjected to criticism and revision, but it is important to at least get started; otherwise, there is a risk that a decade from now analysts or their critics will rightly conclude that the recent analytic research efforts have all been in vain.

Some of the standards I will mention are very obvious but are worth stating if only because they are so often violated in the psychoanalytic research literature; others are more contentious. I will begin by stating some conceptual standards:

1. *A centrality principle*: that the hypothesis being assessed is central to some degree to whatever psychoanalytic theory is being tested. For example, Fisher and Greenberg (1985) claim experimental support for parts of Freud's oedipal theory in that experimental research has confirmed two propositions: (1) During the so-called oedipal period, male and female children are closer to their mother than their father, and (2) Each sex later identifies more with the same-sexed parent. These claims are, indeed, implications of Freud's oedipal theory, but they are clearly not central. If both claims were true but it was false that any child has ever had an unconscious desire to have sex with a parent, and also false that anyone has ever had castration anxiety, then the key parts of Freud's oedipal theory would be wrong. A researcher might sometimes be interested only in confirming relatively minor parts of a psychoanalytic doctrine, but if a particular research method could rarely or never yield cogent evidence for a central thesis, then it would not serve the purpose for which most analysts presumably intend its use.

2. *An identity principle*: that the theoretical hypothesis being tested is a distinctively psychoanalytic principle or, in the case of a therapy, the therapy being studied is actually psychoanalysis. A common defect in the existing set of Freudian experimental studies is that the hypothesis that is said by the experimenter to be confirmed is not distinctive of Freudian theory or any psychoanalytic theory. For example, researchers have commonly cited evidence from the cognitive science literature that unconscious mental events occur, something postulated by Freud, of course, but also by his predecessors, including Herbart, Schopenhauer, and Nietzsche. The idea that Freud's particular theory, the dynamic theory of the unconscious, is true is often confused with the very different hypothesis that unconscious ideas exist. As to the assessment of therapy effects, so-called therapy integrity has been a perennial problem in psychological outcome research. To illustrate, a meta-analytic review of seventy experimental studies of rational emotive therapy (Lyons and Woods 1991) concluded that rational emotive therapy had been shown to be effective, but the author's argument failed because, as they conceded, there was no guarantee that the therapy used in the studies under review was actually rational emotive therapy.

I turn next to some epistemic principles. Here one finds a very large number of "surface" standards that could be developed, most of them dealing with therapy assessment. There are, for example, those dealing with the need to rule out credible placebo and spontaneous remission hypotheses, others that concern therapist training and variance in performance, patient diagnosis, the validity of outcome measures, and so forth. Such principles have been developed and refined in the psychological outcome literature of the last fifty years. Rather than discuss them, I turn briefly to some basic epistemic principles that underlie the surface principles. The following seem to me to be especially pertinent to psychoanalytic research whether it concerns therapy or theory.

3. *Principle of Inference to the Best Explanation*: If *H* is the most likely of several competing explanations of a set of facts *F*, then infer *H* (or believe *H* to some degree or with some degree of confidence) *only if* there is independent evidence that the set of available explanations contains the correct explanation of *F*. Some philosophers do not insist on this extra requirement that there be this independent evidence, but instead substitute something weaker, such as "the competing explanations must be good enough," but as van Fraassen (1989) and others have argued, the resulting principle is illegitimate. Without independent evidence that the set of available explanations contains the true one, there is no reason to think that the best explanation we have come up with is the correct one.

Again, if one looks at the Freudian experimental literature as well as at analytic case studies, this requirement is routinely violated. Analysts have quite often argued that their interpretation of a patient's problem is the correct explanation on the grounds that it is the only one available that accounts for all of the facts. The same sort of argument has been often used in the Freudian experimental literature. As the British psychologist Paul Kline (1986, 230) said in response to my criticisms of certain experiments, Freudian theory provides the only, and therefore the best, explanation of the results of the experiments in question and if a critic is to reasonably question the Freudian interpretations, he or she must provide a more credible one

4. *The differential principle*: Evidence *E* confirms hypothesis *H* only if it does so "differentially," where this means that *E* provides some reason to believe *H* and does not provide equal or better reason to believe some incompatible hypothesis *H1* that is at least as plausible. (Erwin and Siegel 1989)

Some might think that this principle is too obvious to need stating; but some well-known philosophers and analysts have explicitly rejected it. (For a defense of the principle, see Erwin 1996, 44–54, esp. 47).

A different problem with the differential principle is that it might seem uninformative. If we assume that whenever we have a failure to confirm, we fail to show that a proposition is more likely than its negation, and if we count not-p as a rival to p, then all failures to confirm are violations of the differential principle. I agree with this latter point, but not that the principle is uninformative or that it is not worth stating as a separate principle. One of the particular ways of failing to show that p is more plausible than not-p is to neglect to discount a plausible rival to p besides not-p. This failing is quite common in the psychoanalytic literature. An analyst will often construct an interpretation of the patient's behavior but fail to discount an equally plausible rival from commonsense psychology or even a rival psychoanalytic theory; it is also a common error in the Freudian experimental literature. One philosopher, Karen Wilkes (1990, 249), in responding to my critique of certain Freudian experiments, has accused me not only of making a mistake in insisting that credible rivals to Freudian hypotheses be discredited but also of being guilty of "intellectual dishonesty." (For a reply, see Erwin 1996, 51–53.)

One additional principle, discussed by Wesley Salmon and Adolf Grünbaum, concerns causal relevance. The principle is sometimes formulated as stating an ontological condition for X being causally relevant to Y. For reasons discussed in my *A Final Accounting* book (1996, 76–84), I think that the principle should be reformulated as an epistemic principle, at least if it is to be brought to bear on the assessment of methodological strategies. However, it also needs to be reworked to avoid certain types of counter examples. I suggest the following formulation:

5. *The principle of causal relevance*: Except where there is reason to believe that there are countervailing causal factors, to establish that X is generally causally relevant to the occurrence of Y, we need to establish for some reference class c that the frequency of Ys is greater in the subset of c's in which Xs occur than in the subset of c where X does not occur.

PART 2: RESEARCH ON PSYCHOANALYTIC THERAPY

Although there are relatively few experimental studies of any of the newer psychoanalytic theories, there is an extensive experimental literature on Freudian theory (for a partly favorable verdict on these studies, see Kline 1981, and Fisher and Greenberg 1985, 2002; for an essentially skeptical conclusion concerning the same studies, see Erwin 1996, chaps. 4 and 5). In the case of long-term, orthodox psychoanalytic therapy, however, there are no experimental studies at all, and the older nonexperimental studies all have obvious epistemic defects. Referring to these older studies, Fisher and Greenberg claim that some of them support the conclusion that psychoanalytic treatment produced results superior to no treatment, but they also add: "In fact, it is possible to challenge the results of all existing studies because of compromised or contaminated data" (2002, 513). Another writer sympathetic to Freudian theory, Paul Kline, referring to my skeptical arguments about what has been demonstrated with respect to psychoanalytic therapy, writes "I agree with his conclusions, incidentally, that no good evidence exists that it [psychoanalysis] is effective" (Kline 1988, 226; for additional discussion of psychoanalytic therapy and a negative verdict, see Erwin 1996, chap. 6).

The lack of sound outcome studies of psychoanalytic therapy recently assumed greater practical importance when the American Psychiatric Association's committee on Practice Guidelines for the Treatment of Panic Disorder decided not to recommend psychoanalysis as one of the treatments of choice for this disorder, a decision that sent "ripples of dismay and frustration across psychoanalysis" (Tassman 1998). When the president of the American Psychoanalytic Association (Pyles, Fox, and Rosenblitt 1998) complained in the pages of the *American Journal of Psychiatry* about the failure to recommend psychoanalysis, the members of the committee replied that a scientific evaluation of the treatment of panic disorder with psychoanalytic therapies had not yet been undertaken, whereas the effectiveness of medication and cognitive behavior therapy had been demonstrated through a series of controlled studies. Panic disorder is but one type of psychiatric disorder, but both the American Psychiatric Association and the American Psychological Association are committed to the evaluation of treatments for other clinical disorders, using rigorous standards that preclude reliance on uncontrolled case studies. For obvious reasons, controlled studies are difficult to carry out for long-term psychoanalytic treatment, although they can be done for short-term psychoanalytically oriented psychotherapy. To get around this serious problem, some analysts have proposed methods that go beyond case studies without rising to the level of experimental studies.

One such method, used recently by Freedman, Hoffenberg, Vorus, and Frosch (1999), employs the effectiveness questionnaire developed by *Consumer Reports* (*CR*). In 1995, the magazine published the results of a survey that allegedly provided evidence that psy-chotherapy, not necessarily psychoanalysis, is highly effective and that long-term therapy works better than short-term therapy. In fact, the magazine found an even greater ratio of improvement to nonimprove-ment than found in the widely cited, and in my view (1997) deeply flawed, meta-analytic review by Smith, Glass, and Miller (1980): the latter authors claim that eight out of ten clients who undergo psy-chotherapy benefit from it, whereas the *CR* report claims an improve-ment rate of 90 percent.

The *Consumer Reports* survey method has been the subject of much discussion in the psychological literature; that might have hap-pened partly because the person who ran the survey was Martin Seligman, a respected methodologist and recent president of the American Psychological Association.

Seligman draws a distinction, now widely used in psychotherapy outcome research, between two types of research methods, effective-ness methods and efficacy methods, and between two types of studies, effectiveness studies and efficacy studies. Efficacy methods employ experimental techniques, including the random assignment of subjects to treatment and control conditions, the use of standardized therapy manuals, and the use of clinic volunteers with well-diagnosed, uncomplicated disorders. The core problem with efficacy methods is that the experiments resemble real therapy only slightly, thus creating a severe problem of external validity. Real therapy is not of a fixed duration; real patients have multiple problems, and they are not assigned to treatment and control groups.

The effectiveness method, in contrast, is not experimental, but it studies real patients, without a manual, and without a predetermined duration. The main virtue of this sort of method is its realism; in Seligman's words (1996, 1072), it "has no problem of inferential distance because it tests exactly what it wants to generalize to." His position is that the efficacy method and the effectiveness method (that's the nonexperimental one) are both needed and each complements the other. The questionnaire method is a prime example of the effectiveness method.

The idea that we can determine not merely if psychotherapy clients improved but also what caused the improvement by asking clients what they believe does not seem especially plausible on its face, but Seligman is well aware of that fact. Although he does not mention the differential principle by name, he commits himself to observing it when he acknowledges the need to eliminate alternative causal hypotheses (1996). The main alternative hypotheses he tries to discount are that beneficial effects were due mainly to spontaneous remission or that they were caused by placebo factors. In using the questionnaire method, and more generally in using effectiveness methods, neither no-treatment controls nor placebo controls are employed. Seligman argues, however, that so-called internal controls can take their place.

The argument, roughly, is this: Based on the *Consumer Reports* figures, we can infer the following conclusions, which in turn can serve as internal controls: (1) Longer duration of psychotherapy correlates with more improvement; (2) psychotherapy alone does just as well as psychotherapy plus drugs for all disorders, and given the history of placebo controls being inferior to drugs, one can infer that psychotherapy would have outperformed such controls had they been run;

(3) marriage counselors control for many of the nonspecifics of treatment and treat the same sorts of problems of similar severity for the same duration as other mental health professionals, but do significantly worse than their counterparts; and (4) family doctors provide another control in that they too do significantly worse than mental health professionals when treatment continues beyond six months. The most likely explanation of these four facts, the argument continues, is that the psychotherapy was generally effective and the more of it the better; consequently, the *CR* study has provided empirical validation of the effectiveness of psychotherapy as well as supporting a thesis about dosage.

How epistemically useful are Seligman's "internal controls"? As other commentators have pointed out (e.g., Jacobson and Christensen 1996), the first correlation, between longer duration and effectiveness, cannot serve as a control for spontaneous remission. The longer therapy goes on, the greater the opportunity for spontaneous remission; so, even if factors outside the therapy setting are causing improvement, there should be more improvement with longer-term therapies. Another problem is that as people invest more time and money in their therapy, they may well be more motivated to judge that it worked even if in fact it did not.

The second internal control is also useless for several reasons. First, no information was provided about the types of drugs that were used or their dosage; if the drugs were ineffective, perhaps because the dosage was too low, then the psychotherapy's tying the drugs in effectiveness would not translate into effectiveness for the psychotherapy. Furthermore, even if the information about the drugs employed had been supplied, there is a serious methodological problem in relying on drug results as a substitute for placebo controls. Aaron Beck and his colleagues tried this strategy in numerous otherwise well-controlled experimental studies of cognitive therapy for depression in which the cognitive therapy tied or surpassed the drug in every single study, none of which contained a placebo control. When cognitive therapy was finally made to compete with a pill placebo in the collaborative study

(Elkins et al. 1989), the results were very disappointing: the cognitive therapy did no better than the placebo. The methodological problem is this. Drugs for depression, at least, tend to do no better than a credible placebo in two-thirds of all cases (Erwin 1996, 154–55). Without a placebo control, there is no way to know in a particular study if the patient sample consists mainly of those likely to respond to the drug but not a placebo, or equally well to both, or to neither. In the absence of that information, it cannot be determined if a drug used with a given sample of patients worked only for placebo reasons or because of its chemical ingredients.

As to the other two so-called controls, there is no way of knowing whether the clients of the family counselors or the family doctors had as much confidence in their treatment as those going to psychotherapists. So a placebo explanation of the different results is not ruled out. There is also no way of knowing whether the treatment given by the family doctors and family counselors did not have negative effects. There are other problems as well in using groups of these types of practitioners as comparison groups, but suppose they are ignored, and we assume equal credibility of treatments and ignore the possibility of negative effects; at best, we are then left with an unexplained correlation: that the reported results are worse for the family doctors and family counselors. To conclude that the psychotherapy worked and the counseling did not on the grounds that postulating this hypothesis provides the best available explanation for the correlation is to appeal to the illegitimate principle of inference to the best explanation discussed earlier.

Apart from the problem of adequate controls, there are two other serious problems. First, only 13 percent of those receiving the survey responded. There is no way of knowing if the nonresponders did far worse in their treatments than those who responded. Second, because this was a retrospective survey, the results rely on the uncorroborated memories of the respondents.

It should also be noted in passing that the spectacular results—90 percent improvement—were arrived at by lumping together those reporting a great deal of improvement with those who judged that they had improved "somewhat" or a "lot" (Jacobson and Christensen 1996, 1035). If one asks how many said that they had improved a great deal, the answer is: 54 percent, a percentage found in many studies of patients receiving no treatment at all.

In short, the *CR* survey provided evidence that a certain group of those undergoing psychotherapy thought that they had improved to some degree and inferred that their therapy was the cause of their perceived improvement; it did not provide evidence that psychotherapy of any form is effective for any clinical problem.

The failure of the questionnaire method in the *CR* survey does not, by itself, prove that its use by psychoanalysts will also result in failure, but the results do raise legitimate doubts about the idea of relying on so-called internal controls in trying to establish that a therapy works. I will take up this general issue later, but first I want to address some other issues raised by Seligman. One nonsubstantive issue concerns the introduction of the "effectiveness/efficacy" terminology. In outcome studies, whether they be experimental or nonexperimental, researchers are trying to determine whether the therapy being studied caused the specified results, and, consequently, are assessing the effectiveness of a therapy. Why invent new, potentially confusing terminology when there is existing terminology to draw Seligman's distinction: the distinction between an experimental and a naturalistic study?

As to the alleged limitations of experimental studies of psychological therapies, these are all problems of external validity. Seligman does not show that they cannot be dealt with within an experimental paradigm. For example, although this has often not been done, in experimental studies, one can use "real" patients, that is, patients with problems of the same degree of severity as is typical of those treated by therapists, and one can use random assignment not merely in setting up the comparison groups but also in the initial selection of patients. However, the issue here is not whether controlled experiments are sufficient for confirmation of outcome hypotheses but whether they are necessary. I turn now to that issue.

As just noted, the failure of the *CR* survey to support the effectiveness of psychotherapy does not necessarily mean that the analysts
Freedman et al. (1999) also failed in their goal of demonstrating beneficial effects of psychoanalysis. However, the failure of the first
survey is enough to create reasoned suspicion about the second survey.
Freedman and his colleagues modeled their survey on the *CR* study
and even used the same questionnaire. They also claim that their study
provides a replication of the *CR* finding that longer therapy tends to
produce better results; for reasons I gave earlier, this was not demonstrated in the *CR* study.

In the Freedman study, ninety-nine outpatients attending the
IPTAR clinic responded to the *Consumer Reports* questionnaire. On
the basis of the patient responses, the authors claim to have demonstrated that *as a result* of receiving psychoanalysis, the quality of life
of the patients who responded had been enhanced and, further, that
duration and frequency of therapy contributed toward this end. In fact,
the authors close with a bold declaration: "Our empirical findings,
together with those in the evolving literature, establish this as a clinical fact" (Freedman et al. 1999, 770).

What were the findings that are said to justify the authors' claim?
They found an incremental gain in effectiveness scores from six
months to over twenty-four months of treatment. This finding would
appear to at least be consistent with their presumably major hypothesis: that increasing the duration of the therapy incrementally contributes to its effectiveness. This thesis is important to psychoanalysts
who believe that long-term psychoanalysis is better than short-term
therapy.

How was the effectiveness score calculated? Following the practice of the *Consumer Reports* survey, it was composed of three factors:
specific improvement with respect to the problems that led the respondent to therapy, satisfaction with the therapist's treatment, and global
improvement, that is, how respondents felt at the time of the survey
compared with how they felt when they began treatment. Points were
then assigned to each category depending on whether the respondent

rated his or her improvement with respect to each condition as "a lot better," "somewhat better," and so forth. However, this grouping together of the factors creates a problem in interpreting the findings. The third factor is not a measure of therapeutic improvement. If the economy and the stock market were on the upswing when the patients were queried, as in fact both were during the Clinton years, and if their real income was improving, patients might well have responded that they now felt happier compared to when they entered therapy, but such global improvement need not be connected to improvements of the specific problems that led them to undergo therapy. The second factor, liking the therapist's treatment, is also not a measure of therapeutic improvement.

If we rely on composite scores alone, then, there is no way of telling if there was even a therapeutic improvement that correlated with duration of therapy. The authors, however, do provide a breakdown of the composite scores. On the one factor that matters in assessing the effects of the treatment, symptomatic improvement, it *failed* to be significantly related to duration of treatment. In fact, only liking the therapist's treatment was so correlated, and that could just as plausibly be interpreted as the cause of staying in treatment for a longer period rather than being an effect of longer treatment.

There was another finding, however, that is of some interest, although it is not especially pertinent to psychoanalysis as opposed to other psychological treatments. Frequency of treatment, that is, seeing the therapist two or three times a week as opposed to one time, was correlated with greater perceived symptomatic improvement. However, the correlation reached significance with respect to only two out of the five disorders: anxiety and eating disorders. For the most prevalent symptom, depression, there was no significant correlation between frequency of treatment and perceived improvement.

We have, then, at least one significant correlation: a greater perceived improvement in those who came with anxiety or eating disorders if they saw the therapist two or three times a week compared to those who went to therapy once a week. How is this correlation to be explained? The authors discount a cognitive dissonance explanation and favor an explanation that goes like this: increased therapeutic exposure contributes to the experience of greater affective intensity in treatment, which in turn facilitates a perception of the therapist as being optimally responsive, and that in turn contributes to the development of a relatively internal relationship with certain aspects of the therapist (Freedman et al. 1999, 760). But that explanation, even if true, would not explain the correlation unless it is also true that the development of this therapeutic relationship causes an increase in perceived symptomatic improvement, but no evidence was provided for this latter proposition. To argue, as the authors appear to, that their explanation is better than the ones they discount and that therefore there is reason to believe it correct, is to violate the principle stated earlier concerning inference to the best explanation. The most reasonable verdict is that there is no way to tell from the data what explains this single correlation.

There are other problems with the study, some that are likely to reappear when the questionnaire method is used again. First, what I earlier called the "identity" condition may have been violated. The authors label the therapy that was studied "psychoanalytic psychotherapy," but that term is often used to designate psychoanalytically oriented psychotherapy, which is not the same as psychoanalysis. However, this unclarity might be cleared up if the authors were to provide more information.

A more serious problem is that, as in the *CR* report, the authors are relying on the respondents' memory about what first led them to therapy and their subsequent theorizing about the degree to which they had improved. As to the latter matter, if the problem were, say, weight gain, then a client might be able to reliably discern whether improvement had occurred, but if the problem were family disorganization or even stress—two of the other problems that were treated—the client's judgment of improvement might well be unreliable. In any event, there was no independent corroboration of the reported degree of improvement.

As the authors make clear, then, what was being correlated with duration and frequency of treatment was not therapeutic effectiveness but *perceived* effectiveness. The authors conclude that the quality of life of the patients had been improved. Perhaps so, but without a no-treatment and a placebo control, or some alternative way of ruling out a spontaneous remission and placebo explanation, they provide no evidence that the treatment, whether it was psychoanalysis or not, was the cause of such improvement, if any.

A somewhat different use of the questionnaire method can be found in the Stockholm Outcome Project. This study generated a great deal of information and used sophisticated statistical techniques to analyze the data. It is very highly regarded, and in certain respects is indeed impressive.

In carrying out the project, the authors (Sandell et al. 2000) sent questionnaires to 756 patients living in Stockholm County, Sweden, who were in various phases of subsidized psychoanalysis or psychotherapy. The subjects were asked questions before, during, and after the completion of their therapy. Complete data were received from 450 of these people. The main finding was progressive improvement the longer the patients were in treatment, impressively strong, the authors note, among patients in psychoanalysis. The improvement was on self-rating measures of symptom distress and morale. On other measures, the patients did not do well.

Although it is not clear from the abstract, the authors (Sandell et al. 2000) appear to be arguing that the differences in outcomes between the two groups are effects of treatment, that is, the psychoanalytic group did better as time went on than those in the psychotherapy group *because* of the psychoanalytic treatment. The authors say, for example (p. 922), that their methodological challenge was to draw secure conclusions not just about treatment outcomes but about the factors that may have influenced them. However, there is a problem straightaway: the "identity criterion" appears to have been violated. The two treatments being compared were distinguished in terms of session frequency, with a particular application of therapy being counted as psychoanalysis if the patient saw the therapist three to five times a week, although initially, therapist training and licensing were also considered. But even with the addition of these latter factors, there is no guarantee that the analytic process was present, where this includes the presence of free association, resistance, interpretation, and working through (Vaughn and Roose 1995). The authors might not require a presence of these features in order for psychoanalysis to take place; they do conceptualize analysis in terms of internal object relations rather than inter-systematic conflicts. But then it is not clear what they are counting as genuine psychoanalysis. In any event, no attempt was made to construct a therapy manual, something the authors claim to be impossible (p. 922), and then check to see if the therapists in the psychoanalytic group were following the manual and actually practicing psychoanalysis, as opposed to psychoanalytically oriented psychotherapy.

So it is not clear that the comparison was between patients who received psychoanalysis and those who received psychoanalytically oriented psychotherapy. In fact, the authors contend that some therapists in the nonpsychoanalytic group were actually practicing psychoanalysis but were practicing it badly. Consequently, even if there were no other problems, there was no secure basis for concluding that the differences in outcome between the two groups were due to psychoanalysis, because it is not clear that there was a uniform difference in treatment between the two groups; in fact, there was evidence there was not.

But there are other problems. Were the patients randomly assigned to each treatment group? They were not; furthermore, the authors believe that there were systematic differences between the two groups. This matters quite a lot to the assessment of the main positive outcome. At termination, there were no discernable differences between the two groups on symptom remission. The advantage for the psychoanalytic group on this variable emerged only three years after therapy. A lot can happen in three years. Given the differences between the two groups at the onset of the therapy, there is no way to decide if the changes that emerged three years later were due to the therapy or to other factors. One could say that symptom remission is not the right criterion to judge the success of psychoanalysis; many analysts do take this position. Consequently, it could be argued that the more important outcome measure was the SAS score, which is a measure of quantity and quality of social contact, and is said to be correlated with quality of primary object representation. On this more psychoanalytically friendly measure, however, both groups did about the same; furthermore, the results were, according to the authors, "generally unimpressive" (p. 937).

There is an additional problem in measuring the outcomes. By relying on self-ratings, the authors had no way of knowing whether they were generally measuring improvement in symptoms and morale, or merely an improvement in what was *theorized* by the patients to be an improvement. Interviewers also provided ratings of improvement, but the raters were not blind with respect to the kind of treatment the patient had been in.

In brief, there is reasonable doubt as to whether psychoanalysis was even compared to psychotherapy in this study; whether the better improvements on symptom remission three years after therapy completion were not due to differences in the patients when they were initially assigned to one group rather than another, or to unknown factors; and whether and to what degree each patient did improve in morale and symptom remission.

Some of the above problems might be addressed in future uses of the method, but there is still the difficult problem of lack of experimental controls that cannot be resolved without abandoning the methodology. As in the *Consumer Reports* study, there were no controls for either spontaneous remission or placebo factors. In the absence of such controls, I see no argument for saying that patients in the psychoanalytic group improved more than those in the psychotherapy group *because* of the ingredients that constitute psychoanalytic therapy.

In an editorial accompanying the study's publication, an editor notes that the study was not a randomized clinical trial but says that the authors make the case that their innovative and inventive statistical analysis makes the study methodologically equivalent to a randomized clinical trial. I disagree. In the absence of random assignment to treatment and control conditions, objective and valid measures of treatment outcome, evidence that the therapy being evaluated was actually used, and the presence of spontaneous remission and placebo controls, no creative use of statistical analyses can transform the study into the equivalent of a randomized clinical trial of the proper sort.

There remains the issue of whether internal controls, not mere statistical analyses, can sometimes serve as a substitute for genuine controls, even if they failed to do that in any of the studies discussed above. This raises an interesting issue, not merely because of the problems that Seligman points to with respect to many experimental studies, but also because of the obvious problems in using placebo controls in studying long-term therapies. Speaking of "internal controls," however, is just another way of talking about background information, except that some of the information gleaned in a questionnaire study is known before the study commences and some of it is acquired from the study. The question of interest, then, is: Can background information sometimes serve in the absence of controls to establish causal claims about psychological treatments? I think that for *singular* causal judgments about therapy, confirmation is sometimes possible in very special circumstances, such as where there are very dramatic effects and the clinical problem is acute rather than episodic, and other conditions are met (see Erwin 1988 and Kazdin 1981). However, to establish that a therapy is *generally* effective for a given problem, the principle of causal relevance needs to be satisfied; that requires at least a statistical comparison between those who are treated and those who are not. But even that is not enough; there is still the issue of meeting the differential condition, and that requires ruling out a placebo and spontaneous remission hypothesis.

Can background information, then, replace both placebo and spontaneous remission controls, and other features of randomized clinical trials, in establishing that psychoanalysis or other types of psychological therapies are *generally* effective for a given problem? If the epistemic principles I outlined in part I are acceptable—in other words, illicit inferences to the best explanation are to be disallowed, competing credible hypotheses need to be ruled out, and the principle of causal relevance needs to be satisfied—then, the evidence about spontaneous remission and the potency of placebo factors, plus the history of psychotherapy outcome research, makes this quite unlikely. I do not know exactly how to demonstrate this claim, but the recent uses of the questionnaire methodology by psychoanalysts do nothing to undermine it. As a means of demonstrating the effectiveness of psychoanalysis, these newer methodologies do not look promising.

What, then, are psychoanalysts to do in their current predicament? In the United States and elsewhere, they are under intense pressure to demonstrate the effectiveness of their techniques in treating a variety of clinical problems. Some have proposed that, first, the effectiveness of short-term psychoanalytically oriented psychotherapy be experimentally demonstrated, and then it be argued that if this sort of therapy works for a given problem, then psychoanalysis is likely to work as well. It might be possible to work out such a strategy in a satisfactory way, but one cannot reasonably infer from the fact that some psychoanalytically oriented therapy or other is effective for a certain disorder that (long-term) psychoanalysis is also likely to work for the same disorder. There are many different kinds of psychoanalytically oriented short-term therapies; some, such as interpersonal therapy, bear only a superficial resemblance to psychoanalysis. Even for those that use some of the characteristic ingredients of the psychoanalytic process, such as free association and an analysis of transference, it is not enough to argue: If they work, psychoanalysis will work. Whether that hypothetical proposition is true will depend on why they work. If they work but not because of the psychoanalytic ingredients, then their effectiveness may have no implications for psychoanalysis. To make this comparative strategy work, one would have to first demonstrate that a specific set of ingredients that make up the therapeutic package for a particular form of short-term psychoanalytically oriented psychotherapy is causally effective for a certain type of disorder, and perhaps in a certain kind of therapeutic setting, and then argue that the same ingredients are typically present in long-term psychoanalysis.

Another possible experimental strategy might be to try to test the causal efficacy of certain elements of the psychoanalytic process in a nontherapeutic setting. For example, if one could experimentally demonstrate the reliability of free association in detecting unconscious determinants of clinical disorders, that might provide reason to be confident about its use as an investigative tool in a clinical setting. If, however, none of these experimental strategies is feasible, then, for the present at least, I see no solution to the psychoanalytic predicament.

REFERENCES

Compton, A. 1998. "An Investigation of Anxious Thought in Patients with DSM-IV Agoraphobia/Panic Disorder." *Journal of the American Psychoanalytic Association* 46: 691–721.

Eagle, M. 1993. "The Dynamics of Theory Change in Psychoanalysis." In *Philosophical Problems of the Internal and External Worlds: Essays on the Philosophy of Adolf Grünbaum*, edited by J. Earman, A. Janis, G. Massey, and N. Rescher, 373–408. Pittsburgh: University of Pittsburgh Press.

Elkin, I., T. Shea, J. Watkins, S. Imber, S. Sotsky, J. Collins, D. Glass, P. Pilkonis, W. Leber, J. Docherty, A. Fiester, and M. Parloff. 1989. "National Institute of Mental Health Treatment of Depression Collaborative Research Program." *Archives of General Psychiatry* 46: 971–82.

Erwin, E. 1993. "Philosophers on Freudianism: An Examination of Replies to Grünbaum's *Foundations*." In *Philosophical Problems of the Internal and External Worlds: Essays on the Philosophy of Adolf Grünbaum*, edited by J. Earman, A. Janis, G. Massey, and N. Rescher, 409–60. Pittsburgh: University of Pittsburgh Press.

———. 1996. *A Final Accounting: Philosophical and Empirical Issues in Freudian Psychology*. Cambridge, MA: MIT Press.

———. 1997. *Philosophy and Psychotherapy: Razing the Troubles of the Brain*. London: Sage.

Erwin, E., and H. Siegel. 1989. "Is Confirmation Differential?" *British Journal for the Philosophy of Science* 40: 105–19.

Fisher, S., and R. Greenberg. 1985. *The Scientific Credibility of Freud's Theories and Therapy*. 1977; New York: Columbia University Press.

———. 2002. "Scientific Tests of Freud's Theories and Therapy." In *The Freud Encyclopedia: Theory, Therapy, and Culture*, edited by E. Erwin. New York: Routledge.

Freedman, N., J. Hoffenberg, N. Vorus, and A. Frosch. 1999. "The Effectiveness of Psychoanalytic Psychotherapy: The Role of Treatment Duration, Frequency of Sessions, and the Therapeutic Relationship." *Journal of the American Psychoanalytic Association* 47: 741–72.

Grünbaum, A. 1984. *The Foundations of Psychoanalysis: A Philosophical Critique*. Berkeley: University of California Press.

———. 2002. "Critique of Psychoanalysis." In *The Freud Encyclopedia: Theory, Therapy, and Culture*, edited by E. Erwin. New York: Routledge.

Hanly, C. 1992. "Inductive Reasoning in Clinical Psychoanalysis." *International Journal of Psychoanalysis* 73: 293–301.

Jacobson, N., and A. Christensen. 1996. "Studying the Effects of Psychotherapy: How Well Can Clinical Trials Do the Job?" *American Psychologist* 51: 1031–39.

Kazdin, A. 1981. "Drawing Valid Inferences from Case Studies." *Journal of Consulting and Clinical Psychology* 49: 183–92.

Kline, P. 1981. *Fact and Fantasy in Freudian Theory*. 1972; New York: Methuen.

———. 1988. "Kline Replies to Erwin." In *Mind, Psychoanalysis, and Science*, edited by P. Clark and C. Wright. New York: Basil Blackwell.

Lyons, L., and P. Woods. 1991. "The Efficacy of Rational-Emotive Therapy: A Quantitative Review of the Outcome Research." *Clinical Psychology Review* 11: 357–69.

Sandell, R., J. Blomberg, A. Lazar, J. Carlsson, J. Broberg, and J. Schubert. 2000. "Varieties of Long-Term Outcome among Patients in Psychoanalysis and Long-Term Psychotherapy: A Review of Findings in the Stockholm Outcome of Psychoanalysis and Psychotherapy Project (STOPP)." *International Journal of Psychoanalysis* 81: 921–42.

Seligman, M. 1995. "The Effectiveness of Psychotherapy: The *Consumer Reports* Study." *American Psychologist* 50: 965–74.

Smith, M., G. Glass, and T. Miller. 1980. *The Benefits of Psychotherapy.* Baltimore, MD: Johns Hopkins University Press.

Tasman, A. 1998. "Opinion." *Journal of the American Psychoanalytic Association* 46: 669–72.

Van Fraassen, B. 1989. *Laws and Symmetry.* Oxford: Oxford University Press.

Vaughn, S., and S. Roose. 1995. "The Analytic Process: Clinical and Research Definitions." *International Journal of Psychoanalysis* 76: 343–56.

Wilkes, K. 1990. "Analyzing Freud." *Philosophical Quarterly* 40: 241–54.

7.

HOW CAN WE KNOW WHAT MADE THE RATMAN SICK?

Singular Causes and
Population Probabilities

NANCY CARTWRIGHT

M uch of Adolf Grünbaum's work on psychoanalytic theory over the years has been concerned with causal claims. These can be claims either about the aetiology of psychic disorders or about the efficacy of various aspects of psychoanalytic treatment; and they can be either singular causal claims intended to describe a single individual, such as "Anna O's neurotic symptoms were cured by Breuer's inducement of cathartic recall,"[1] or generic claims meant to be true in general (what I call "causal laws"), like "The adult male disposition of castration anxiety is attributable to oedipal childhood events prior to age 6."[2]

I am here going to discuss singular causal claims and how to support them. Grünbaum and I both,[3] I gather, accept what is conventionally regarded as best practice in so-called "quasi-experiments"—so long as all the precautions are tended to. I have in mind here methods like pretest-posttest control group designs, Solomon four groups designs, simple ANOVA, or Latin square designs.

In all these cases we look for a difference in the outcome between when the individual is "treated" with the cause and when not. The methods in my list are increasingly difficult to apply, and correlatively, increasingly more powerful. The simpler designs are more apt to go astray because they are less good at guarding against reasons for the difference in outcome *other than* the treatment. This is the same idea that Grünbaum uses in his criticisms of many of the psychoanalytic claims about individual etiologies: the reasons employed in arriving at a causal conclusion are not good enough to ensure that the putative cause made a difference to the outcome.

I do not agree with Grünbaums's insistence that the cause must make a difference in the single case. The cause must *contribute* to the effect. But that is no guarantee that it will make a difference. The idea that it must make a difference lies at the core of counterfactual accounts of causality; and these notoriously give wrong verdicts in cases of preemption. The president would nevertheless have died if the trainee assassin had not fired because, in that case, the experienced assassin standing beside the trainee would have shot.

There are of course ways to adjust our semantics for either counterfactuals or causal claims to deal with cases of preemption. For instance, withdraw the claim that causes make a difference to their effects and replace it by a two-stage analysis. One factor causes another if there is a chain of causal dependence between the cause and the effect, where the counterfactual connection does hold for each link. We may dispute whether this tactic does the job since it depends on the empirical hypothesis that between a cause and its effect there is always a chain of events where each is so close—"close enough"—to the next that preemption is impossible and the first will make a difference to the second. But whether it works or not, it does not rescue the claim that a cause must make a difference to its effect but rather gives up on it. I maintain that that is all to the good. For the claim mistakes a "sometimes" test, or symptom, of causation for a necessary condition: "sometimes," perhaps often, causes contribute to their effects in such a way as to make a difference. But it is not necessary that they do so.

Even should we adopt the assumption that causes make a difference, we still face enormous difficulties in putting it to use. That will be my topic here. How can we tell whether one factor makes a difference or not to another, outside of one of the really good single case designs? Grünbaum offers a necessary condition, which I will describe in a moment. A necessary condition can of course rule out singular causal claims even though it is not enough to admit them. Here I am more cautious even than Grünbaum. I do not know of any general criteria I am happy with that will rule out singular claims as false, let alone some criteria that will rule them in as true. Outside of a good quasi-experiment, unless we are in a situation where we have a great deal of background knowledge to deploy and are not relying on general criteria (i.e., on ones that are valid in every case),[4] the verdict must be left open.

The problem of "other reasons" for a positive outcome is very familiar in the case of treatments. The patient got better, but was it the treatment that did it as opposed to "spontaneous remission"? Lots of conditions do just go away "on their own." Or might it instead be a "placebo effect"—the result of confidence and expectation that does not depend on the specific character of the treatment but might accompany any treatment?[5] These are just the problems Grünbaum raises. For instance, Anna O, it seemed, lost her neurotic symptoms under Breuer's treatment of cathartic recall. But what convincing reason is there to think the treatment caused the relief?[6]

It is similar for claims about the sources of an individual patient's neuroses. Was the Ratman's obsession provoked by his encounter with the Czech captain? Was its base cause certain repressed (or imagined) childhood experiences? Was Anna O's inability to drink water due to her disgust at seeing a dog lapping water from a friend's glass? Much of the argument in favor of these hypotheses depends, Grünbaum notes, on "thematic affinities"; and thematic affinities are not enough!

In explaining why thematic affinities are not enough, Grünbaum considers a couple of analogies. He asks what licenses a tourist to make the causal inference that shapes in the sand that look very much like human footprints were caused by human feet walking there. He says,

> To draw the inference, the tourist avails himself of a crucial piece of additional information . . . footlike beach formations in the sand

never or hardly ever result from the mere collocation of sand particles under the action of air, such as gusts of wind. Indeed, the additional evidence is that, within the class of beaches, the incursion of
a pedestrian onto the beach *makes the difference* between the
absence and presence of the footlike formations.[7]

He also contrasts two dreams about houses. The first is Agnes's
dream of a house that looks just like Frank Lloyd Wright's Falling
Water. She has the dream just after she has visited it for the first time,
never having seen or heard about it before then. The second is his own
dream about houses after a typical day on which he has passed a lot of
houses. In the first we have reason to postulate a causal connection
because "Agnes's visit *made a difference* to her having that dream."[8]
The contrary is the case with Grünbaum's own dream:

> To put it more precisely, seeing a house on the day before a dream
> does not divide the class of the day's waking experiences on the
> prior days into two subclasses, such that the probabilities (or fre-
> quencies) of the appearance of a house in the next dream *differ* as
> between the two subclasses. On the other hand, in Agnes's life, such
> a division does occur, with ensuing *different* probabilities of
> dreaming about that house.[9]

More generally,

> If X is to be causally relevant to Y in reference class C, X must *par-
> tition* C into *two* subclasses in which the probabilities or incidences
> of Y are different from one another.[10]

He repeats this general demand verbatim in his 1992 "postscript"
to his criticism of the way the case study method is used in psychoanalysis in *Validation,* and he notes there that this criterion is defended
by Wesley Salmon in his work on statistical relevance as a mark of
causation. It is this condition I want to focus on.

The first thing I want to note is that this is, as I said, offered as a necessary condition. Showing that X increases the probability of Y in the right reference class C shows that X can cause Y; on my account of causal laws[11] it even shows that some individual Xs do cause Ys in C. But that does not show that for the case at hand X did cause Y. A lot of other factors that can cause Y might be present as well and they might have been responsible for Y. Or Y might have, in a sense, had no cause at all, as in the examples of spontaneous remission—Y's occurrence was the outcome of the natural development of the system. The requisite increase in probability of Y on X is not enough to show that Grünbaum's own condition—that X made a difference to Y in the case at hand—is met.

This might be overlooked if we focus only on the two analogues Grünbaum mentions. For in each case the other kinds of factors that raise the probability of the effect are extremely unlikely to have occurred. For instance, the prints in the sand could be made by a hand-held mold as part of an elaborate hoax, or a line of children in a strange sand castle competition. If these had occurred, then more needs to be done to show that the putative cause is really the one that brought about the effect (or, where the effect is cumulative, that it contributed).

Let us turn now to the condition itself. I have been reading this in the usual way, as a condition intended to bear immediately on *generic* causal claims, not singular ones, on what I call "causal laws." My paradigms are "Aspirins relieve headaches" and "Inverting and sparking a population of atoms causes lasing." For generic claims, I want to argue, the condition may be a sufficient condition, but it is not necessary.

Whether it will be a sufficient condition in a reference class depends very much on what that reference class is like. We are all familiar with the problem of spurious correlation. One factor raises the probability of a second, but it does not cause it. Rather, they are both joint effects of a common cause. A common tactic here is to demand reference classes in which all common causes are held fixed. But that is not enough. Even in a reference class like that, if the members of the class are also preselected for a given effect, causes of that effect can be correlated or anticorrelated, without causing each other. Also, if causes can act purely probabilistically, then joint effects of a common cause can be correlated even in a population where the cause is held fixed. There are also those strange quantum mechanical cases, where distant factors are correlated even—so the standard story goes—without any causal explanation.

I think the best we can do here is to employ a process of elimination. Causes can increase the probability of their effects. But there are also many other possible explanations for such an increase in probability. If we have a reference class in which all other explanations for the increase in probability have been eliminated, then it must be due to the fact that the one factor causes the other.

Next, why do I maintain that the condition is not necessary? There are two well-known reasons. For the first, let us consider a version of Wesley Salmon's example,[12] which was central to his work on statistical relevance that Grünbaum cites. We consider an experimental situation in which by the random flip of a fair coin either a strong radioactive element, E_1, or a weak one, E_2, will be introduced into an empty box. We look for the presence of a radioactive particle, α, later. The probability that E_1 produces an α particle in the designated time is .9; for E_2, the probability of α is .1. Since the probability is .5 for either element to be in the box, the overall probability for α is .5. Notice that the probability of α with E_2 in the box is lower than this. But when E_2 is in the box and an α particle appears, it is certainly E_2 that causes it. This led Salmon to claim that causes may reduce the probability of their effects as well as raising it.

A slight change in the example leads to a more radical conclusion, however. Imagine we have available a large number of different radioactive elements; two—E_3 and E_4—happen to have the same probability for producing α particles. Repeat the experiment now with E_3 and E_4. If E_3 is in the box when an α particle appears, definitely E_3 causes it and similarly with E_4. Yet in this case the probability stays the same. The cause makes no difference to the probability of the effect.

Salmon's example is a case of Simpson's paradox, which I rediscovered years ago in criticizing Patrick Suppes's use of increase in

probability as a necessary condition for causality.[13] Suppose we parti-
tion a population into cells according to different values of a variable
Z. Then X can increase the probability of Y in every cell of the parti-
tion and yet not do so in the population as a whole. In order for this to
happen X and Y must be probabilistically dependent on Z. In my
amended version of Salmon's example X = radioactive element 1, Y =
presence of decay particle, and Z = presence of radioactive element 2.
X increases the probability of Y given Z and it also increases it given
–Z; but it does not do so in the population as a whole. That is because,
as we have seen, the presence of the one element is correlated with the
absence of the other.

One proposal conventionally offered in solution to this problem is
the one I described above in talking about sufficient conditions for
causal claims: insist that the reference class for testing "X causes Y"
should be homogeneous with respect to all other causes of Y other than
X. In a reference class like that, nothing should be correlated (or, so it is
supposed) with X, and hence Simpson's paradox cannot make problems. I do not know what Grünbaum thinks of this solution. He might
have intended this as a constraint on the reference class C all along. He
probably discusses this somewhere that I missed. Here is a remark I did
find, in a discussion of a similar problem in a singular case:

> But it is *utterly chimerical* to predicate a research design on a situa-
> tion in which two people differ *only* with respect to the property that
> the investigator *conjecturally* deems relevant to the outcome![14]

Of course the quote only says that we will not be likely to ensure
such a research design; it does not tell us whether our condition is a
necessary condition if we did have such a design.

Besides the strategy of demanding an increase in probability only
in reference classes that are homogeneous with respect to all other
causal factors, there is a second strategy to deal with Simpson's
paradox—simply deny the phenomenon. I suppose this is not a crazy
proposal because very clever people like Clark Glymour, Peter Spirtes,

Judea Pearl, and a host of followers insist on it. It is a major assumption in the currently fashionable Bayes-nets methods for causal discovery and causal inference. The chief argument offered in its support is one about mathematical spaces. Notice that in the Salmon example, only some specific arrangements of probabilities will make $P(Y/X)=P(Y/-X)$. This is characteristic of Simpson's paradox situations. I can construct millions of them, but always one has to get the numbers to balance out correctly. This is reflected in a theorem: if you put a Lebesgue measure over n-tuples of numbers, the set of n-tuples that gives rise to Simpson's paradox has measure zero. From this it is concluded that these situations never occur in nature. Not only is this a bad inference, the conclusion seems palpably untrue as well. I shall return briefly to this point in a moment.

There is a second well-known reason why a cause may not increase the probability of its effect: a particular cause may itself have different capacities with respect to the same effect and these capacities may balance out. This is especially likely to happen in systems we design ourselves, either consciously or by trial and error. One and the same cause can have opposing tendencies—both to enhance an effect and to retard it. In any case, where these tendencies just balance out, we will see no increase in the probability of the effect on the cause. Consider, for instance, a certain brand of nondrowsy decongestant: it does not put you to sleep. Nevertheless the patented chemical in it is a powerful soporific. The decongestant does not induce drowsiness because the chemical is always packaged with an equally potent stimulant.[15]

For cases like this where there is no change in the probability of the effect whether the cause is present or not, it may nevertheless be extremely useful to know the specific causal facts. Armed with the information about the Janus-faced nature of a cause, we may, for instance, be able to place some block in one of its pathways and thus be left with only the result of the tendency in the other direction.

Those who deny that Simpson's paradox situations ever occur generally also deny that cancellations of opposing tendencies will ever lead to equality of the conditional probabilities. The primary argument is the theorem as before: the numbers that afford exact cancellation form a set of measure zero. Again, this argument seems both invalid and unsound. Exact cancellations are often just what we try to achieve. For the decongestants, for example, they are supposed merely to decongest—they are supposed *neither* to put one to sleep nor to stimulate one. So exact cancellation matters.

Economic methodologist Kevin Hoover also takes this kind of cancellation to be common:

> Spirtes et al. (1993, 95) acknowledge the possibility that particular parameter values might result in [exact cancellation], but they dismiss their importance as having "measure zero." But this will not do for macroeconomics. It fails to account for the fact that in macroeconomic and other control contexts, the policymaker aims to set parameter values in just such a way to make this supposedly measure-zero situation occur. To the degree that policy is successful, such situations are common, not infinitely rare.[16]

It would also be a mistake to put too much emphasis on questions of whether *exact* cancellations occur regularly or not for a reason central to the genuine practical considerations that underwrite Grünbaum's concerns about psychoanalytic claims. We are not primarily concerned with pure abstract philosophy—is increase of probability a necessary condition for causality or not? Rather we are interested in real methodologies that can be used to give reasonable confidence in claims that support often-costly strategies for relief from very severe problems.

Our tools of statistical estimation are never, even in principle, good enough to settle a claim about exact equality of conditional probabilities. I think we have good empirical evidence that both near-Simpson's paradox situations and near-cancellations occur regularly and we know that our best methods will frequently estimate conditional probabilities as equal. If we adopt increase in probability as a necessary condition for a generic causal relation, we will give wrong verdicts about these. And this really matters because in both cases there may be large numbers of individuals for whom the cause has a profound effect.

This is my case against increase in probability as a necessary condition for generic level causation. Perhaps, however, Grünbaum might want to use it directly as a test for singular causation. Maybe not, though. I myself have never been comfortable with talk of single-case probabilities —and I think I learned this suspicion from Grünbaum himself! They at any rate just push the problem back a step—how *for the single case* can we establish probability claims in a psychoanalytic setting?

Also I suspect we would frequently get the wrong verdicts. Consider one of those cases of opposing tendencies that cancel, where the cancellation in probabilities takes place not because different members of the population experience opposing outcomes, but rather because the opposing tendencies balance out in each individual. Gerhard Hesslow's example of birth control pills and thrombosis is, I believe, supposed to work like this.[17] The pills cause thrombosis; they also prevent pregnancy, which itself is a cause of thrombosis. If the two tendencies cancel for an individual, say me, then I suppose we might find, for single case probabilities, P(N gets thrombosis/N takes pills) = P(N gets thrombosis/N does not take pills). Still we would want to know *both* causal facts because we may be able to find a way to block the deleterious pathway and leave me with only the beneficial effects (thereby changing the conditional probabilities).

The more usual way to use probabilities to treat single cases is in a two-step process. We use the probabilities to establish generic causal claims, then—like Donald Davidson—we insist that all admissible singular claims fall under a generic law. So (I) if we take increase in probability as a necessary condition for a generic causal truth, "Cs cause Es" and (II) we take as a necessary condition for "this C caused this E" that it be generically true that Cs cause Es, then we would have as a result that increase in probability of E given C is a necessary condition for the related singular causal claim. I have already explained why I do not accept the first step. I am also suspicious of the second, whether or not we accept the first. I particularly worry about the second in medical and psychiatric settings of the kind that Grünbaum is concerned with.

Let us begin anecdotally. I offer Emily a diet regime to lose weight. She says, "That kind of crash diet never works." I reply, "It works for me." And I do so with good reason. I am fairly careful about these matters. I have tried this diet frequently. It always works. I know about the possibility of spontaneous weight loss and about the placebo effect, about long-term versus short-term outcomes, and so forth, and I have evidence these are not a problem. Now there *may* be some description, D, of me that fixes *in a law-like way* the efficacy of this diet for people who satisfy D. In that case we would have a reference class, picked out in a non-question-begging way, in which P(weight loss/diet) > P(weight loss/no diet). But there may not be such a description. There may simply be individual variation. To insist that there is always such a description is to let a big—and insecure—metaphysical assumption guide our methodology. That I think is a wrong thing to do.

It is however widely assumed. For instance consider Hersen and Barlow's text, still in use, *Single Case Experimental Designs*. Here they endorse this view, citing in turn the earlier text of M. Sidman on experimental data in psychology:

> Physics assumes that variability is imposed by error of measurement or other identifiable factors. Experimental efforts are then directed to discovering and eliminating as many sources of variability as possible. . . . Sidman proposes that basic researches in psychology adopt this strategy. Rather than assuming that variability is intrinsic to the organism, one should make every effort to discover sources of behavioral variability among organisms.[18]

Here of course we do not see the strong metaphysical position—variability is impossible, but merely a methodological injunction. But we need to be cautious even about methodological injunctions. We want to look for sources of variations because that knowledge could be powerful. But every hunt is costly, especially if we do not have good starting ideas of where to look. And in the case of human beings—who are not after all electrons—we cannot rule out the possibility of intrinsic variability. So we may be hunting for what in fact can never be found. In cases like this, we need to weigh the costs and the benefits, the probabilities of success and the probabilities of failure from following the advice.

The question is a live one, particularly now regarding pharmaceuticals. Some drugs may work very well for some people but for others may worsen the very condition they are supposed to treat. Ideally we would like a testable description for those in the two categories; and we would like to ensure that within these two categories, the treatment passes all the tests for generic-level causes. But such testable descriptions are often not available, and we must even leave open the possibility that they simply do not exist. The good news is that sometimes the bad effects set in gradually and in some known pattern so that patients can be monitored and taken off the treatment if it is proving harmful.

One possible example that has recently received notoriety is Prozac, which is used to treat depressives. It seems, however, sometimes to induce akathesia, a kind of restlessness that can lead to suicide or even the killing of others. So far nobody knows how to predict who will fall into the category where Prozac increase the chance of suicide (or even if there is such a proper category). The controversy has arisen in part because many previous antidepressants were recommended to be taken with tranquilizers; this was not so for Prozac—except in Germany; and now there is the suspicion that a number of people have killed themselves and others on account of taking the antidepressant intended, inter alia, to prevent suicide.

There is a mire of legal, social, medical, moral, and methodological problems here. It is commonplace that standard clinical trial procedures, even when the trial population is large and compliant, do not tell us when treatments have opposing effects. I have not found anything yet in Grünbaum's own writing that looks at the methodology in cases like this. One hope I have in raising these issues is that we can encourage him to turn to this tangle of problems to help sort them out, as he has so many other issues of importance to human welfare.

NOTES

Research for this paper was aided by the AHRB project, *Causality: Metaphysics and Methods*, and was supported by a grant from the Latsis Foundation. I would like to thank both. I first became seized with the importance and excitement of philosophy of science from Adolf Grünbaum as well as with a respect for careful thought and serious dialogue, which I am afraid I have never been able to emulate sufficiently; and I am deeply grateful to him, as well as for his support and friendship over many years.

1. A. Grünbaum, *Validation in the Clinical Theory of Psychoanalysis, A Study in the Philosophy of Psychoanalysis* (Madison, CT: International Universities Press, 1993), pp. 238ff.

2. Ibid., p. 176.

3. See by comparison A. Grünbaum's endorsement that "the A-B-A does succeed in discrediting spontaneous remission." Ibid., p. 237.

4. I do not mean here that the criterion can actually be applied in every case, but rather that its deliverances are universally reliable if properly applied.

5. To be more careful, we should follow Grünbaum's definition: "I speak of a treatment Gain as a *placebo effect* with respect to a particular target disorder, therapeutic *theory*, and type of patient, just when that positive effect is produced by treatment factors *other than* those designated as the efficacious ones by the therapeutic theory." Grünbaum, *Validation in the Clinical Theory of Psychoanalysis*, p. 189.

6. See by means of comparison ibid., pp. 237ff.

7. A. Grünbaum, "'Meaning' Connections and Causal Connections in the Human Sciences," *Journal of the American Psychoanalytic Association* 38 (1990): 559–77, esp. 568.

8. Ibid., p. 570.

9. Ibid.

10. Ibid., p. 571.

11. See by means of comparison N. Cartwright, *Nature's Capacities and Their Measurement* (Oxford: Oxford University Press, 1989).

12. W. Salmon, *Statistical Explanation and Statistical Relevance* (Pittsburgh: University of Pittsburgh Press, 1971).

13. N. Cartwright, *How the Laws of Physics Lie* (Oxford: Oxford University Press, 1983).

14. Grünbaum, *Validation in the Clinical Theory of Psychoanalysis*, p. 242.

15. Thanks to Lisa Lloyd for this example.

16. K. Hoover, *Causality in Macroeconomics* (Cambridge: Cambridge University Press, 2001), p. 17.

17. G. Hesslow, "Discussion: Two Notes on the Probabilistic Theory of Causality," *Philosophy of Science* 43, no. 2 (1976): 290–92.

18. D. H. Barlow and M. Hersen, *Single-Case Experimental Designs* (Oxford: Pergamon Press, 1981), p. 35.

References

Barlow, David H., and Michel Hersen. *Single-Case Experimental Designs.* Oxford: Pergamon Press, 1981.

Cartwright, Nancy. *How the Laws of Physics Lie.* Oxford: Oxford University Press, 1983.

———. *Nature's Capacities and Their Measurement.* Oxford: Oxford University Press, 1989.

Grünbaum, Adolf. "'Meaning' Connections and Causal Connections in the Human Sciences." *Journal of the American Psychoanalytic Association* 38 (1990): 559–77.

———. *Validation in the Clinical Theory of Psychoanalysis, A Study in the Philosophy of Psychoanalysis.* Madison, CT: International Universities Press, 1993.

Hesslow, Gerhard. "Discussion: Two Notes on the Probabilistic Theory of Causality." *Philosophy of Science* 43, no. 2 (1976): 290–92.

Hoover, Kevin. *Causality in Macroeconomics.* Cambridge: Cambridge University Press, 2001.

Salmon, Wesley. *Statistical Explanation and Statistical Relevance.* Pittsburgh: University of Pittsburgh Press, 1971.

BIBLIOGRAPHY OF ADOLF GRÜNBAUM

GRÜNBAUM WEB SITE:
WWW.PITT.EDU/~GRUNBAUM

[1]   (1942) (With R. S. Cohen) *An Outline of Trigonometry (Plane and Spherical)*. Boston: Hymarx Student Outlines.

[2]   (1948) Review of "The Logic of the Sciences and the Humanities," by F. S. C. Northrop. *Yale Law Journal* 57, no. 7 (June): 1332–38.

[3]   (1950) "Realism and Neo-Kantianism in Professor Margenau's Philosophy of Quantum Mechanics." *Philosophy of Science* 17, no. 1 (January): 26–34.

[4]   (1950) "Relativity and the Atomicity of Becoming." *Review of Metaphysics* 4, no. 2 (December): 143–86. This article contains some serious distorting misprints.

[5]   (1951) "Some Recent Writings in the Philosophy of Mathematics." *Review of Metaphysics* 5, no. 2 (December): 281–92.

[6]   (1951) *The Philosophy of Continuity*. PhD thesis, Yale University. See also [148].

[7]   (1952) "A Consistent Conception of the Extended Linear Continuum as an Aggregate of Unextended Elements." *Philosophy of Science* 19, no. 4 (October): 288–306.

[8]   (1952) "Causality and the Science of Human Behavior." *American Scientist* 40, no. 4 (October): 665–76. See also [9], [12], [43], [47], [64], [90], [91], [96], [136], [137], [189], [214], and [257].

[9]   (1952) "Causality and the Science of Human Behavior." In Bobbs-Merrill Reprint Series in the Social Sciences, Item P-639, pp. 665–76. This is a reprint of the article in *American Scientist* (1952) [8]. See also [12], [43], [47], [64], [90], [91], [96], [136], [137], [189], [214], and [257].

[10]   (1952) "Messrs. Black and Taylor on Temporal Paradoxes." *Analysis* 12 (June): 144–48.

[11]   (1952) "Some Highlights of Modern Cosmology and Cosmogony." *Review of Metaphysics* 5, no. 3 (March): 481–98.

[12]   (1953) "Causality and the Science of Human Behavior." In *Readings in the Philosophy of Science*, edited by H. Feigl and M. Brodbeck. New York: Appleton-Century-Crofts, pp. 766–78. This is a reprint of the article in *American Scientist* (1952) [8]. See also [9], [43], [47], [64], [90], [91], [96], [136], [137], [189], [214], and [257].

[13]   (1953) "Comments on Power and Science." *Review of Metaphysics* 6, no. 3 (March): 478–80.

[14]   (1953) "Critique of H. Dingle's Objection to the Bondi-Gold Cosmology." *Scientific American* 189, no. 6 (December): 6–8.

[15]   (1953) "Relativity, Causality and Weiss's Theory of Relations." *Review of Metaphysics* 7, no. 1 (September): 115–23.

[16] (1953) Review of *The Rise of Scientific Philosophy*, by Hans Reichenbach. *Scripta Mathematica* 19, no. 1 (March): 48–54.

[17] (1953) "Some Remarks on Professor Ushenko's Interpretation of Causal Law." *Journal of Philosophy* 50, no. 4 (February 12): 115–20.

[18] (1953) "Whitehead's Method of Extensive Abstraction." *British Journal for the Philosophy of Science* 4, no. 15: 215–26.

[19] (1954) "E. A. Milne's Scales of Time." *British Journal for the Philosophy of Science* 4, no. 16: 329–30.

[20] (1954) "Operationism and Relativity." *Scientific Monthly* 79, no. 4 (October): 228–31. See also [41] and [58].

[21] (1954) Review of *Readings in the Philosophy of Science*, edited by H. Feigl and M. Brodbeck. *American Journal of Physics* 22, no. 7 (October): 498–99.

[22] (1954) Review of *Histoire de la Mécanique*, by R. Dugas. *Philosophy and Phenomenological Research* 15, no. 1 (September): 119–21.

[23] (1954) Review of *Positivism, A Study in Human Understanding*, by R. von Mises. *Scripta Mathematica* 20, no. 1–2 (March–June): 75–76.

[24] (1954) "Science and Ideology." *Scientific Monthly* 79, no. 1 (July): 13–19. See also [25].

[25] (1954) "Science and Ideology." *Humanist* 14, no. 4 (July–August): 161–73. This is a version of the article in *Scientific Monthly* (1954) [24].

[26]   (1954) "The Clock Paradox in the Special Theory of Relativity." *Philosophy of Science* 21, no. 3 (July): 249–53. See also [30] and [32].

[27]   (1955) "Logical and Philosophical Foundations of the Special Theory of Relativity." *American Journal of Physics* 23, no. 7 (October): 450–64. See also [45], [51], and [307].

[28]   (1955) "Modern Science and Refutation of the Paradoxes of Zeno." *Scientific Monthly* 81, no. 5 (November): 234–39. Reprinted in W. C. Salmon, ed. *Zeno's Paradoxes.* See also [94], [105], and [123].

[29]   (1955) "Reply to Dr. Leaf." *Philosophy of Science* 22, no. 1 (January): 53.

[30]   (1955) "Reply to Dr. Törnebohm's Comments on 'The Clock Paradox in the Special Theory of Relativity.'" *Philosophy of Science* 22, no. 3 (July): 233. See also [26] and [32].

[31]   (1955) Review of *Sovereign Reason*, by E. Nagel. *Science* 121, no. 3155 (June 17): 862.

[32]   (1955) "The Clock Paradox in the Special Theory of Relativity." In *Proceedings of the 2nd International Congress of the IUHPS*, vol. 3. Neuchâtel, Switzerland: Editions du Griffon, pp. 55–60. This is a version of the article in *Philosophy of Science* (1954) [26]. See also [30].

[33]   (1955) "Time and Entropy." *American Scientist* 43, no. 4 (October): 550–72. See also [39].

[34]   (1955) "Time and Ethics." *Scientific Monthly* 80, no. 3 (March): 200–201.

[35] (1956) "Historical Determinism, Social Activism, and Predictions in the Social Sciences." *British Journal for the Philosophy of Science* 7, no. 27: 236–40. See also [303].

[36] (1956) "Relativity of Simultaneity within a Single Galilean Frame: A Rejoinder." *American Journal of Physics* 24, no. 8 (November): 588–90.

[37] (1956) Review of *The Direction of Time*, by H. Reichenbach. *American Scientist* 44 (October): 294A–300A.

[38] (1957) "Complementarity in Quantum Physics and Its Philosophical Generalization." *Journal of Philosophy* 54, no. 23 (November 7): 713–27. This paper was presented in the Symposium on Determinism in the Light of Recent Physics at the Meeting of the American Philosophical Association, Eastern Division, on December 29, 1957.

[39] (1957) "Das Zeitproblem." *Archiv für Philosophie* 7, no. 3/4: 165–208. This is a German translation of "Time and Entropy" in *American Scientist* (1955) [33].

[40] (1957) (With E. L. Hill) "Irreversible Processes in Physical Theory." *Nature* 179 (June 22): 1296–97.

[41] (1957) "Operationism and Relativity." In *The Validation of Scientific Theories*, edited by P. G. Frank. Boston: Beacon Press, pp. 84–94. This is a version of the article in *Scientific Monthly* (1954) [20]. See also [58].

[42] (1957) "Remarks Concerning Moon and Spencer's 'On the Establishment of a Universal Time.'" *Philosophy of Science* 24, no. 1 (January): 77–78.

[43]   (1957) "Science and Man." In *Philosophic Problems: An Introductory Book of Readings*, edited by M. Mandelbaum, F. W. Gramlich, and A. R. Anderson. New York: Macmillan, pp. 328–38. This is a version of the article in *American Scientist* (1952) [8]. See also [9], [12], [47], [64], [90], [91], [96], [137], [189], [214], and [257].

[44]   (1957) "The Philosophical Retention of Absolute Space in Einstein's General Theory of Relativity." *Philosophical Review* 66, no. 4 (October): 525–34. See also [88].

[45]   (1958) "Fundamental Philosophical Issues in the Special Theory of Relativity." In *Kritik und Fortbildung der Relativitätstheorie*, edited by K. Sapper. Graz, Austria: Akademische Druck u. Verlagsanstalt, pp. 1–26. This is a revised version of the article in *American Journal of Physics* (1955) [27]. See also [51] and [307].

[46]   (1958) "Philosophical Principles in the Special Theory of Relativity" (in Hebrew). *IYYUN* 9, no. 4 (October): 249–62.

[47]   (1959) "Causality and the Science of Human Behavior." In *Contemporary Readings in General Psychology*, edited by R. S. Daniel. Boston: Houghton Mifflin, pp. 328–36. This is a version of the article in *American Scientist* (1952) [8]. See also [9], [12], [43], [64], [90], [91], [96], [137], [189], [214], and [257].

[48]   (1959) "Conventionalism in Geometry." In *Proceedings of the Symposium on the Axiomatic Method, with Special Reference to Geometry and Physics*, edited by L. Henkin, P. Suppes, and A. Tarski. Studies in Logic and the Foundations of Mathematics. Amsterdam, Netherlands: North-Holland Publishing, pp. 204–22.

[49]   (1959) "Remarks on Dr. Kubie's Views." In *Psychoanalysis, Scientific Method, and Philosophy*, edited by S. Hook. New York: New York University Press, p. 225.

[50]   (1959) "The Falsifiability of the Lorentz-Fitzgerald Contraction Hypothesis." *British Journal for the Philosophy of Science* 10, no. 37: 48–50. See also [54].

[51]   (1960) "Logical and Philosophical Foundations of the Special Theory of Relativity." In *Philosophy of Science*, edited by A. Danto and S. Morgenbesser. New York: Meridian Books, pp. 399–434. This is a version of the article in *American Journal of Physics* (1955) [27]. See also [45] and [307].

[52]   (1960) "The Duhemian Argument." *Philosophy of Science* 27, no. 1 (January): 75–87. See also [53] and [158].

[53]   (1960) "The Duhemian Argument." In Bobbs-Merrill Reprint Series in Philosophy, Item Phil-90, pp. 75–87. This is a reprint of the article in *Philosophy of Science* (1960) [52]. See also [158].

[54]   (1960) "The Falsifiability of the Lorentz-Fitzgerald Contraction Hypothesis: A Rejoinder to Professor Dingle." *British Journal for the Philosophy of Science* 11, no. 42: 143–45. See also [50].

[55]   (1960) "The Role of *A Priori* Elements in Physical Theory." In *The Nature of Physical Knowledge*, edited by L. W. Friedrich. Bloomington: Indiana University Press, pp. 109–28.

[56]   (1961) "A Rejoinder to Feyerabend's Comments on 'Law and Convention in Physical Theory.'" In *Current Issues in the Philosophy of Science*, edited by H. Feigl and G. Maxwell. New York: Holt, Rinehart and Winston, pp. 161–68. See also [57].

[57]   (1961) "Law and Convention in Physical Theory." In *Current Issues in the Philosophy of Science*, edited by H. Feigl and G. Maxwell. New York: Holt, Rinehart and Winston, pp. 140–55. See also [56].

[58]   (1961) "Operationism and Relativity." In *The Validation of Scientific Theories*, edited by P. G. Frank. New York: Collier Books, pp. 83–92. This is a reprint of the article in *Scientific Monthly* (1954) [20]. See also [41].

[59]   (1961) "Professor Dingle on Falsifiability: A Second Rejoinder." *British Journal for the Philosophy of Science* 12, no. 46 (August): 153–57.

[60]   (1961) "The Genesis of the Special Theory of Relativity." In *Current Issues in the Philosophy of Science*, edited by H. Feigl and G. Maxwell. New York: Holt, Rinehart and Winston, pp. 43–53. See also [107] and [139].

[61]   (1962) "Discussion: The Structure of Science." *Philosophy of Science* 29, no. 3 (July): 294–305.

[62]   (1962) "Geometry, Chronometry, and Empiricism." In *Scientific Explanation, Space, and Time*, edited by H. Feigl and G. Maxwell. Minnesota Studies in the Philosophy of Science, vol. 3. Minneapolis: University of Minnesota Press, pp. 405–526. See also [82].

[63]   (1962) Prologue and epilogue in P. W. Bridgman (posthumous). *A Sophisticate's Primer of Relativity*. Middletown, CT: Wesleyan University Press, pp. vii–viii, 165–91. See also [75].

[64]   (1962) "Science and Man." *Perspectives in Biology and Medicine* 5, no. 4 (Summer): 483–502. This is a version of the article in *American Scientist* (1952) [8]. See also [9], [12], [43], [47], [90], [91], [96], [137], [189], [214], and [257].

[65]   (1962) "Temporally Asymmetric Principles, Parity between Explanation and Prediction, and Mechanism versus Teleology." *Phi-*

losophy of Science 29, no. 2 (April): 146–70. See also [78], [79], and [113].

[66]   (1962) "The Falsifiability of Theories: Total or Partial? A Contemporary Evaluation of the Duhem-Quine Thesis." *Synthese* 14, no. 1 (March): 17–34. For W. V. O. Quine's comment on this paper, see (1976) W. V. O. Quine. "A Comment on Grünbaum's Claim." In *Can Theories Be Refuted?* edited by S. G. Harding. Dordrecht, Netherlands: D. Reidel Publishing, p. 132. See also [80], [92], and [213].

[67]   (1962) "The Nature of Time." In *Frontiers of Science and Philosophy*, edited by R. G. Colodny. University of Pittsburgh Series in the Philosophy of Science, vol. 1. Pittsburgh, PA: University of Pittsburgh Press, pp. 147–88. See also [68].

[68]   (1962) "The Nature of Time." In *Frontiers of Science and Philosophy*, edited by R. G. Colodny. London: Allen and Unwin, pp. 147–88. This book is a British edition of [67].

[69]   (1962) "The Relevance of Philosophy to the History of the Special Theory of Relativity." *Journal of Philosophy* 59, no. 21 (October 11): 561–74.

[70]   (1962) "The Special Theory of Relativity as a Case Study of the Importance of the Philosophy of Science for the History of Science." *Annali di Matematica* 57, no. 4: 257–82. See also [81].

[71]   (1962) "Whitehead's Philosophy of Science." *Philosophical Review* 71, no. 2 (April): 218–29.

[72]   (1963) "Carnap's Views on the Foundations of Geometry." In *The Philosophy of Rudolf Carnap*, edited by P. A. Schilpp. Library of Living Philosophers, vol. 11. La Salle, IL: Open Court Publishing, pp. 599–684. See also [89] and [143].

[73]   (1963) "Comments on Professor Roger Buck's Paper 'Reflexive Predictions.'" *Philosophy of Science* 30, no. 4 (October): 370–72. See also [112].

[74]   (1963) *Philosophical Problems of Space and Time*. New York: Alfred A. Knopf. A British edition was published by London: Routledge & Kegan Paul (1964). See also [114], [140], [165], and [209].

[75]   (1963) Prologue and epilogue in P. W. Bridgman (posthumous). *A Sophisticate's Primer of Relativity*. London: Routledge & Kegan Paul, pp. vii–viii; 165–91. This is a British edition of [63].

[76]   (1963) "Space." Unpublished, but see [77] and [97].

[77]   (1963) "Space and Time." In *The Voice of America Forum Lectures*. Philosophy of Science Series, vol. 12. Washington, DC: US Information Agency. This was a lecture broadcast by Voice of America on July 1, 1963. See also [76] and [97].

[78]   (1963) "Temporally Asymmetric Principles, Parity between Explanation and Prediction, and Mechanism versus Teleology." In *Philosophy of Science*, edited by B. Baumrin. The Delaware Seminar, vol. 1, 1961–1962. New York: Interscience Publishers, pp. 57–96. This is a version of the article in *Philosophy of Science* (1962) [65]. See also [79] and [113].

[79]   (1963) "Temporally Asymmetric Principles, Parity between Explanation and Prediction, and Mechanism versus Teleology." In *Induction: Some Current Issues*, edited by H. E. Kyburg Jr. and E. Nagel. Middletown, CT: Wesleyan University Press, pp. 114–49. This is a version of the article in *Philosophy of Science* (1962) [65]. See also [78] and [113].

[80] (1963) "The Falsifiability of Theories: Total or Partial? A Contemporary Evaluation of the Duhem-Quine Thesis." In *Proceedings of the Boston Colloquium for the Philosophy of Science, 1961/1962,* edited by M. W. Wartofsky. Boston Studies in the Philosophy of Science, vol. 1. Dordrecht, Netherlands: D. Reidel Publishing, pp. 178–95. This is a version of the article in *Synthese* (1962) [66]. See also [92] and [213].

[81] (1963) "The Special Theory of Relativity as a Case Study of the Importance of the Philosophy of Science for the History of Science." In *Philosophy of Science,* edited by B. Baumrin. The Delaware Seminar, vol. 2, 1962–1963. New York: Interscience Publishers, pp. 171–204. This is a version of the article in *Annali di Matematica* (1962) [70].

[82] (1964) "Geometrie, Zeitmessung und Empirismus." *Archiv für Philosophie* 12, no. 3/4: 179–303. This is a German translation of "Geometry, Chronometry, and Empiricism," in *Scientific Explanation, Space, and Time* (1962) [62].

[83] (1964) "Is a Universal Nocturnal Expansion Falsifiable or Physically Vacuous?" *Philosophical Studies* 15, no. 5 (October): 71–79. Errata in this article are listed in (1965) *Philosophical Studies* 16, no. 3 (April): 47–48.

[84] (1964) "Popper on Irreversibility." In *The Critical Approach to Science and Philosophy: Essays in Honor of Karl R. Popper,* book 1, edited by M. Bunge. London: Macmillan, pp. 316–31.

[85] (1964) "Questions for Brand Blanshard." In *Philosophical Interrogations,* edited by S. Rome and B. Rome. New York: Holt, Rinehart and Winston, p. 205.

[86] (1964) "The Anisotropy of Time." *Monist* 48, no. 2 (April): 219–47. See also [98] and [122].

[87]   (1964) "The Bearing of Philosophy on the History of Science."
Science 143, no. 3613 (March 27): 1406–12.

[88]   (1964) "The Philosophical Retention of Absolute Space in
Einstein's General Theory of Relativity." In *Problems of Space and
Time*, edited by J. J. C. Smart. New York: Macmillan, pp. 313–17. This
is a version of the article in *Philosophical Review* (1957) [44].

[89]   (1964) "Time, Irreversible Processes, and the Physical Status
of Becoming." In *Problems of Space and Time*, edited by J. J. C.
Smart. New York: Macmillan, pp. 397–425.

[90]   (1966) "Causality and the Science of Human Behavior." In
Control of Human Behavior, edited by R. Ulrich, T. Stachnik, and J.
Mabry. Glenview, IL: Scott, Foresman, pp. 3–10. This is a version of
the article in *American Scientist* (1952) [8]. See also [9], [12], [43],
[47], [64], [91], [96], [137], [189], [214], and [257].

[91]   (1966) "Science, Causality, and Man." In *Value and Man:
Readings in Philosophy*, edited by L. Z. Hammer. New York:
McGraw-Hill, pp. 55–66. This is a version of the article in *American
Scientist* (1952) [8]. See also [9], [12], [43], [47], [64], [90], [96],
[137], [189], [214], and [257].

[92]   (1966) "The Falsifiability of a Component of a Theoretical
System." In *Mind, Matter, and Method: Essays in Philosophy and Sci-
ence in Honor of Herbert Feigl*, edited by P. K. Feyerabend and G.
Maxwell. Minneapolis: University of Minnesota Press, pp. 273–305.
This is a version of the article in *Synthese* (1962) [66]. See also [80]
and [213].

[93]   (1967) *Modern Science and Zeno's Paradoxes*. Middletown,
CT: Wesleyan University Press. See also [104].

[94]   (1967) "Modern Science and Zeno's Paradoxes of Motion." In *The Philosophy of Time: A Collection of Essays*, edited by R. M. Gale. New York: Anchor Doubleday Books, pp. 422–94. See also [28], [105], and [123].

[95]   (1967) "Relativity Theory, Philosophical Significance of." In *The Encyclopedia of Philosophy*, vol. 7, edited by P. Edwards. New York: Macmillan, pp. 133–40.

[96]   (1967) "Science and Man." In *Philosophic Problems: An Introductory Book of Readings*, 2nd ed., edited by M. Mandelbaum, F. W. Gramlich, A. R. Anderson, and J. B. Schneewind. New York: Macmillan, pp. 448–62. This is a version of the article in *American Scientist* (1952) [8]. See also [9], [12], [43], [47], [64], [90], [91], [137], [189], [214], and [257].

[97]   (1967) "Space and Time." In *Philosophy of Science Today*, edited by S. Morgenbesser. New York: Basic Books, pp. 125–35. This is a version of the lecture in *The Voice of America Forum Lectures* (1963) [77]. See also [76].

[98]   (1967) "The Anisotropy of Time." In *The Nature of Time*, edited by T. Gold and D. L. Schumacher. Ithaca, NY: Cornell University Press, pp. 149–77, 245–47; discussion on pp. 177–86. This is a revision of the article in *Monist* (1964) [86] and [122].

[99]   (1967) "The Denial of Absolute Space and the Hypothesis of a Universal Nocturnal Expansion: A Rejoinder to George Schlesinger." *Australasian Journal of Philosophy* 45, no. 1 (May): 61–91.

[100]   (1967) "The Status of Temporal Becoming." *Annals of the New York Academy of Sciences* 138 (February 6): 374–95. See also [101], [108], and [132].

[101]   (1967) "The Status of Temporal Becoming." In *The Philosophy of Time: A Collection of Essays*, edited by R. M. Gale. New York: Anchor Doubleday Books, pp. 322–53. This is a version of the essay in *Annals of the New York Academy of Sciences* (1967) [100]. See also [108] and [132].

[102]   (1968) "Are 'Infinity Machines' Paradoxical?" *Science* 159, no. 3810 (January 5): 396–406.

[103]   (1968) *Geometry and Chronometry in Philosophical Perspective*. Minneapolis: University of Minnesota Press. See also [119] for a reprint of chapter 3 of this book.

[104]   (1968) *Modern Science and Zeno's Paradoxes*. London: George Allen and Unwin. This is a second edition of the book in [93].

[105]   (1968) "Modern Science and Zeno's Paradoxes of Motion." In *The Philosophy of Time: A Collection of Essays*, edited by R. M. Gale. London: Macmillan, pp. 422–501. This is a revised British edition of the essay in [94]. See also [28] and [123].

[106]   (1968) "Spatial and Temporal Congruence in Physics: Newton"; "Has the General Theory of Relativity Repudiated Absolute Space?"; and "Is There a 'Flow' of Time or Temporal 'Becoming'?" In *Philosophy of Science: An Introduction*, edited by P. R. Durbin. New York: McGraw-Hill, pp. 142–44; 151–57; 189–94.

[107]   (1968) "The Genesis of the Special Theory of Relativity." In *Relativity Theory, Its Origins and Impact on Modern Thought*, edited by L. P. Williams. New York: John Wiley & Sons, pp. 107–14. This is a version of the article in *Current Issues in the Philosophy of Science* (1961) [60]. See also [139].

[108] (1968) "The Status of Temporal Becoming." In *The Philosophy of Time: A Collection of Essays*, edited by R. M. Gale. London: Macmillan, pp. 322–54. This is a revised British edition of the essay in *Annals of the New York Academy of Sciences* (1967) [100]. See also [101] and [132].

[109] (1969) "Are Physical Events Themselves Transiently Past, Present and Future? A Reply to H. A. C. Dobbs." *British Journal for the Philosophy of Science* 20, no. 2 (August): 145–53.

[110] (1969) "Can an Infinitude of Operations Be Performed in a Finite Time?" *British Journal for the Philosophy of Science* 20, no. 3 (October): 203–18.

[111] (1969) "Can We Ascertain the Falsity of a Scientific Hypothesis?" *Studium Generale* 22, no. 11 (November 17): 1061–93. See also [129] and [154].

[112] (1969) "Comments on Professor Roger Buck's Paper 'Reflexive Predictions.'" In *The Nature and Scope of Social Science*, edited by L. I. Krimmerman. New York: Appleton-Century-Crofts, pp. 163–65. This is a version of the article in *Philosophy of Science* (1963) [73].

[113] (1969) "Explanation and Prediction are Symmetrical." In *The Nature and Scope of Social Science*, edited by L. I. Krimerman. New York: Appleton-Century-Crofts, pp. 126–32. This is an excerpt from the article in *Philosophy of Science* (1962) [65]. See also [78] and [79].

[114] (1969) *Filosofski problemy prostranstva i vremeni.* Moscow: Progress Publishers. This is a Russian translation, with revisions and additions, of *Philosophical Problems of Space and Time* (1963) [74]. See also [140], [165], and [209].

[115]   (1969) "Free Will and Laws of Human Behavior." *L'Age de la Science* no. 2 (April/June): 105–27. Errata in this article are listed in (1970) *L'Age de la Science* no. 4. See also [118], [126], [130], [135], [136], [144], [149], and [163].

[116]   (1969) (With W. C. Salmon) Introduction to a panel discussion of simultaneity by slow clock transport in the special and general theories of relativity. "The Context of These Essays." *Philosophy of Science* 36, no. 1 (March): 1–4.

[117]   (1969) "Simultaneity by Slow Clock Transport in the Special Theory of Relativity." *Philosophy of Science* 36, no. 1 (March): 5–43. See also [124].

[118]   (1969) "Libero arbitrio e norme del comportamento umano." In *Le Implicazioni Politiche della Scienza*, edited by R. Campa. Rome, Italy: Edizioni della Nuova Antologia, pp. 87–123. This is an Italian translation of "Free Will and Laws of Human Behavior," in *L'Age de la Science* (1969) [115]. See also [126], [130], [135], [136], [144], [149], and [163].

[119]   (1968) "Reply to Hilary Putnam's 'An Examination of Grünbaum's Philosophy of Geometry.'" In *Proceedings of the Boston Colloquium for the Philosophy of Science*, edited by R. S. Cohen and M. W. Wartofsky. Boston Studies in the Philosophy of Science, vol. 5. Dordrecht, Netherlands: D. Reidel Publishing, pp. 1–150. This is a reprint of chapter 3 of *Geometry and Chronometry in Philosophical Perspective* (1968) [103].

[120]   (1969) "The Meaning of Time." In *Essays in Honor of Carl G. Hempel: A Tribute on the Occasion of His Sixty-fifth Birthday*, edited by N. Rescher et al. Dordrecht, Netherlands: D. Reidel Publishing, pp. 147–77. See also [131], [159], and [188].

[121] (1970) (With M. Strauss) "Comments on Professor Schmutzer's Paper 'New Approach to Interpretation Problems of General Relativity by Means of the Splitting-Up-Formalism of Space-Time.'" In *Induction, Physics, and Ethics: Proceedings and Discussions of the 1968 Salzburg Colloquium in the Philosophy of Science*, edited by P. Weingartner and G. Zecha. Dordrecht, Netherlands: D. Reidel Publishing, pp. 137–39.

[122] (1970) "Die Anisotropie der Zeit." In *Erkenntnisprobleme der Naturwissenschaften*, edited by L. Krüger. Köln, Germany: Verlag Kiepenheuer & Witsch, pp. 476–508. This is a German translation of "The Anisotropy of Time." in *Monist* (1964) [86] and [98].

[123] (1970) "Modern Science and Zeno's Paradoxes of Motion." In *Zeno's Paradoxes*, edited by W. C. Salmon. New York: Bobbs-Merrill, pp. 200–50. This is a version of the essay in R. M. Gale, ed. *The Philosophy of Time: A Collection of Essays* (1967) [94]. This book also contains a reprint of the article from *Scientific Monthly* (1955) [28] and selections from (1968) [104]. See also [105].

[124] (1970) "Simultaneity by Slow Clock Transport in the Special Theory of Relativity." In *Induction, Physics, and Ethics: Proceedings and Discussions of the 1968 Salzburg Colloquium in the Philosophy of Science*, edited by P. Weingartner and G. Zecha. Dordrecht, Netherlands: D. Reidel Publishing, pp. 140–66; discussion remarks on pp. 167–79. This is a version of the article in *Philosophy of Science* (1969) [117].

[125] (1970) "Space, Time and Falsifiability, Critical Exposition and Reply to 'A Panel Discussion of Grünbaum's Philosophy of Science.'" *Philosophy of Science* 37, no. 4 (December): 469–588.

[126] (1970) "Svoboda voli i zakony chelovecheskogo povedeniya." *Voprosi Filosofii*, no. 6: 62–74. This is a Russian translation of "Free

Will and Laws of Human Behavior," in *L'Age de la Science* (1969) [115]. See also [118], [130], [135], [136], [144], [149], and [163].

[127] (1970) "Zeno's Metrical Paradox of Extension." In *Zeno's Paradoxes*, edited by W. C. Salmon. New York: Bobbs-Merrill, pp. 176–99.

[128] (1971) "Are Spatial and Temporal Congruence Conventional?" Letter to the editor. *General Relativity and Gravitation* 2, no. 3: 281–84.

[129] (1971) "Can We Ascertain the Falsity of a Scientific Hypothesis?" In *Observation and Theory in Science*, edited by M. Mandelbaum. Baltimore, MD: Johns Hopkins University Press, pp. 69–129. This is a version of the article in *Studium Generale* (1969) [111]. See also [154].

[130] (1971) "Free Will and Laws of Human Behavior." *American Philosophical Quarterly* 8, no. 4 (October): 299–317. This is a version of the article in *L'Age de la Science* (1969) [115]. See also [118], [126], [135], [136], [144], [149], and [163].

[131] (1971) "The Meaning of Time." In *Basic Issues in the Philosophy of Time*, edited by E. Freeman and W. Sellars. La Salle, IL: Open Court Publishing, pp. 195–228. This is a version of the essay in *Essays in Honor of Carl G. Hempel: A Tribute on the Occasion of His Sixty-fifth Birthday* (1969) [120]. See also [159] and [188].

[132] (1971) "The Status of Temporal Becoming." In *Time in Science and Philosophy, An International Study of Some Current Problems*, edited by J. Zeman. Prague, Czechoslovakia: Academia Publishing House, pp. 67–87. This is a version of the essay in *Annals of the New York Academy of Sciences* (1967) [100]. See also [101] and [108].

[133] (1971) "Why I Am Afraid of Absolute Space." *Australasian Journal of Philosophy* 49, no. 1 (May): 96.

[134] (1972) "Abelson on Feigl's Mind-Body Identity Thesis." *Philosophical Studies* 23, nos. 1–2 (February): 119–21.

[135] (1972) "El libre albedrío y las leyes de la conducta humana." In *La ciencia de la conducta*, edited by G. Fernández Pardo and L. F. S. Natalicio. Mexico: Editorial Trillas, S. A., pp. 229–56. This is a Spanish translation of "Free Will and Laws of Human Behavior," in *L'Age de la Science* (1969) [115]. It is replete with errors. See also [118], [126], [130], [136], [144], [149], and [163].

[136] (1972) "Free Will and Laws of Human Behavior." In *New Readings in Philosophical Analysis*, edited by H. Feigl, W. Sellars, and K. Lehrer. New York: Appleton-Century-Crofts, pp. 605–27. This is a revised version of the article in *L'Age de la Science* (1969) [115]. See also [118], [126], [130], [135], [144], [149], and [163].

[137] (1973) "Causality and the Science of Human Behavior." In *Patterns of Psychology: Issues and Prospects*, edited by A. C. Kamil and N. R. Simonson. Boston: Little, Brown, pp. 18–25. This is a version of the article in *American Scientist* (1952) [8]. See also [9], [12], [43], [47], [64], [90], [91], [96], [189], [214], and [257].

[138] (1973) "Geometrodynamics and Ontology." *Journal of Philosophy* 70, no. 21 (December 6): 775–800.

[139] (1973) "La Genesis de la Teoría Especial de la Relatividad." In *La Teoría de la Relatividad*, edited by L. P. Williams. Madrid, Spain: Alianza Editorial, pp. 119–25. This is a Spanish translation of "The Genesis of the Special Theory of Relativity," in *Current Issues in the Philosophy of Science* (1961) [60]. See also [107].

[140]   (1973) *Philosophical Problems of Space and Time.* Boston Studies in the Philosophy of Science, vol. 12. Dordrecht, Netherlands: D. Reidel Publishing. This is the 2nd ed. of [74]. See also [114], [165], and [209].

[141]   (1973) "Reply to J. Q. Adams' 'Grünbaum's Solution to Zeno's Paradoxes.'" *Philosophia* (Israel) 3, no. 1 (January): 51–57.

[142]   (1973) "The Ontology of the Curvature of Empty Space in the Geometrodynamics of Clifford and Wheeler." In *Space, Time and Geometry*, edited by P. Suppes. Dordrecht, Netherlands: D. Reidel Publishing, pp. 268–95.

[143]   (1974) "Carnap's Views on the Foundations of Geometry." In *La filosofia di Rudolf Carnap*. Milan, Italy: Il Saggiatore, pp. 583–658. This is an Italian translation of the essay in *The Philosophy of Rudolf Carnap* (1963) [72]. See also [89].

[144]   (1974) "Free Will and Laws of Human Behavior." In *Psychoanalysis and Contemporary Science: An Annual of Integrative and Interdisciplinary Studies*, vol. 3, edited by L. Goldberger and V. H. Rosen. New York: International Universities Press, pp. 3–39. This is a version of the article in *L'Age de la Science* (1969) [115]. See also [118], [126], [130], [135], [136], [149], and [163].

[145]   (1974) "Is the Coarse-Grained Entropy of Classical Statistical Mechanics an Anthropomorphism?" In *Modern Developments in Thermodynamics*, edited by B. Gal-Or. New York/Jerusalem: John Wiley & Sons/Israel Universities Press, pp. 413–28. See also [150] and [156].

[146]   (1974) "Popper's Views on the Arrow of Time." In *The Philosophy of Karl Popper*, book 2, edited by P. A. Schilpp. Library of Living Philosophers, vol. 14. La Salle, IL: Open Court Publishing, pp. 775–97.

[147] (1974) "Space, Time, and Matter: The Foundations of Geometrodynamics." Introductory remarks in *Proceedings of the 1972 Biennial Meeting of the Philosophy of Science Association*, edited by K. F. Schaffner and R. S. Cohen. Boston Studies in the Philosophy of Science, vol. 20. Dordrecht, Netherlands: D. Reidel Publishing, pp. 3–5.

[148] (1974) *The Philosophy of Continuity.* (February 1951) PhD thesis, Yale University. Ann Arbor, MI: University Microfilms. This is a microfilm version of the thesis in [6].

[149] (1975) "In Defense of Determinism." In *Philosophy Now*, 2nd ed., edited by P. R. Struhl and K. J. Struhl. New York: Random House, pp. 211–28. This is a version of "Free Will and Laws of Human Behavior," in *L'Age de la Science* (1969) [115]. See also [118], [126], [130], [135], [136], [144], and [163].

[150] (1975) "Is the Coarse-Grained Entropy of Classical Statistical Mechanics an Anthropomorphism?" In *Entropy and Information in Science and Philosophy*, edited by L. Kubát and J. Zeman. Prague, Czechoslovakia: Academia Publishing House, pp. 173–86. This is a version of the article in *Modern Developments in Thermodynamics* (1974) [145]. See also [156].

[151] (1976) "Can a Theory Answer More Questions Than One of Its Rivals?" *British Journal for the Philosophy of Science* 27, no. 1 (March): 1–23. See also [359].

[152] (1976) "Is the Method of Bold Conjectures and Attempted Refutations *Justifiably* the Method of Science?" *British Journal for the Philosophy of Science* 27, no. 2 (June): 105–36.

[153] (1976) "Is Falsifiability the Touchstone of Scientific Rationality? Karl Popper versus Inductivism." In *Essays in Memory of Imre*

Lakatos, edited by R. S. Cohen, P. K. Feyerbend, and M. W. Wartofsky. Boston Studies in the Philosophy of Science, vol. 39. Dordrecht, Netherlands: D. Reidel Publishing, pp. 213–52. See also [195].

[154] (1976) "Is It *Never* Possible to Falsify a Hypothesis Irrevocably?" In *Can Theories Be Refuted? Essays on the Duhem-Quine Thesis*, edited by S. G. Harding. Dordrecht, Netherlands: D. Reidel Publishing, pp. 260–88. This is a version of the article in *Studium Generale* (1969) [111]. See also [129].

[155] (1976) "Is Preacceleration of Particles in Dirac's Electrodynamics a Case of Backward Causation? The Myth of Retrocausation in Classical Electrodynamics." *Philosophy of Science* 43, no. 2 (June): 165–201. See also [161], [167], [174], [175], and [178].

[156] (1976) "Is the Coarse-Grained Entropy of Classical Statistical Mechanics an Anthropomorphism?" In *Vistas in Physical Reality, Festschrift for Henry Margenau*, edited by E. Laszlo and E. B. Sellon. New York: Plenum Press, pp. 11–29. This is a version of the essay in *Modern Developments in Thermodynamics* (1974) [145]. See also [150].

[157] (1976) "*Ad Hoc* Auxiliary Hypotheses and Falsificationism." *British Journal for the Philosophy of Science* 27, no. 4 (December): 329–62.

[158] (1976) "The Duhemian Argument." In *Can Theories Be Refuted? Essays on the Duhem-Quine Thesis*, edited by S. G. Harding. Dordrecht, Netherlands: D. Reidel Publishing, pp. 116–31. This is a version of the article in *Philosophy of Science* (1960) [52]. See also [53].

[159] (1976) "The Exclusion of Becoming from the Physical World." In *The Concepts of Space and Time: Their Structure and Their*

Development, edited by M. Čapek. Boston Studies in the Philosophy of Science, vol. 22. Dordrecht, Netherlands: D. Reidel Publishing, pp. 471–500. This is a version of the essay in *Essays in Honor of Carl G. Hempel: A Tribute on the Occasion of His Sixty-fifth Birthday* (1969) [120]. See also [131] and [188].

[160] (1977) "Absolute and Relational Theories of Space and Space-Time." In *Foundations of Space-Time Theories*, edited by J. Earman, C. Glymour, and J. Stachel. Minnesota Studies in the Philosophy of Science, vol. 8. Minneapolis: University of Minnesota Press, pp. 303–73.

[161] (1977) (With A. I. Janis) Comments and criticism. "Is There Backward Causation in Classical Electrodynamics?" *Journal of Philosophy* 74, no. 8 (August): 475–82. See also [155], [167], [174], [175], and [178].

[162] (1977) "How Scientific Is Psychoanalysis?" In *Science and Psychotherapy*, 2nd ed., vol. 2, edited by R. Stern, L. S. Horowitz, and J. Lynes. New York: Haven Publishing, pp. 219–54.

[163] (1977) "Is Freedom a Form of Determinism?" In *Discovering Philosophy*, edited by M. Lipman. Englewood Cliffs, NJ: Prentice-Hall, pp. 421–31. This is a version of the article in *L'Age de la Science* (1969) [115]. See also [118], [126], [130], [135], [136], [144], and [149].

[164] (1977) "Is Psychoanalysis a Pseudo-Science? Karl Popper versus Sigmund Freud (Part 1)." *Zeitschrift für Philosophische Forschung* 31, no. 3 (July–September): 333–53. See [170] for the second installment of this article.

[165] (1977) "Remarks on Miller's Review of *Philosophical Problems of Space and Time*." *Isis* 68, no. 243: 447–48. See also [74], [114], [140], and [209].

[166]   (1977) (With A. I. Janis) "The Geometry of the Rotating Disk in the Special Theory of Relativity." *Synthèse* 34: 281–99. See also [179].

[167]   (1978) (With A. I. Janis) "Can the Effect Precede Its Cause in Classical Electrodynamics?" *American Journal of Physics* 46, no. 4 (April): 337–41. See also [155], [161], [174], [175], and [178].

[168]   (1978) "Hans Reichenbach's Definitive Influence on Me." In *Hans Reichenbach Selected Writings: 1909—1953*, vol. 1, edited by M. Reichenbach and R. S. Cohen. Vienna Circle Collection, vol. 4. Dordrecht, Netherlands: D. Reidel Publishing, pp. 65–67.

[169]   (1978) "How Is Science Distinguished from Pseudo-Science? Two Views." *Proceedings of the Pennsylvania Academy of Science* 52: 96–98.

[170]   (1978) "Is Psychoanalysis a Pseudo-Science? (Part 2)." *Zeitschrift für Philosophische Forschung* 32, no. 1 (January–March): 49–69. This is the second installment of "Is Psychoanalysis a Pseudo-Science? Karl Popper versus Sigmund Freud (Part 1)," in *Zeitschrift für Philosophische Forschung* (1977) [164].

[171]   (1978) "Poincaré's Thesis that Any and All Stellar Parallax Findings are Compatible with the Euclideanism of the Pertinent Astronomical 3-Space." *Studies in History and Philosophy of Science* 9, no. 4: 313–18.

[172]   (1978) "Popper versus Inductivism." In *Progress and Rationality in Science*, edited by G. Radnitzky and G. Andersson. Boston Studies in the Philosophy of Science, vol. 58. Dordrecht, Netherlands: D. Reidel Publishing, pp. 117–42. See also [182].

[173]   (1978) "Psychological Explanations for the Rejection or Acceptance of Scientific Theories." *Humanities in Society* 1, no. 4 (Fall): 293–304. See also [180], [185], and [186].

[174] (1978) "The Myth of Retrocausation in Classical Electrodynamics." *Epistemologia* 1, no. 2 (July–December): 353–96. This is a version of the article in *Philosophy of Science* (1976) [155]. See also [161], [167], [175], and [178].

[175] (1979) "Der Mythos kausaler Rückkopplung in der klassischen Elektrodynamik." *Allgemeine Zeitschrift für Philosophie* 4, no. 1: 1–39. This is a German translation of "Is Preacceleration of Particles in Dirac's Electrodynamics a Case of Backward Causation? The Myth of Retrocausation in Classical Electrodynamics," in *Philosophy of Science* (1976) [155]. See also [161], [167], [174], and [178].

[176] (1979) "Epistemological Liabilities of the Clinical Appraisal of Psychoanalytic Theory." *Psychoanalysis and Contemporary Thought* 2, no. 4: 451–526. See also [181] and [229].

[177] (1979) "Is Freudian Psychoanalytic Theory Pseudo-Scientific by Karl Popper's Criterion of Demarcation?" *American Philosophical Quarterly* 16, no. 2 (April): 131–41. See also [196].

[178] (1979) (With A. I. Janis) "Retrocausation and the Formal Assimilation of Classical Electrodynamics to Newtonian Mechanics: A Reply to Nissim-Sabat's 'On Grünbaum and Retrocausation.'" *Philosophy of Science* 46, no. 1 (March): 136–60. See also [155], [161], [167], [174], and [175].

[179] (1979) (With A. I. Janis) "The Geometry of the Rotating Disk in the Special Theory of Relativity." In *Hans Reichenbach: Logical Empiricist*, edited by W. C. Salmon. Synthèse Library, vol. 132. Dordrecht, Netherlands: D. Reidel Publishing, pp. 321–39. This is a version of the article in *Synthèse* (1977) [166].

[180] (1979) "The Role of Psychological Explanations of the Rejection or Acceptance of Scientific Theories." In *Perspectives in Meta-*

science: A Festschrift for Håkan Törnebohm, edited by J. Bärmark. Göteborg, Sweden: University of Göteborg Press, pp. 95–115. This is a version of the essay in *Humanities in Society* (1978) [173]. See also [185] and [186].

[181] (1980) "Epistemological Liabilities of the Clinical Appraisal of Psychoanalytic Theory." *Noûs* 14, no. 3 (September): 307–85. This is an enlarged version of the essay in *Psychoanalysis and Contemporary Thought* (1979) [176]. See also [229].

[182] (1980) "Popper und der Induktivismus." In *Fortschritt und Rationalität der Wissenschaft*, edited by G. Radnitzky and G. Andersson. Tübingen, Germany: J. C. B. Mohr, pp. 129–56. This is a German translation of "Popper versus Inductivism," in *Progress and Rationality in Science* (1978) [172].

[183] (1980) Preface in H. Mehlberg, *Time, Causality, and the Quantum Theory*, vol. 1. Boston Studies in the Philosophy of Science, vol. 19. Dordrecht, Netherlands: D. Reidel Publishing, pp. xiii–xiv.

[184] (1980) "Psychoanalysis." *Commentary* 70, no. 5 (November): 21–23.

[185] (1980) "The Role of Psychological Explanations of the Rejection or Acceptance of Scientific Theories." In *Science, Pseudo-Science and Society*, edited by M. P. Hanen, M. J. Osler, and R. G. Weyant. Waterloo, ON, Canada: Wilfrid Laurier University Press, pp. 29–53. This is a version of the essay in *Humanities in Society* (1978) [173]. See also [180] and [186].

[186] (1980) "The Role of Psychological Explanations of the Rejection or Acceptance of Scientific Theories." In *Science and Social Structure: A Festschrift for Robert K. Merton*, edited by T. F. Gieryn. April 24. *Transactions of the New York Academy of Sciences*, series 2,

vol. 39, pp. 75–90. This is a version of the article in *Humanities in Society* (1978) [173]. See also [180] and [185].

[187]   (1980) (With A. I. Janis) "The Rotating Disk: Reply to Grøn." *Foundations of Physics* 10, nos. 5–6 (June): 495–98.

[188]   (1980) "Τὶ νἰημα του χρἰνου." *Deucalion*, no. 32 (December): 343–82. This is a Greek translation of "The Meaning of Time," in *Essays in Honor of Carl G. Hempel: A Tribute on the Occasion of His Sixty-fifth Birthday* (1969) [120]. See also [131] and [159].

[189]   (1981) "Causality and the Science of Human Behavior." In *The Individual and the Universe: An Introduction to Philosophy*, edited by O. A. Johnson. New York: Holt, Rinehart and Winston, pp. 148–57. This is a version of the article in *American Scientist* (1952) [8]. See also [9], [12], [43], [47], [64], [90], [91], [96], [137], [214], and [257].

[190]   (1981) "How Valid Is Psychoanalysis? An Exchange." *New York Review of Books* 28, no. 3 (March 5): 40.

[191]   (1981) "The Placebo Concept." *Behaviour Research and Therapy* 19, no. 2: 157–67.

[192]   (1982) "Can Psychoanalytic Theory Be Cogently Tested 'On the Couch'? (Part 1)." *Psychoanalysis and Contemporary Thought* 5, no. 2: 155–255. See also [193] and [203].

[193]   (1982) "Can Psychoanalytic Theory Be Cogently Tested 'On the Couch'? (Part 2)." *Psychoanalysis and Contemporary Thought* 5, no. 3: 311–436. This is the second installment of the article in *Psychoanalysis and Contemporary Thought* (1982) [192]. See also [203].

[194] (1983) Comments and criticism. "Is Object-Relations Theory Better Founded Than Orthodox Psychoanalysis? A Reply to Jane Flax." *Journal of Philosophy* 80, no. 1 (January): 46–51.

[195] (1983) "Es la falsabilidad la piedra de toque de la racionalidad científica? Karl Popper contra el inductivismo." *Cuadernos de Crítica* 22: 5–63. This is a Spanish translation of "Is Falsifiability the Touch-stone of Scientific Rationality? Karl Popper versus Inductivism," in *Essays in Memory of Imre Lakatos* (1976) [153].

[196] (1983) "Es la teoria psicoanalítica freudiana pseudocientífica con el criterio de demarcación de Karl Popper?" *Teorema* (Madrid) 13, nos. 1–2: 179–99. This is a Spanish translation of "Is Freudian Psychoanalytic Theory Pseudo-Scientific by Karl Popper's Criterion of Demarcation?" in *American Philosophical Quarterly* (1979) [177].

[197] (1983) "Freud's Theory: The Perspective of a Philosopher of Science." Presidential address to the American Philosophical Association (Eastern Division) on December 28, 1982. *Proceedings and Addresses of the American Philosophical Association* 57, no. 1 (September): 5–31. See also [205], [211], [216], [237], [263], and [267].

[198] (1983) "Logical Foundations of Psychoanalytic Theory." *Erkenntnis* 19, no. 1–3 (May): 109–52. See also [199], [200], and [277].

[199] (1983) "Logical Foundations of Psychoanalytic Theory." In Theories of Personality and Psychopathology, 3rd ed., edited by T. Millon. New York: Holt, Rinehart and Winston, pp. 193–216. This is a reprint of the article in Erkenntnis (1983) [198]. See also [200] and [277].

[200] (1983) "Logical Foundations of Psychoanalytic Theory." In *Methodology, Epistemology, and Philosophy of Science: Essays in*

Honour of Wolfgang Stegmüller on the Occasion of His 60th Birthday,
June 3rd 1983, edited by C. G. Hempel, H. Putnam, and W. K. Essler.
Dordrecht, Netherlands: D. Reidel Publishing, pp. 109–52. This is a
reprint of the article in *Erkenntnis* (1983) [198]. See also [199] and
[277].

[201] (1983) "Psychoanalysis and Hermeneutics: An Ill-Conceived
Marriage." *Dialectics and Humanism* 10, no. 1: 135–45.

[202] (1983) "Retrospective versus Prospective Testing of Aetiolog-
ical Hypotheses in Freudian Theory." In *Testing Scientific Theories,*
edited by J. Earman. Minnesota Studies in the Philosophy of Science,
vol. 10. Minneapolis: University of Minnesota Press, pp. 315–47.

[203] (1983) "The Foundations of Psychoanalysis." In *Mind and*
Medicine: Problems of Explanation and Evaluation in Psychiatry and
the Biomedical Sciences, edited by L. Laudan. Pittsburgh Series in the
Philosophy and History of Science, vol. 8. Berkeley: University of
California Press, pp. 143–309. This is a revised version of the articles
in *Psychoanalysis and Contemporary Thought* (1982) [192] and [193].

[204] (1984) "Explication and Implications of the Placebo Concept."
In *Rationality in Science and Politics: Festschrift for Gerard Rad-*
nitzky, edited by G. Andersson. Boston Studies in the Philosophy of
Science, vol. 79. Dordrecht, Netherlands: D. Reidel Publishing, pp.
131–58. See also [210], [219], [243], [261], [264], and [292].

[205] (1984) "Freudova teorija: pogled filozofa znanosti." *Filozofska*
Istrazivanja (Zagreb, Yugoslavia) 10, no. 3: 295–309. This is an
abridged version/Serbo-Croatian translation of "Freud's Theory: The
Perspective of a Philosopher of Science," in *Proceedings and*
Addresses of the American Philosophical Association (1983) [197].
See also [211], [216], [237], [263], and [267].

[206]   (1984) *The Foundations of Psychoanalysis: A Philosophical Critique*. Berkeley: University of California Press. See also [218], [228], [230], [232], [233], [256], [261], [299], [305], and [308].

[207]   (1984) "The Hermeneutic Construal of Psychoanalytic Theory and Therapy: An Ill-Conceived Paradigm for the Human Sciences." In *Science and Reality: Recent Work in the Philosophy of Science. Essays in Honor of Ernan McMullin*, edited by J. T. Cushing, C. F. Delaney, and G. M. Gutting. Notre Dame, IN: University of Notre Dame Press, pp. 54–82.

[208]   (1985) "A warrant for psychoanalysis?" *Times Higher Education Supplement* (London) (June 21): 14.

[209]   (1985) Excerpt from *Philosophical Problems of Space and Time*. In *Ideas of Time*, edited by Z. Popov and Z. Bojadgiev. Sofia, Bulgaria: Science and Art Publishing House, pp. 462–76. This is a Bulgarian translation from the book in [140]. See also [74], [114], and [165].

[210]   (1985) "Explication and Implications of the Placebo Concept." In *Placebo: Theory, Research, and Mechanisms*, edited by L. White, B. Tursky, and G. E. Schwartz. New York: Guilford Press, pp. 9–36. This is a slightly different version of the article in *Rationality in Science and Politics: Festschrift for Gerard Radnitzky* (1984) [204]. See also [219], [243], [261], [264], and [292].

[211]   (1985) "La teoria di Freud nella prospettiva di un filosofo della scienza." In *L'anima e il compasso*, edited by P. Repetti. Rome, Italy: Edizioni Theoria, pp. 87–138. This is an Italian translation of "Freud's Theory: The Perspective of a Philosopher of Science," in *Proceedings and Addresses of the American Philosophical Association* (1983) [197]. See also [205], [216], [237], [263], and [267].

[212] (1985) "Some Reflections on the Intellectual Contributions of German Jewry." In *Jews in Germany under Prussian Rule*. A historical exhibition at the University of Pittsburgh, originally compiled by the German *"Bildarchiv Preussischer Kulturbesitz, Berlin."* Pittsburgh: University of Pittsburgh, University Center for International Studies, pp. 1–15.

[213] (1985) "The Falsifiability of Theories: Total or Partial? A Contemporary Evaluation of the Duhem-Quine Thesis." In *A Portrait of Twenty-five Years, Boston Colloquium for the Philosophy of Science 1960–1985*, edited by R. S. Cohen and M. W. Wartofsky. Dordrecht, Netherlands: D. Reidel Publishing, pp. 1–18. This is a reprint of the article in *Synthèse* (1962) [66]. See also [80] and [92].

[214] (1986) "Causality and the Science of Human Behavior." In *The Tradition of Philosophy*, edited by H. Hall and N. E. Bowie. Belmont, CA: Wadsworth Publishing, pp. 307–13. This is a version of the article in *American Scientist* (1952) [8]. See also [9], [12], [43], [47], [64], [90], [91], [96], [137], [189], and [257].

[215] (1986) "Homenaje a Alberto Coffa." *Revista Latinoamericana de Filosofía* (Argentina) 12, no. 2 (July): 242–45. This is a Spanish translation of "Alberto Coffa: In Memoriam," an oral presentation given at Indiana University, Bloomington, IN, January 19, 1985. See also [231].

[216] (1986) "Ist die Psychoanalyse rational oder kausalistisch begründet?" In *Zur Kritik der wissenschaftlichen Rationalität*, edited by H. Lenk. Freiburg, Germany: Verlag Karl Alber GmbH, pp. 351–94. This is a somewhat shortened German translation of "Freud's Theory: The Perspective of a Philosopher of Science," in *Proceedings and Addresses of the American Philosophical Association* (1983) [197]. See also [205], [211], [237], [263], and [267].

[217]   (1986) "L'impresa psicoanalitica: una valutazione." *Nuova Civiltà delle Macchine* 4, nos. 3/4, 15/16: 123–30. This is an Italian translation of "The Psychoanalytic Enterprise: An Assessment," an oral presentation given at the International Congress, Rationality in Science and in Politics, Locarno, Switzerland, June 14, 1986.

[218]   (1986) Précis of *The Foundations of Psychoanalysis: A Philosophical Critique* and "Author's Response to 41 Reviewers: 'Is Freud's Theory Well-Founded?'" *Behavioral & Brain Sciences* 9, no. 2 (June): 217–28, 266–84. See also [206], [228], [230], [233], [256], [261], [299], [305], and [308].

[219]   (1986) "The Placebo Concept in Medicine and Psychiatry." *Psychological Medicine* (England) 16: 19–38. This is an enlarged, revised version of the essay in *Rationality in Science and Politics: Festschrift for Gerard Radnitzky* (1984) [204]. See also [210], [243], [261], [264], and [292].

[220]   (1986) "What Are the Clinical Credentials of the Psychoanalytic Compromise Model of Neurotic Symptoms?" In *Contemporary Directions in Psychopathology Toward the DSM-IV*, edited by T. Millon and G. L. Klerman. New York: Guilford Press, pp. 193–214.

[221]   (1987) "God and the Holocaust." *Free Inquiry* 8, no. 1 (Winter 1987/1988): 23.

[222]   (1987) *Psychoanalyse in wissenschaftstheoretischer Sicht— Zum Werk Sigmund Freuds und seiner Rezeption*. Konstanz, Germany: Universitätsverlag Konstanz GmbH. This is the text of six lectures inaugurating the series of Konstanz Dialogues presented at the University of Konstanz, Germany, June 21–July 7, 1983.

[223]   (1987) "Psychoanalysis and Theism." *Monist* 70, no. 2 (April): 152–92. A revised version of this essay appeared as chapter 7 in *Vali-*

dation in the Clinical Theory of Psychoanalysis (1993) [283]. See also [262].

[224] (1987) "The Place of Secular Humanism in Current American Political Culture." *Vital Speeches of the Day* 54, no. 2 (November 1): 42–47. See also [225] and [239].

[225] (1987) "Secular Humanism in American Political Culture." *Free Inquiry* 8, no. 1 (Winter 1987/1988): 21–25. This is a truncated version of the article in *Vital Speeches of the Day* (1987) [224]. See also [239].

[226] (1988) "Are Hidden Motives in Psychoanalysis Reasons but Not Causes of Human Conduct?" In *Hermeneutics and Psychological Theory: Interpretive Perspectives on Personality, Psychotherapy, and Psychopathology*, edited by S. B. Messer, L. A. Sass, and R. L. Woolfolk. Rutgers Symposia on Applied Psychology, vol. 2. New Brunswick, NJ: Rutgers University Press, pp. 149–67.

[227] (1988) "A Rejoinder to Richard Bernstein." In *Hermeneutics and Psychological Theory: Interpretive Perspectives on Personality, Psychotherapy, and Psychopathology*, edited by S. B. Messer, L. A. Sass, and R. L. Woolfolk. Rutgers Symposia on Applied Psychology, vol. 2. New Brunswick, NJ: Rutgers University Press, pp. 175–81.

[228] (1988) *Die Grundlagen der Psychoanalyse, Eine Philosophische Kritik*. Stuttgart, Germany: Philipp Reclam jun. GmbH. This is a considerably enlarged and revised German version of *The Foundations of Psychoanalysis: A Philosophical Critique* (1984) [206]. See also [218], [230], [232], [233], [256], [261], [299], [305], and [308].

[229] (1988) "Epistemological Liabilities of the Clinical Appraisal of Psychoanalytic Theory." In *La Verifica Empirica in Psicoanalisi:*

Itinerari Teorici e Paradigmi di Ricerca, edited by N. Dazzi and M. Conte. Bologna, Italy: Il Mulino. This is an Italian translation of the essay in *Psychoanalysis and Contemporary Thought* (1979) [176]. See also [181].

[230]   (1988) *I Fondamenti della Psicoanalisi: Una Critica Filosofica.* Milan, Italy: Il Saggiatore (Mondadori). This is an Italian translation of *The Foundations of Psychoanalysis: A Philosophical Critique* (1984) [206]. See also [218], [228], [232], [233], [256], [261], [299], [305], and [308].

[231]   (1988) "In Memoriam: J. Alberto Coffa." In *Limitations of Deductivism*, edited by A. Grünbaum and W. C. Salmon. Berkeley: University of California Press, pp. xvii–xx. This is a transcription of the oral presentation given at Indiana University, Bloomington, IN, January 19, 1985 [215].

[232]   (1988) "*Précis* of *The Foundations of Psychoanalysis: A Philosophical Critique.*" In *Mind, Psychoanalysis and Science*, edited by P. C. Clark and C. Wright. Oxford: Basil Blackwell, pp. 3–32. This is a reprint of the précis in *Behavioral & Brain Sciences* (1986) [218]. See also [206], [228], [230], [233], [256], [261], [299], [305], and [308].

[233]   (1988) *Psicoanalisi: Obiezioni e risposte.* Rome, Italy: Armando Editore. This is an Italian translation of the précis of *The Foundations of Psychoanalysis: A Philosophical Critique* (1986) [218] with an introduction by M. Pera. See also [206], [228], [230], [232], [256], [261], [299], [305], and [308].

[234]   (1988) "The Role of the Case Study Method in the Foundations of Psychoanalysis." In *Die Philosophen und Freud*, edited by H. Vetter and L. Nagl. Wiener Reihe, Themen der Philosophie, vol. 3. Vienna, Austria: R. Oldenbourg Verlag, pp. 134–74. See also [235] and [261].

[235] (1988) "The Role of the Case Study Method in the Foundations of Psychoanalysis." *Canadian Journal of Philosophy* 18, no. 4 (December): 623–58. This is a reprint of the article in *Die Philosophen und Freud* (1988) [234]. See also [261].

[236] (1988) (With W. C. Salmon, eds.) *The Limitations of Deductivism.* Berkeley: University of California Press.

[237] (1989) "Freud's Theory: The Perspective of a Philosopher of Science." In *Freud and the Impact of Psychoanalysis*, edited by L. Spurling. Sigmund Freud, Critical Assessments, vol. 4. London: Routledge, pp. 168–98. This is a reprint of the address in *Proceedings and Addresses of the American Philosophical Association* (1983) [197]. See also [205], [211], [216], [263], and [267].

[238] (1989) "La Cattiva Scienza." *Prometeo* 7, no. 26 (June): 10–17.

[239] (1989) "Secular Humanism in American Political Culture." In *On the Barricades: Religion and Free Inquiry in Conflict*, edited by R. Basil, M. B. Gehrman, and T. Madigan. Amherst, NY: Prometheus Books, pp. 52–61. This is a truncated version of the article in *Vital Speeches of the Day* (1987) [224]. See also [225].

[240] (1989) "Soviet Atheism and Psychoanalysis under Perestroika." *Free Inquiry* 9, no. 2 (Spring): 52–53.

[241] (1989) "The Degeneration of Popper's Theory of Dermarcation." In *Freedom and Rationality: Essays in Honor of John Watkins*, edited by F. D'Agostino and I. C. Jarvie. Boston Studies in the Philosophy of Science, vol. 117. Dordrecht, Netherlands: Kluwer Academic Publishers, pp. 141–61. See also [242] and [358].

[242]   (1989) "The Degeneration of Popper's Theory of Dermarca-tion." *Epistemologia* 12, no. 2 (July–December): 235–60. This is a reprint of the essay in *Freedom and Rationality: Essays in Honor of John Watkins* (1989) [241]. See also [358].

[243]   (1989) "The Placebo Concept in Medicine and Psychiatry." In *Non-Specific Aspects of Treatment*, edited by M. Shepherd and N. Sartorius. World Health Organization. Toronto, ON, Canada: Hans Huber Publishers, pp. 7–38. This is a reprint of the article in *Psychological Medicine* (England) (1986) [219]. See also [204], [210], [261], [264], and [292].

[244]   (1989) "The Pseudo-Problem of Creation in Physical Cosmology." *Philosophy of Science* 56, no. 3 (September): 373–94. See also [245], [246], [253], [260], [326], [332], and [333].

[245]   (1989) "The Pseudo-Problem of Creation in Physical Cosmology." *Epistemologia* 12, no. 1: 3–31. This is a reprint of the article in *Philosophy of Science* (1989) [244]. See also [246], [253], [260], [326], [332], and [333].

[246]   (1989) "The Pseudo-Problem of Creation in Physical Cosmology." *Free Inquiry* 9, no. 4 (Fall): 48–57. This is a version of the article in *Philosophy of Science* (1989) [244]. See also [245], [253], [260], [326], [332], and [333].

[247]   (1989) "Why Thematic Kinships between Events Do NOT Attest Their Causal Linkage." In *An Intimate Relation: Studies in the History and Philosophy of Science. A Festschrift for Robert E. Butts*, edited by J. R. Brown and J. Mittelstrass. Boston Studies in the Philosophy of Science, vol. 116. Dordrecht, Netherlands: Kluwer Academic Publishers: 477–94. See also [248] and [255].

[248] (1989) "Affinità tematiche come connessioni causali: riflessioni su una fallacia logica." *Iride* 2: 5–21. This is an Italian translation of "Why Thematic Kinships between Events Do NOT Attest Their Causal Linkage," in *An Intimate Relation: Studies in the History and Philosophy of Science. A Festschrift for Robert E. Butts* (1989) [247]. See also [255].

[249] (1990) "Creation in Physical Cosmology: Pseudo-Problem or Superior Truth?" *Nuova Civiltà delle Macchine* 8, no. 4: 114–23. An Italian translation of this article, "La Creazione nella Cosmologia Fisica: Pseudo-Problem o una Verità Superiore?" appears on pp. 13–23.

[250] (1990) "'Meaning' Connections and Causal Connections in the Human Sciences: The Poverty of Hermeneutic Philosophy." *Journal of the American Psychoanalytic Association* 38, no. 3 (September): 559–77. See also [252] and [295].

[251] (1990) "Pseudo-Creation of the Big Bang." *Nature* 344, no. 6269 (April 26): 821–22. See also [269] and [271].

[252] (1990) "'Sinn'- Zusammenhänge und Kausalbeziehungen in den Humanwissenschaften: Die Armut der hermeneutischen Philosophie." In *Philosophie und Psychoanalyse. Symposium der Wiener Festwochen*, edited by L. Nagl, H. Vetter, and H. Leupold Löwenthal. Frankfurt, Germany: Nexus Verlag GmbH, pp. 121–36. This is a German translation of "'Meaning' Connections and Causal Connections in the Human Sciences: The Poverty of Hermeneutic Philosophy," in *Journal of the American Psychoanalytic Association* (1990) [250]. See also [295].

[253] (1990) "The Pseudo-Problem of Creation in Physical Cosmology." In *Physical Cosmology and Philosophy*, edited by J. Leslie. Philosophical Issues Series, edited by P. Edwards. New York: Macmillan,

428 BIBLIOGRAPHY OF ADOLF GRÜNBAUM

pp. 92–112. This is a reprint of the article in *Philosophy of Science* (1989) [244]. See also [245], [246], [260], [326], [332], and [333].

[254] (1990) "The Psychoanalytic Enterprise in Scientific Perspective." In *Scientific Theories*, edited by C. W. Savage. Minnesota Studies in the Philosophy of Science, vol. 14. Minneapolis: University of Minnesota Press, pp. 41–58. See also [273].

[255] (1990) "Why Thematic Kinships between Events Do NOT Attest Their Causal Linkage." *Epistemologia* 13, no. 2 (July–December): 187–207. This is a reprint of the essay in *An Intimate Relation: Studies in the History and Philosophy of Science* (1989) [247]. See also [248].

[256] (1991) "Author's Response, Etiology and Therapy in Psychoanalytic Theory." *Behavioral & Brain Sciences* 14, no. 4 (December): 729–32. This is a continuation of commentary and response in *Behavioral & Brain Sciences* (1986) [218]. See also [206], [228], [230], [232], [233], [261], [299], [305], and [308].

[257] (1991) "Causality and the Science of Human Behavior." In *Readings in Introductory Psychology*, edited by J. Robertson. Acton, MA: Copley Publishing Group, pp. 36–42. This is a reprint of the article in *American Scientist* (1952) [8]. See also [9], [12], [43], [47], [64], [90], [91], [96], [137], [189], and [214].

[258] (1991) "Creation as a Pseudo-Explanation in Current Physical Cosmology." *Erkenntnis* 35 (July): 233–54. See also [259] and [329].

[259] (1991) "Creation as a Pseudo-Explanation in Current Physical Cosmology." In *Erkenntnis Orientated*, edited by W. Spohn. Boston: Kluwer Academic Publishers, pp. 233–54. This reprint of the article in *Erkenntnis* (1991) [258] is part of a special volume in honor of Rudolf Carnap and Hans Reichenbach. See also [329].

[260]   (1991) "Die Schöpfung als Scheinproblem der physikalischen Kosmologie." In *Wege der Vernunft: Festschrift zum Siebzigsten Geburtstag von Hans Albert,* edited by A. Bohnen and A. Musgrave. Tübingen, Germany: J. C. B. Mohr (Paul Siebeck), pp. 164–91. This is a German translation of a revised version of "The Pseudo-Problem of Creation in Physical Cosmology," in *Philosophy of Science* (1989) [244]. See also [245], [246], [253], [326], [332], and [333].

[261]   (1991) *Kritische Betrachtungen zur Psychoanalyse,* edited by A. Grünbaum. Heidelberg, Germany: Springer-Verlag. This book contains a German translation of the précis of*The Foundations of Psychoanalysis: A Philosophical Critique* titled "Eine zusammenfassende Darstellung von *Die Grundlagen Der Psychoanalyse: Eine philosophische Kritik,*" pp. 3–31 (1986) [218]. See also [206], [228], [230], [232], [233], [256], [299], [305], and [308]. It also contains the German translations of the following articles: "The Placebo Concept in Medicine and Psychiatry," titled "Der Plazebobegriff in Medizin und Psychiatrie," pp. 326–57 (1986) [219], see also [204], [210], [243], [264], and [292]; and "The Role of the Case Study Method in the Foundations of Psychoanalysis," titled "Die Rolle der Fallstudienmethode in den Grundlagen der Psychoanalyse," pp. 289–325 (1988) [234], see also [235]; as well as M. N. Eagle's review article on [206] in *Philosophy of Science* 53 (1986), pp. 65–88, and B. von Eckardt's "Adolf Grünbaum's Psychoanalytische Erkenntnistheorie" (1985).

[262]   (1991) *Psicoanalisi e Teismo.* Naples, Italy: Edizioni Scientifiche Italiane. This is an Italian translation of "Psychoanalysis and Theism," *Monist* (1987) [223] with an introduction by A. Pagnini. See also [275] and [283].

[263]   (1991) "Teorija Freida i filosofija nauki." *Voprosi Filosofii,* no. 4: 90–106. This is a Russian translation of a slightly shortened version of "Freud's Theory: The Perspective of a Philosopher of Science," in *Proceedings and Addresses of the American Philosophical Association* (1983) [197]. See also [205], [211], [216], [237], and [267].

[264]   (1991) "The Placebo Concept in Medicine and Psychiatry." In *Thinking Clearly about Psychology*, vol. 1, *Matters of Public Interest: Essays in Honor of Paul E. Meehl*, edited by D. Ciccetti and W. M. Grove. Minneapolis: University of Minnesota Press, pp. 140–70. This is a version of the essay in *Rationality in Science and Politics: Festschrift for Gerard Radnitzky* (1984) [204]. See also [210], [219], [243], [261], and [292].

[265]   (1992) "Agassi-Grünbaum Exchange on Popper and Psycho-analysis." *Popper Newsletter* 4, nos. 1 and 2 (March): 12–14.

[266]   (1992) "Big Bang Doesn't Equate with Start of Time." Letter to the editor, *New York Times*, May 13, p. A14.

[267]   (1992) "Freud's Theory: The Perspective of a Philosopher of Science." In *The Restoration of Dialogue: Philosophy, Psychology and Clinical Practice*, edited by R. B. Miller. Washington, DC: American Psychological Association Press, pp. 366–87. This is a reprint of the Address in *Proceedings and Addresses of the American Philosophical Association* (1983) [197]. See also [205], [211], [216], [237], and [263].

[268]   (1992) "In Defense of Secular Humanism." *Free Inquiry* 12, no. 4 (Fall): 30–39. See also [282] and [288].

[269]   (1992) "Perché il Big Bang non è la creazione." *Il Sole-24 Ore* 20 (May 3): 24. This is an Italian translation of "Pseudo-Creation of the 'Big Bang,'" in *Nature* (1990) [251]. See also [271].

[270]   (1992) "Pluralism and the Church." Letter to the editor, *Commentary* 94, no. 4 (October): 7–8.

[271]   (1992) "Pseudo-Creation of the 'Big Bang.'" *Free Inquiry* 13, no. 1 (Winter 1992/1993): 14–15. This is a reprint of the article in *Nature* (1990) [251]. See also [269].

[272] (1992) Review of *Motivation and Explanation: An Essay on Freud's Philosophy of Science*, by Nigel MacKay. *Isis* 83, no. 1 (March): 175–76.

[273] (1992) "The Psychoanalytic Enterprise in Scientific Perspective." In *Interpreting the World: Science and Society*, edited by W. R. Shea and A. Spadafora. Canton, MA: Science History Publications, pp. 145–66. This is a reprint of the essay in *Scientific Theories* (1990) [254].

[274] (1992) "This 'Bang' Wasn't as Big as You Make It." Letter to the editor, *Pittsburgh Post-Gazette*, April 29, p. 12. A misprint "correction" appeared in the (1992) April 30 issue, p. 2.

[275] (1992) "Two Major Difficulties for Freud's Theory of Dreams." In *Freud and the History of Psychoanalysis*, edited by T. Gelfand and J. Kerr. Hillsdale, NJ: Analytic Press, pp. 193–213. A revised version of this essay appeared as chapter 10 in *Validation in the Clinical Theory of Psychoanalysis* (1993) [283]. See also [223] and [262].

[276] (1993) "A New Critique of Freud's Theory of Dreams." In *Scientific Philosophy: Origins and Developments*, edited by F. Stadler. Dordrecht, Netherlands: Kluwer Academic Publishers, pp. 169–91.

[277] (1993) "A pszichoanalízis logikai alapjai." In *Filozófusok Freudról És a Pszichoanalízisrol*, edited by C. Szummer and F. Eros. Budapest, Hungary: Cserépfalvi Kiadása, pp. 139–58. This is a Hungarian translation of "Logical Foundations of Psychoanalytic Theory," in *Erkenntnis* (1983) [198]. See also [199] and [200].

[278] (1993) "Creation in Cosmology." in *Encyclopedia of Cosmology: Scientific, Philosophical and Historical Foundations of Modern Cosmology*, edited by N. S. Hetherington. New York: Garland Publishing, pp. 126–36.

[279]   (1993) *La Psychanalyse À l'Épreuve*. Collection Tiré à Part. Paris, France: Éditions de l'Éclat. See also [304] and [308].

[280]   (1993) "Memory Playing False." Letter to the editor, *Times Literary Supplement* (London) (November 19): 17.

[281]   (1993) "Narlikar's 'Creation' of the Big Bang Universe Was a Mere Origination." *Philosophy of Science* 60, no. 4 (December): 638–46. See also [353].

[282]   (1993) "Seymour Cain's Jeremiad: A Rejoinder." *Free Inquiry* 14, no. 1 (Winter 1993/1994): 58–62. This issue also includes S. Cain, "In Response to Grünbaum's Defense," a critique of "In Defense of Secular Humanism." See also [268] and [288].

[283]   (1993) *Validation in the Clinical Theory of Psychoanalysis, A Study in the Philosophy of Psychoanalysis*. Madison, CT: International Universities Press, pp. 257–309. See also [223], [262], and [275].

[284]   (1994) "A Brief Appraisal of Freud's Dream Theory." *Dreaming* 4, no. 1 (March): 80–82. See also [309].

[285]   (1994) Autobiographical statement in *Wesleyan University 50th Reunion, Class of '44, June 2–5, 1994*. Middletown, CT: Wesleyan University Press, pp. 29–34.

[286]   (1994) "Does Psychoanalysis Have a Future? Doubtful." Harvard Mental Health Letter 11, no. 4 (October): 3–6. This is an invited "Insights" article.

[287]   (1994) "'Freud's Permanent Revolution': An Exchange," a response to T. Nagel. *New York Review of Books* 41, no. 14 (August 11): 54–55.

[288] (1994) "In Defense of Secular Humanism." In *Challenges to the Enlightenment: In Defense of Reason and Science*, edited by P. Kurtz and T. J. Madigan. Amherst, NY: Prometheus Books, pp. 102–29. This is a reprint of the article in *Free Inquiry* (1992) [268]. See also [282].

[289] (1994) Letter to the editor. *Psychoanalytic Books* 5, no. 1: 154–67. This is a reply to L. Berger's review of *Validation in the Clinical Theory of Psychoanalysis* (1993) [283]. See also [223], [262], and [275].

[290] (1994) "Psychoanalytic Theory and Therapy After 100 Years, and Its Future." Keynote address presented at the awards ceremony of the Prix Latsis Universitaires. Geneva, Switzerland: Fondation Latsis Internationale, pp. 1–34. Later, considerably modified versions of this address are [312], [313], [315], [316], [317], [318], [323], [327], [328], [335], [345], [347], [352], [354], [371], and [373].

[291] (1994) "Some Comments on William Craig's 'Creation and Big Bang Cosmology.'" *Philosophia Naturalis* 31, no. 2: 225–36. See also [331].

[292] (1994) "The Placebo Concept in Medicine and Psychiatry." In *Philosophical Psychopathology*, edited by G. Graham and G. L. Stephens. Cambridge, MA: MIT Press, pp. 285–324. This is a reprint of the article in [219]. See also [210], [243], [261], and [264].

[293] (1995) "A Short Reminiscence of F. S. C. Northrop." In *An Introduction to the Philosophical Works of F. S. C. Northrop*, edited by F. Seddon. Problems in Contemporary Philosophy, vol. 27. Lewiston, NY: Edwin Mellen Press, pp. viii–x.

[294] (1995) "Le Erronee Interpretazioni Teologiche della Cosmologia Fisica Recente." In *Il Tempo nella Scienza e nella Filosofia,* edited by E. Agazzi. Proceedings of the Conference Il Tempo nella Scienza e nella Filosofia in Naples, October 12–14, 1992. Naples, Italy: Guida Editori, pp. 167–211.

[295] (1995) "Les carences de la philosophie herméneutique de la psychanalyse." *Psychothérapies* 15, no. 2: 55–64. This is a French translation/revised version of "'Meaning' Connections and Causal Connections in Human Sciences: The Poverty of Hermeneutic Philosophy," in *Journal of the American Psychoanalytic Association* (1990) [250]. See also [252].

[296] (1995) "Origin versus Creation in Physical Cosmology." In *Physik, Philosophie und die Einheit der Wissenschaften: Festschrift for Erhard Scheibe*, edited by Krüger and B. Falkenburg. Heidelberg, Germany: Spektrum Akademischer Verlag, pp. 221–54. See also [297].

[297] (1995) "Proishojzdenie protiv tvorenija v fizicheskoi kosmologii." *Voprosi Filosofii*, no. 2: 48–60. This is a Russian translation of "Origin versus Creation in Physical Cosmology," in *Physik, Philosophie und die Einheit der Wissenschaften: Festschrift for Erhard Scheibe* (1995) [296].

[298] (1995) "The Poverty of Theistic Morality." In *Science, Mind and Art: Essays on Science and the Humanistic Understanding in Art, Epistemology, Religion and Ethics, in Honor of Robert S. Cohen*, edited by K. Gavroglu, J. Stachel, and M. W. Wartofsky. Boston Studies in the Philosophy of Science, vol. 165. Dordrecht, Netherlands: Kluwer Academic Publishers, pp. 203–42. See also [344], [362], [372], and [374].

[299] (1996) *A pszichoanalízis alapjai*. Budapest, Hungary: Osiris Kiadó. This is a poor Hungarian translation of *The Foundations of Psychoanalysis: A Philosophical Critique* (1984) [206]. See also [218], [228], [230], [232], [233], [256], [261], [305], and [308].

[300] (1996) "Empirical Evaluations of Theoretical Explanations of Psychotherapeutic Efficacy: A Reply to John D. Greenwood." *Philosophy of Science* 63, no. 4 (December): 622–41.

[301] (1996) "Energy Conservation and Theological Misinterpretations of Current Physical Cosmology." In *Henri Poincaré: Science and Philosophy*, edited by J-L.Greffe, G. Heinzmann, and K. Lorenz. Congrès International, Nancy, France, 1994. Berlin, Germany/Paris, France: Akademie Verlag/Albert Blanchard, pp. 209–30. See also [310], [325], [334], and [357].

[302] (1996) "Freud, in Full View." *Pittsburgh Post-Gazette*, January 7, pp. B1, B4. This article pertains to the controversy occasioned by the 1998/1999 Library of Congress Exhibition *Freud: Conflict and Culture*.

[303] (1996) "Historical Determinism, Social Activism, and Predictions in the Social Sciences." In *Logical Empiricism and the Special Sciences*, vol. 4, edited by S. Sarkar. Science and Philosophy in the Twentieth Century. New York: Garland Publishing, pp. 252–56. This is a reprint of the article in *British Journal for the Philosophy of Science* (1956) [35].

[304] (1996) "Is Psychoanalysis Viable?" In *The Philosophy of Psychology*, edited by W. O'Donohue and R. F. Kitchener. London: Sage Publications, pp. 281–90. A French translation of this essay appeared as the introduction in *La Psychanalyse Á l'Épreuve* (1993) [279]. A Japanese translation appeared as part of a prefatory chapter in *The Foundations of Psychoanalysis* (1996) [308].

[305] (1996) *Les Fondements de la Psychanalyse*. Paris, France: Press Universitaires de France. This is a French translation of a revised edition of *The Foundations of Psychoanalysis: A Philosophical Critique* (1984) [206]. The introduction is a translation of "Is Psychoanalysis Viable?" in *The Philosophy of Psychology* (1996) [304]. See also [218], [228], [230], [232], [233], [256], [261], [299], and [308].

[306]   (1996) Letter to the editor. *New Republic*, January 29, p. 5. This is a reply to J. Lear, "The Shrink Is In."

[307]   (1996) "Logical and Philosophical Foundations of the Special Theory of Relativity." In *Logical Empiricism and the Special Sciences*, edited by S. Sarkar. Science and Philosophy in the Twentieth Century, vol. 4. New York: Garland Publishing, pp. 98–112. This is a reprint of the article in *American Journal of Physics* (1955) [27]. See also [45] and [51].

[308]   (1996) *The Foundations of Psychoanalysis: A Philosophical Critique* (in Japanese). Tokyo, Japan: Sangyo-Tosho Publishing. This is a Japanese translation of *The Foundations of Psychoanalysis: A Philosophical Critique* (1984) [206], equipped with a new prefatory chapter and a modified version and translation of "Is Psychoanalysis Viable?" in *The Philosophy of Psychology* (1996) [304], [279]. See also [218], [228], [230], [232], [233], [256], [261], [299], and [305].

[309]   (1996) "The Liabilities of Freudian Psychoanalysis." In *Introduction to the Philosophy of Science*, edited by A. Zucker. Upper Saddle River, NJ: Prentice-Hall, pp. 311–20. This is a reprint of the article in *Dreaming* (1994) [284].

[310]   (1996) "Theological Misinterpretations of Current Physical Cosmology." *Foundations of Physics* 26, no. 4 (April): 523–43. See also [301], [325], [334], and [357].

[311]   (1997) "Is the Concept of 'Psychic Reality' a Theoretical Advance?" *Psychoanalysis and Contemporary Thought* 20, no. 2: 245–67. This is an invited paper presented at the meeting of the International Psychoanalytic Association in San Francisco, CA, August 1, 1995.

[312] (1997) "One Hundred Years of Psychoanalytic Theory and Therapy: Retrospect and Prospect." In *Mindscapes: Philosophy, Science, and the Mind,* edited by M. Carrier and P. Machamer. Pittsburgh, PA/Konstanz, Germany: University of Pittsburgh Press/University of Konstanz Press, pp. 323–60. This is a paper presented at the third biennial meeting of the Pittsburgh-Konstanz Colloquium in Philosophy of Science in Konstanz, May 1995. It is a modified version of the keynote address presented at the awards ceremony of the Prix Latsis Universitaires (1994) [290]. Later considerably modified versions of this address are [313], [315], [316], [317], [318], [323], [327], [328], [335], [345], [347], [352],[354], [371], and [373].

[313] (1997) "Psychoanalytic Theory and Therapy After 100 Years, and Its Future." In*Proceedings of the 40th Congress of the Deutsche Gesellschaft für Psychologie* (German Psychological Society), München, 1996, edited by M. Mandl. Göttingen, Germany: Hogrefe, pp. 59–71. This is a modified version of the keynote address presented at the awards ceremony of the Prix Latsis Universitaires (1994) [290], [312]. Later considerably modified versions of this address are [315], [316], [317], [318], [323], [327], [328], [335], [345], [347], [352], [354], [371], and [373].

[314] (1997) "Robert E. Butts, In Memoriam." *Erkenntnis* 47: 1–2.

[315] (1997) "Sto let psihoanaliza: itogi i perspectivy." *Voprosi Filosofii,* no. 7: 85–98. This is a Russian translation of "A Century of Psychoanalysis," a modified version of the keynote address presented at the awards ceremony of the Prix Latsis Universitaires (1994) [290], [312], [313]. Later considerably modified versions of this address are [316], [317], [318], [323], [327], [328], [335], [345], [347], [352], [354], [371], and [373].

[316] (1997) "Un siècle de psychanalyse: critique rétrospective et perspective." *L'unebévue* no. 10: 25–38. This French translation of "A

Century of Psychoanalysis: Critical Retrospect and Prospect" is a modified version of the keynote address presented at the awards ceremony of the Prix Latsis Universitaires (1994) [290], [312], [313], [315]. Later considerably modified versions of this address are [317], [318], [323], [328], [335], [345], [347], [352], [354], [371], and [373].

[317]   (1998) "A Century of Psychoanalysis: Critical Retrospect and Prospect." In *Freud: Conflict and Culture*, edited by M. S. Roth. New York: Alfred A. Knopf, pp. 183–95. This is a modified version of the keynote address presented at the awards ceremony of the Prix Latsis Universitaires (1994) [290], [312], [313], [315], [316]. Later considerably modified versions of this address are [318], [323], [327], [328], [335], [345], [347], [352], [354], [371], and [373].

[318]   (1998) "A Century of Psychoanalysis: Critical Retrospect & Prospect." Newsletter of *University of Pittsburgh, University Center for International Studies, Center for West European Studies* (February): 1, 7–8. This is a drastically shortened version of the essay in *Freud: Conflict and Culture* (1998) [290], [312], [313], [315], [316], [317], [323], [327], [328], [335], [345], [347], [352], [354], [371], and [373].

[319]   (1998) "Freud in Conflict." Letter to the editor. *New York Times Book Review*, December 27, p. BR2.

[320]   (1998) "Freud's Theorie der Wunscherfüllung beim Traum: Ein Artefakt seines widerlegten neurobiologischen Traummodells." In *Über das Wünschen: Ein seelisches und poetisches Phänomen wird erkundet*, edited by B. Boothe, R. Wepfer, and A. von Wyl. Göttingen, Germany: Vandenhoeck & Ruprecht, pp. 148–69.

[321]   (1998) Letter to the editor. *University Times* 31, no. 6 (November 12): 2–3.

[322] (1998) "Made-to-Order Evidence." In *Unauthorized Freud: Doubters Confront a Legend*, edited by F. C. Crews. New York: Viking, pp. 76–84.

[323] (1998) "Sto lat psychoanalizy: krytyczna retrospekcja i perspektywy." Translated by Z. Rosinska and L. Zielinski. *Swiat Psychoanalizy* no. 2: 47–63. This is a Polish translation/version of "A Century of Psychoanalysis: Critical Retrospect and Prospect," in *Freud: Conflict and Culture* (1998) [317]. See also [290], [312], [313], [315], [316], [317], [318], [327], [328], [335], [345], [347], [352], [354], [371], and [373].

[324] (1998) "The Failure of Freud's Explanation of Counter-Wish Dreams." In *Philosophy and the Many Faces of Science*, edited by D. Anapolitanos, A. Baltas, and S. Tsinorema. Proceedings of the International Athens Philosophy of Science Conference, May 1992. Lanham, MD: Rowman & Littlefield Publishers, pp. 273–85.

[325] (1998) "Theological Misinterpretations of Current Physical Cosmology." *Philo* 1, no. 1 (Spring/Summer): 15–34. This is a substantially revised version of the article in *Foundations of Physics* (1996) [310]. See also [301], [334], and [357].

[326] (1998) "The Pseudo-Problem of Creation in Physical Cosmology." In *Modern Cosmology and Philosophy*, edited by J. Leslie. Amherst, NY: Prometheus Books, pp. 98–118. This is a reprint of the article in *Philosophy of Science* (1989) [244]. See also [245], [246], [253], [260], [332], and [333].

[327] (1998) "Un secolo di psicoanalisi: bilancio e prospettive." *KOS*, no. 152 (May): 26–31. This is an Italian translation of "A Century of Psychoanalysis: Critical Retrospect and Prospect," in *Freud: Conflict and Culture* (1998) [317]. See also [290], [312], [313], [315], [316], [317], [318], [323], [328], [335], [345], [347], [352], [354], [371], and [373].

[328]   (1999) "A Century of Psychoanalysis: Critical Retrospect & Prospect." POL.it, Psychiatry On Line. http://www.pol-it.org/ital/ 9grunb-i.htm or http://www.pol-it.org//ital/9grunb-i.htm. This is a reprint of the article in the newsletter of *University of Pittsburgh, University Center for International Studies, Center for West European Studies* (1998) [318]. See also [290], [312], [313], [315], [316], [317], [323], [327], [328], [335], [345], [347], [352], [354], [371], and [373].

[329]   (1999) "Creation as a Pseudo-Explanation in Current Physical Cosmology." Secular Web. http://www.infidels.org/library/modern/ adolf_grunbaum/explanation.html. This is a reprint of the article in Erkenntnis (1991) [258]. See also [259].

[330]   (1999) "My Exodus to Secular Humanism." *Free Inquiry* 19, no. 4 (Fall): 25.

[331]   (1999) "Some Comments on William Craig's 'Creation and Big Bang Cosmology.'" Secular Web. http://www.infidels.org/library/ modern/adolf_grunbaum/comments.html. This is a reprint of the article in *Philosophia Naturalis* (1994) [291].

[332]   (1999) "The Hermeneutic versus the Scientific Conception of Psychoanalysis: An Unsuccessful Effort to Chart a *Via Media* for the Human Sciences." In *Einstein Meets Magritte, an Interdisciplinary Reflection: The White Book of Einstein Meets Magritte*, edited by D. Aerts, J. Broekaert, and E. Mathijs. Dordrecht, Netherlands: Kluwer Academic Publishers, pp. 219–39. See also [340], [343], [356], and [360].

[333]   (1999) "The Pseudo-Problem of Creation in Physical Cosmology." Secular Web. http://www.infidels.org/library/modern/adolf _grunbaum/problem.html. This is a reprint of the article in *Philosophy of Science* (1989) [244]. See also [245], [246], [253], [260], [326], and [332].

[334] (1999) "Theological Misinterpretations of Current Physical Cosmology." Secular Web. http://www.infidels.org/library/modern/adolf_grunbaum/theological.html. This is a reprint of the article in *Philo* (1998) [325]. See also [301], [310], and [357].

[335] (1999) "Un secolo di psicoanalisi: bilancio e prospettive." POL.it, Psychiatry On Line. http://www.pol-it.org/ital/9grunbau.htm or http://www.pol-it.org//ital/9grunbau.htm. This is a reprint of the Italian translation of "A Century of Psychoanalysis: Critical Retrospect and Prospect," in *KOS* (1998) [327]. See also [290], [312], [313], [315], [316], [317], [323], [328], [345], [347], [352], [354], [371], and [373].

[336] (2000) "A New Critique of Theological Interpretations of Physical Cosmology." *British Journal for the Philosophy of Science* 51 (March): 1–43. See also [346] and [355].

[337] (2000) "Does Freudian Theory Resolve 'The Paradoxes of Irrationality'?" In *Proceedings of the 20th World Congress of Philosophy*, edited by B. Elevitch. Philosophy of Mind and Philosophy of Psychology, vol. 9. Bowling Green, OH: Philosophy Documentation Center, pp. 203–18. See also [338], [341], and [349].

[338] (2000) "Does Freudian Theory Resolve 'The Paradoxes of Irrationality'?" POL.it, Psychiatry On Line. Introduction, http://www.pol-it.org/ital/adolf.htm; chapter 2, http://www.pol-it.org/ital/adolf2.htm; chapter 3, http://www.pol-it.org/ital/adolf3.htm; and endnotes, http://www.pol-it.org/ital/adolf4.htm. This is a reprint of the paper in *Proceedings of the 20th World Congress of Philosophy* (2000) [337]. See also [341] and [349].

[339] (2000) "Ein Jahrhundert Psychoanalyse: Ein kritischer Rückblick—ein kritischer Ausblick." *Forum der Psychoanalyse* 16: 285–96. This is a German translation of "A Century of Psychoanalysis: Critical Retrospect and Prospect," in *Freud: Conflict and Culture* (1998) [317]. Errata in this article are listed in (2000) *Forum der Psychoanalyse* 17, no. 2 (June). See also [318], [323], [327], [328], and [335].

[340]   (2000) "Ermeneutica versus Concezione Scientific della Psicoanalisi: Uno Sforzo Inutile di Stabilire Una *Via Media* per le Scienze Umane." POL.it, Psychiatry On Line. Introduction, http://www.pol-it.org/ital/grunba_it.htm; chapter 2, http://www.pol-it.org/ital/grunba 2_it.htm; chapter 3, http://www.pol-it.org/ital/grunba 3_it.htm; chapter 4, http://www.pol-it.org/ ital/grunba 4_it.htm; and chapter 5, http://www.pol-it.org/ital/grunba 5_it.htm. This is an Italian translation of "The Hermeneutic versus the Scientific Conception of Psychoanalysis: An Unsuccessful Effort to Chart a *Via Media* for the Human Sciences," in *Einstein Meets Magritte, an Interdisciplinary Reflection: The White Book of Einstein Meets Magritte* (1999) [332]. See also [356] and [360].

[341]   (2000) "La Teoria Freudiana Risolve 'I Paradossi dell'Irrazionalita'?" POL.it, Psychiatry On Line. Chapter 1, http://www.pol-it.org/ital/adolf_it.htm; chapter 2, http://www.pol-it.org/ital/adolf2_it.htm; chapter 3, http://www.pol-it.org/ital/adolf3_it.htm; and endnotes, http://www.pol-it.org/ital/adolf4_it.htm. This is an Italian translation of "Does Freudian Theory Resolve 'The Paradoxes of Irrationality'?" in *Proceedings of the 20th World Congress of Philosophy* (2000) [337]. See also [338] and [349].

[342]   (2000) "Reply to Symposium on the Grünbaum Debate." *Psychoanalytic Dialogues* 10, no. 2 (March): 335–42.

[343]   (2000) "The Hermeneutic versus the Scientific Conception of Psychoanalysis: An Unsuccessful Effort to Chart a *Via Media* for the Human Sciences." POL.it, Psychiatry On Line. Chapter 1, http://www.pol-it.org//ital/grunba.htm; chapter 2, http://www.pol-it.org//ital/grunba2.htm;

chapter 3, http://www.pol-it.org//ital/grunba3.htm; chapter 4, http://www .pol-it.org//ital/grunba4.htm; chapter 5, http://www.pol-it.org//ital/grunba 5.htm; and endnotes, http://www.pol-it.org//ital/grunba6.htm. This is a reprint of the article in *Einstein Meets Magritte, an Interdisciplinary Reflection: The White Book of Einstein Meets Magritte* (1999) [332]. See also [340], [356], and [360].

[344] (2000) "The Poverty of Theistic Morality." Secular Web. http://www.infidels.org/library/modern/adolf_grunbaum/poverty.htm. This is a reprint of the essay in *Science, Mind and Art: Essays on Science and the Humanistic Understanding in Art, Epistemology, Religion and Ethics, in Honor of Robert S. Cohen* (1995) [298], [362], [372], and [374].

[345] (2001) "A Century of Psychoanalysis: Critical Retrospect and Prospect." *International Forum of Psychoanalysis* 10, no. 2 (June): 105–12. This is a reprint of the article in *Freud: Conflict and Culture* (1998) [317]. See also [290], [312], [313], [315], [316], [323], [327], [328], [335], [347], [352], [354], [371], and [373].

[346] (2001) "A New Critique of Theological Interpretations of Physical Cosmology." In *John von Neumann and the Foundations of Quantum Physics, Vienna Circle Institute Yearbook 2000*, no. 8, edited by M. Rédei and M. Stöltzner. Boston: Kluwer Academic Publishers, pp. 269–88. This is a lecture presented at the Institute Vienna Circle, Vienna, Austria, June 7, 1999. It is a much condensed and revised version of the article in *British Journal for the Philosophy of Science* (2000) [336] and [355].

[347] (2001) "Critique of Freud's Notions of Mental Illness." In *Life—Interpretation and the Sense of Illness within the Human Condition*, edited by A-T. Tymieniecka and E. Agazzi. Analecta Husserliana, the Yearbook of Phenomenological Research, vol. 72. Dordrecht, Netherlands: Kluwer Academic Publishers, pp. 57–70. This is a version of the essay in *Freud: Conflict and Culture* (1998) [317]. See also [290], [312], [313], [315], [316], [323], [327], [328], [335], [345], [352], [354], [371], and [373].

[348]   (2001) "David Malament and the Conventionality of Simultaneity: A Reply." PhilSci Archive. http://philsci-archive.pitt.edu/documents/disk0/00/00/01/84/index.html. This is a prepublished essay and will appear as a chapter in A. Grünbaum, *Philosophy of Science in Action*, vol. 1, to be published by New York: Oxford University Press. See [372].

[349]   (2001) "Does Freudian Theory Resolve 'The Paradoxes of Irrationality'?" *Philosophy and Phenomenological Research* 62, no. 1 (January): 129–43. This is a reprint of the paper in *Proceedings of the 20th World Congress of Philosophy* (2000) [337]. See also [338] and [341].

[350]   (2001) "Freud and the Interpreters." Letter to the editor, *New York Review of Books* 47, no. 1 (January 11): 56.

[351]   (2001) *La Mia Odissea Dalla Filosofia Alla Psicoanalisi.* Translated by A. Carotenuto. Rome, Italy: Di Renzo Editore.

[352]   (2001) "L'inconscient à l'épreuve" (The Unconscious under Scrutiny). Translated by A. Barberousse. *Sciences et Avenir* (The Sciences and the Future) (July/August): 42–49. This is a shortened version of the essay in *Mindscapes: Philosophy, Science, and the Mind* (1997) [312]. It is a modified version of the keynote address presented at the awards ceremony of the Prix Latsis Universitaires (1994) [290], [312], [313], [315], [316], [317], [318], [323], [327], [328], [335], and [345]. A later considerably modified version of this address is [354]. See also [371] and [373].

[353] (2001) "Narlikar's 'Creation' of the Big Bang Universe Was a Mere Origination." Secular Web. http://www.infidels.org/library/modern/adolf_grunbaum/narlikar.htm. This is a reprint of the article in *Philosophy of Science* (1993) [281].

[354] (2002) "Critique of Psychoanalysis." In *The Freud Encyclopedia: Theory, Therapy, and Culture*, edited by E. Erwin. New York/London: Routledge, pp. 117–36. This is a revised, enlarged version of the essay in *Mindscapes: Philosophy, Science, and the Mind* (1997) [312]. It is a much modified version of the keynote address presented at the awards ceremony of the Prix Latsis Universitaires (1994) [290], [313], [315], [316], [317], [318], [323], [327], [328], [335], [345], and [352]. See also [361], [371], [373], and [380].

[355] (2002) "Novaya kritika teologicheskih interpretatsii fizicheskoi kosmologii." *Voprosi Filosofii*, no. 4: 67–88. This is a Russian translation of "A New Critique of Theological Interpretations of Physical Cosmology," in *British Journal for the Philosophy of Science* (2000) [336]. See also [346].

[356] (2003) "The Poverty of the Semiotic Turn in Psychoanalytic Theory and Therapy." In *Between Suspicion and Sympathy: Paul Ricoeur's Unstable Equilibrium, a Festschrift in Honor of Paul Ricoeur's 90th Birthday*, edited by A. Wiercinski. International Institute for Hermeneutics, Hermeneutic Series, vol. 3. Toronto, ON, Canada: Hermeneutic Press, pp. 602–19. This is a somewhat enlarged version of "The Hermeneutic versus the Scientific Conception of Psychoanalysis: An Unsuccessful Effort to Chart a *Via Media* for the Human Sciences," in *Einstein Meets Magritte, an Interdisciplinary Reflection: The White Book of Einstein Meets Magritte* (1999) [332]. See also [340], [343], and [360].

[357] (2003) "Theological Misinterpretations of Current Physical Cosmology." In *The Existence of God*, edited by R. M. Gale and A. R.

Pruss. Hants, UK: Dartmouth Publishing, pp. 327–46. This is a reprint of the article in *Philo* (1998) [334]. See also [151], [301], [310], and [325].

[358]   (2004) "The Degeneration of Popper's Theory of Demarcation." In *Karl Popper: Critical Assessments of Leading Philosophers*, vol. 2, *Philosophy of Science 1*, edited by A. O'Hear. Part 3, "Demarcation." London: Routledge, pp. 392–411. This is a reprint of the article in *Epistemologia* (1989) [242]. See also [241] and [242].

[359]   (2004) "Can a Theory Answer More Questions Than One of Its Rivals?" In *Karl Popper: Critical Assessments of Leading Philosophers*, vol. 3, *Philosophy of Science 2*, edited by A. O'Hear. Part 4: "Verisimilitude." London: Routledge, pp. 55–77. This is a reprint of the article in *British Journal for the Philosophy of Science* (1976) [151]. See also [241] and [242].

[360]   (2004) "The Hermeneutic versus the Scientific Conception of Psychoanalysis." In *Psychoanalysis at the Limit: Epistemology, Mind, and the Question of Science*, edited by J. Mills. New York: State University of New York Press, pp. 139–60. This is a reprint of the article in *Einstein Meets Magritte, an Interdisciplinary Reflection: The White Book of Einstein Meets Magritte* (1999) [332]. See also [340], [343], and [356].

[361]   (2004) "Critique of Psychoanalysis." In *Who Owns Psychoanalysis?* edited by A. Casement. London: Karnac Books, pp. 263–305. This is a reprint of the essay in *The Freud Encyclopedia: Theory, Therapy, and Culture* (2002) [354]. See also [371], [373], and [380].

[362]   (2004) "Das Elend der Theistischen Moral." In *Moral als Gabe: Zur Ambivalenz von Moral und Religion*, edited by B. Boothe and P. Stoellger. Würzburg, Germany: Königshausen und Neumann, pp. 143–75. This is a German translation of an enlarged version of

"The Poverty of Theistic Morality," in *Science, Mind and Art: Essays on Science and the Humanistic Understanding in Art, Epistemology, Religion and Ethics, in Honor of Robert S. Cohen* (1995) [298]. See also [344], [362], [372], and [374].

[363]   (2004) *Podstawy psychonanlizy. Krytyka filozoficzna.* Translated by E. Olender-Dmowska. Edited by Z. Rosinska. Poland: Universitas Publishing House. This is a Polish translation of *The Foundations of Psychoanalysis: A Philosophical Critique* (1984) [206]. See also [218], [228], [230], [232], [233], [256], [261], [299], [305], and [308].

[364]   (2004) "Why Is There Something Rather Than Nothing? An Ill-Conceived Question Whose Theistic Answer Fails." In *Knowledge and Belief—Wissen und Glauben*, edited by Löffler and P. Weingartner. Proceedings of the 26th International Wittgenstein Symposium, Kirchberg, Austria, August 2003. Vienna, Austria: Osterreichischer Bundesverlag/Holder-Pichler-Tempsky, pp. 287–97.

[365]   (2004) "Nishcheta teisticheskoi kosmologii." Translated by D. G. Lakhuti. *Voprosy Filosofii*, no. 8: 99–114; no. 9: 149–62; no. 10: 114–24. This is the Russian translation of an *earlier version* of "The Poverty of Theistic Cosmology" 2004 [367], which occupied fifty-four consecutive printed pages as the lead article in the *British Journal for the Philosophy of Science* (2004) [367]. It is to be noted that, because of its great length, this essay was spread over *three successive issues* of *Voprosy Filosofii*.

[366]   (2004) "Quelques Objections Fondamentales." *Le Nouvel Observateur*. Hors-série, October/November, p. 11. This is a French translation of "Is Freudian Psychoanalysis Moribund?" See also [378].

[367]   (2004) "The Poverty of Theistic Cosmology." *British Journal for the Philosophy of Science* 55, no. 4: 561–614. See also [365].

[368] (2004) "Wesley Salmon's Intellectual Odyssey and Achievements." Proceedings of PSA 2002, the Symposium on Wesley C. Salmon, 1925–2001: A Symposium Honoring His Contributions to the Philosophy of Science. *Philosophy of Science* 71, no. 5: 922–25.

[369] (2005) "Does Leibniz's Principle of Sufficient Reason License His Primordial Existential Question 'Why Is There Something Contingent Rather than Nothing Contingent'?" In *Homo Sapiens und Homo Faber: Zur Epistemischen und Technischen Rationalität in Antike und Gegenwart—Festschrift für Jürgen Mittelstrass*, edited by G. Wolters and M. Carrier. Berlin: Walter de Gruyter, pp. 147–56.

[370] (2005) "Theological Misinterpretations of Current Physical Cosmology." In *Metaphysics: Classic and Contemporary Readings*, 2nd ed., edited by L. N. Oaklander and R. C. Hoy. Belmont, CA: Thomson Wadsworth Publishing, pp. 525–39 [357]. See also [301], [325], and [334].

[371] (2005) "Critique of Psychoanalysis." In *Handbook of Personology and Psychopathology: Festschrift for Ted Millon*, edited by S. Strack. New York, John Wiley & Sons, pp. 73–99; this is a reprint of (2002) [354]. See also [361], [373], and [380].

[372] (2005) "Das Elend der Theistischen Moral." In *Brauchen wir Gott? Moderne Texte zur Religionskritik*, edited by E. Dahl. Stuttgart, Germany: S. Hirzel, pp. 125–57. This German translation of "The Poverty of Theistic Morality" is a reprint of the essay in *Moral als Gabe: Zur Ambivalenz von Moral und Religion* (2004) [362]. See also [298], [344], and [348].

[373] (2005) "Critique of Psychoanalysis." In *The Freud Wars: An Introduction to the Philosophy of Psychoanalysis*, edited by L. Gomez. London/New York: Routledge, pp. 109–37. See also [354], [361], [371], and [380].

[374] (2005) "The Poverty of Theistic Morality." *Free Inquiry* 25, no. 6 (October/November): 41. This is a condensed version of the essay that first appeared in *Science, Mind and Art: Essays on Science and the Humanistic Understanding in Art, Epistemology, Religion and Ethics, in Honor of Robert S. Cohen* (1995) [298]. See also [344] and [362].

[375] (2005) "Freud, Sigmund." In *Encyclopedia of Philosophy*, 2nd ed., vol. 3, edited by D. Borchert. Farmington Hills, MI: Macmillan Reference, Gale Group/Thomson, pp. 736–48.

[376] (2005) "Answers to Questions about Formal Methods." In *Formal Philosophy*, edited by V. F. Hendricks and J. Symons. New York/London: Automatic Press/VIP, p. 75.

[377] (2005) "Rejoinder to Richard Swinburne's 'Second Reply to Grünbaum.'" *British Journal for the Philosophy of Science* 56, no. 4: 927–38.

[378] (2006) "Otzjivaet li freydistski psychoanliz svoi vek?" *Voprosy Filosofii*, no. 1: 173–74. This is a Russian translation of "Is Freudian Psychoanalysis Moribund?" which first appeared as "Quelques Objections Fondamentales," in *Le Nouvel Observateur* (2004) [366].

[379] (2006) "Is Sigmund Freud's Psychoanalytic Edifice Relevant to the 21st Century?" *Psychoanalytic Psychology* (150th Freud Anniversary Issue) 23, no. 2: 257–84. See also [380].

[380] (2007) "Kritika psychoanaliza." *Voprosi Filosofii* 12, no. 3: 105–29. This is a Russian reprint of "Critique of Psychoanalysis," in *The Freud Encyclopedia: Theory, Therapy, and Culture* (2002) [354]. See also [361], [371], [373], and [379].

[381] (2007) "The Reception of my Freud-Critique in the Psychoanalytic Literature (Part 1)." *Psychoanalytic Psychology* 24, no. 3: 545–76. This is the first of two installments.

[382] (2007) "Why Is There a Universe AT ALL, Rather Than Just Nothing?" In *The Routledge Companion to the Philosophy of Religion*, edited by C. Meister and P. Copan. London/New York: Routledge, pp. 441–51. See also [385], [389], [390], [393], and [395].

[383] (2007) "Is Simplicity Evidence of Truth?" In *Philosophy of Science*, edited by A. O'Hear. London: Cambridge University Press, pp. 261–75. Also published as supplement 61 to subscribers of the journal *Philosophy*. This is the text of a lecture delivered at All Souls College, Oxford University, March 9, 2006, and at the Royal Institute of Philosophy, London, UK, March 10, 2006. See also [386].

[384] (2007) "Critique of Psychoanalysis." In *Filosofia, scienza e bioetica nel dibattito contemporaneo*, edited by F. Minazzi. Studi internazionali in onore di Evandro Agazzi. Rome: Instituto Poligrafico e Zecca dello Stato, pp. 547–73. This is a reprint of the essay in *The Freud Encyclopedia: Theory, Therapy, and Culture* (2002) [354]. See also [361], [371], [373], and [379].

[385] (2008) "Why Is There a Universe AT ALL, Rather Than Just Nothing? (Part 1)." *Free Inquiry* 28, no. 4 (June/July): 32–35; (Part 2), *Free Inquiry* 28, no. 5 (August/September): 37–41. This is a variant of the presidential address given at the 13th quadrennial World Congress of the Division of Logic, Methodology, and Philosophy of Science (DLMPS) of the International Union of History and Philosophy of Science (IUHPS), held at Tsinghua University, Beijing, China, August 9, 2007. A much earlier version of this paper was published under the same title in *The Routledge Companion to the Philosophy of Religion* (2007). See also [382], [389], [390], [393], and [395].

[386] (2008) "Is Simplicity Evidence of Truth?" *American Philosophical Quarterly* 45, no. 2: 179–89. This article is a significantly revised and expanded version of an earlier paper that was published under the same title in *Philosophy of Science* (2007) [383].

[387] (2008) "The Anisotropy of Time." In *Philosophy of Time: Critical Concepts in Philosophy*, edited by N. Oaklander. London: Routledge. See also [98].

[388] (2009) "Popper's Fundamental Misdiagnosis of the Scientific Defects of Freudian Psychoanalysis, and of Their Bearing on the Theory of Demarcation." *Psychoanalytic Psychology* 25, no. 4 (October): 574–89. A slightly longer version appears in the spring of 2009. See [394].

[389] (2008) "Why Is There a Universe, AT ALL, Rather Than Just Nothing?" Part 2. *Free Inquiry* 28, no. 5 (August/September 2008): 37–41. This is the second half of a variant of Grünbaum's presidential address of August 9, 2007, at the 13th Quadrennial World Congress of the Division of Logic Methodology, and Philosophy of Science (DLMPS) of the International Union of History and Philosophy of Science (IUHPS), held at Tsinghua University, Beijing, China. See also [382], [385], [390], [393], and [395].

[390] (2008) "Why Is There a Universe, AT ALL, Rather Than Just Nothing?" *World Philosophy*, vol. 5. This is a Chinese translation of the presidential address delivered by Grünbaum at the 13th Quadrennial World Congress of the Division of Logic Methodology, and Philosophy of Science (DLMPS) of the International Union of History and Philosophy of Science (IUHPS) at Tsinghua University, Beijing, China, on August 9, 2007. See also [382], [385], [389], [393], and [395].

[391] (2009) "Autobiographical-Philosophical Narrative." In the present volume: *Philosophy of Physics and Psychology: Essays in*

Honor of Adolf Grünbaum, edited by A. Jokic. Proceedings of the international conference "The Adolf Grünbaum Symposium in Honor of the Works of Professor Adolf Grünbaum," Santa Barbara, CA, October 2002. Amherst, NY: Prometheus Books. (This volume also contains a reprinting of item [367] above.)

[392]   (2009) "The Poverty of Theistic Cosmology," in the present volume: *Philosophy of Physics and Psychology: Essays in Honor of Adolf Grünbaum*, edited by A. Jokic. Proceedings of the international conference, "The Adolf Grünbaum Symposium in Honor of the Works of Professor Adolf Grünbaum," Santa Barbara, CA, October 2002. Amherst, NY: Prometheus Books. See also [367].

[393]   (2009) "Why Is There a Universe, AT ALL, Rather Than Just Nothing?" In *Logic, Methodology and Philosophy of Science: Proceedings of the Thirteenth International Congress*, edited by C. Glymour, Wang Wei, and D. Westerstahl. London: Kings College Publications. This article was the presidential address at the Congress, delivered by Grünbaum at Tsinghua University, Beijing, China, on August 9, 2007. See also [382], [385], [389], [390], and [395].

[394]   (2009) "Popper's Fundamental Misdiagnosis of the Scientific Defects of Freudian Psychoanalysis, and of Their Bearing on the Theory of Demarcation." In *Rethinking Popper*, edited by Z. Parusnikova and R. S. Cohen, from proceedings of a conference held in Prague, Czech Republic, on September 11, 2007, at the Institute of Philosophy, Academy of Sciences of the Czech Republic, Boston Studies in the Philosophy of Science, vol. 272, Springer, pp. 117–34. See also [388].

[395]   (2009) "Why Is There a Universe AT ALL, Rather Than Just Nothing?" *Ontology Studies* 9. Proceedings of the VIII International Ontology Congress held in San Sebastian, Spain, September 29–October 3, 2008. See also [382], [385], [389], [390], and [393].

[396] (2009) "Psychoanalyse: Wissenschaft, Weltanschauung, Religion," a German translation of "Psychoanalysis: Science, World-View, Religion," which is the lead chapter in the opening section, "Perspektiven der Wissenschaft" [Scientific Perspectives], in *Sigmund Freud in der Sicht der psychotherapeutischen Schulen: Gestern und Heute* [Sigmund Freud in the Light of Psychotherapeutic Schools–Yesterday and Today], edited by A. Leitner and H. G. Petzold. Vienna, Austria: Krammer Verlag.

[397] (2009) "David Malament and the Conventionality of Simultaneity: A Reply." In Peter Mittelstaedt *Festschrift* issue of *Foundations of Physics*. See also [348].

(Forthcoming) "The Reception of My Freud-Critique in the Psychoanalytic Literature." Part II, *Psychoanalytic Psychology.* See [381] Part I.

(Forthcoming) (title to be determined). In *Living Legacies Speak*, edited by A. Mahrer, a special issue of the *Journal of Contemporary Psychotherapy.*

(Forthcoming) "Why Is There a Universe AT ALL, Rather Than Just Nothing?" In *The Future of Naturalism*, edited by J. Shook and P. Kurtz. Amherst, NY: Prometheus Books. This article was first published in *Free Inquiry* as Part 1, June/July, 2008 and Part 2, August/ September 2008. See also [382], [385], [389], [390], [393], and [395].

(Forthcoming) "Sigmund Freud." In *History of Western Philosophy of Religion*, edited by Graham Oppy and Nick Trakakis. England: Acumen Publishing Co.

(Forthcoming) "Freud's Theory: The Perspective of a Philosopher of Science." In Centenary Celebration volume of the *Presidential Addresses of The American Philosophical Association,*

454 BIBLIOGRAPHY OF ADOLF GRÜNBAUM

1900–2000, edited by R. T. Hull. This is a much enlarged, revised version of the author's presidential address in *Proceedings and Addresses of The American Philosophical Association* (1983) [197]. See also [205], [211], [216], [237], [263], and [267].

(Forthcoming) *5 Questions in Philosophy of Science*. Interview answers by Grünbaum in this collection edited by R. Rosenberger.

(Forthcoming) *Philosophy of Science in Action*. Vols. 1 and 2. New York: Oxford University Press.

INDEX

A. A. Knopf, 127

Abraham, 16–17

"Absolute and Relational Theories of Space and Space-Time" (conference), 39

absolute simultaneity. *See* simultaneity

Académie Internationale de Philosphie des Sciences, 145

Academy of Social Sciences, Institute of Atheism, 45, 47

active diffeomorphisms. *See* diffeomorphisms

"Adolffest" (University of Pittsburgh), 138, 142

"Adolf Grünbaum: Psychoanalytic Epistemology" (von Eckhardt), 95–96

"Adolf Grünbaum on Religion, Cosmology, and Morals" (Murphy), 235–55

"Adolf Grünbaum on Space and Time" (Tooley), 259–85

Adolf Grünbaum Philosophical Reading Room, University of Pittsburgh, 138, 146

Adolf Grünbaum Symposium (Santa Barbara, CA), 9–10, 146–47

After Virtue (MacIntyre), 249

Agassi, Joseph, 81

Ağca, Mehmet Ali, 19–20

agnosticism, 23–24, 201

Albert, David, 331

Albert, Robert C., 63–64

Allen, Woody, 16

All Souls College, Oxford University, 121

American Academy of Arts and Sciences, 37, 145

American Association for the Advancement of Science, 44, 145

American Journal of Psychiatry, 118, 359

American Philosophical Association, Eastern Division, 62, 73–79, 145

American Philosophical Quarterly (journal), 121

American Psychiatric Association, 359–60

American Psychoanalytic Association, 118, 359

American Psychological Association, 91, 96, 126, 360

Anderson, Alan, 61

Andrew Mellon Educational and Charitable Trust, 64

Andrew Mellon Professorships (University of Pittsburgh), 52–53, 64, 66, 105, 121, 133, 143, 146

455